# Fundamentals of Thermal Spraying

This book discusses the concepts and uses of thermal spraying including starting powder, spraying parameters, diagnostics, coating deposition, evolved microstructure and resulting properties complemented with several case studies to associate the learnings with applied concepts. The major parts of the instrumentation, the spraying gun, which is the fundamental aspect of different thermal spraying conditions are also discussed. Solved examples, numerical problems and descriptive questions are included for self-assessment at the end of every chapter.

The book:

- Discusses all aspects from starting powder, spraying parameters, diagnostics and coating deposition;
- Explores schematics to highlight the conceptual notes;
- Includes multiple case studies from domains including aerospace, biomedical, manufacturing, wettability and others to highlight the practical application of thermally sprayed coatings;
- Covers classification of thermal spray techniques; and
- Contains solved examples, numerical problems and descriptive questions for self-assessment.

This book is aimed at senior undergraduates and graduates in materials science and engineering.

# Fundamentals of Thermal Spraying

Edited by Ariharan S, Rubia Hassan,
Alok Bhadauria, Ashutosh Tiwari, Ritik Tandon,
Anup Kumar Keshri and Kantesh Balani

CRC Press is an imprint of the
Taylor & Francis Group, an **informa** business

First edition published 2023
by CRC Press
6000 Broken Sound Parkway NW, Suite 300, Boca Raton, FL 33487–2742

and by CRC Press
4 Park Square, Milton Park, Abingdon, Oxon, OX14 4RN

*CRC Press is an imprint of Taylor & Francis Group, LLC*



ISBN: 978-1-032-34400-3 (hbk)
ISBN: 978-1-032-34402-7 (pbk)
ISBN: 978-1-003-32196-5 (ebk)

DOI: 10.1201/9781003321965

Typeset in Times
by Apex CoVantage, LLC

# Contents

# Editor Biographies

**Ariharan S** earned his doctorate in Materials Science and Engineering from Indian Institute of Technology (IIT) Kanpur, India, in 2018. He is recipient of the Nanomaterials and Energy Prize 2019 by the journal *Nanomaterials and Energy*. He was awarded a National Post-Doctoral Fellowship 2018, SERB, Government of India, during February 2019–August 2021 at IIT Madras. Currently, he is a post-doctoral researcher at Alexander Dubček University of Trenčín, Slovakia. He has published more than 25 research articles (*h*-index of 13, *i*10-index of 13) with 450+ citations in peer-reviewed journals and conference proceedings.

**Rubia Hassan** earned a PhD from the Department of Materials Science and Engineering at IIT Kanpur, India, in February 2022. She is a recipient of the prestigious Prime Ministers Research Fellowship. She is a recipient of the Bogineni Chenchu Raman Naidu Gold Medal and the Cadence Gold Medal, recognizing her academic excellence. She has co-authored eight research papers published in reputed international journals.

**Alok Bhadauria** is a post-doctoral fellow in the Materials Science and Engineering Department at IIT Kanpur, India. He earned a PhD from the Metallurgical and Materials Engineering Department at IIT Kharagpur, India. He has published 13 technical papers in peer-reviewed journals and is currently working in alumina and yttria-stabilized zirconia (YSZ)-based thermal barrier coatings for high-temperature gas-turbine engine blades.

**Ashutosh Tiwari** is Dean (PG and Research) and Registrar at Rajkiya Engineering College Banda, Uttar Pradesh, India. He is a member of the Indian Nuclear Society, a life member of the Indian Association of Physics Teachers, the Indian Society of Technical Education, and the Indian Science Congress. He bagged the DST TARE project and has received various fellowships, such as the IGCAR Post-Doctoral Fellowship, the Dr D.S. Kothari Post-Doctoral Fellowship of UGC, an institute post-doctoral fellowship at IIT Kanpur, and SRF and JRF from CSIR New Delhi. He has published a book on nuclear science. He has 11 years of teaching experience and has published 25+ papers in various international journals. His research interests are safety analysis of power plants, thermal barrier coatings and development of virtual labs.

**Ritik Tandon** studied Chemical Engineering at Purdue University, West Lafayette, Indiana, USA. He is involved with Anod Plasma Spray Limited, Kanpur, India, in the area of repair, refurbishment and coating of turbine components for end use in power generation. His firm offers services in industrial, hydro and thermal sectors. His company enjoys an excellent reputation with esteemed customers like NTPC, NHPC, BHEL and RIL, and has been associated with Mitsubishi Heavy Industries in repair and coating of gas turbine components for more than 25 years.

**Anup Kumar Keshri** is an Associate Professor in the Department of Metallurgical and Materials Engineering at IIT Patna, India. He earned his PhD degree in Materials Science and Engineering from Florida International University (FIU), Miami, Florida, USA, in July 2010. He has published 75+ papers in peer-reviewed journals and has an *h*-index of 28 (2600+ citations), which strongly endorses his research productivity. He has filed five Indian patents, two of which have been granted. He is a recipient of awards including a Research Stay Grant by the Humboldt Foundation, a Dissertation Year Fellowship (2009–2010) from FIU, the Arthur E. Focke Leadership Award by ASM Foundation, and delegate of President's Council of Student Advisors (PCSA) formed by the American Ceramic Society (ACerS). He also serves as a reviewer for several journals in the areas of coatings and thermal spray.

**Kantesh Balani** is Full Professor in the Department of Materials Science and Engineering, IIT Kanpur, India, since November 2018. He earned his doctorate in Mechanical Engineering from FIU in 2007. He has published over 170 articles (*h*-index of 40, *i*10-index of 120+) with 7500+ citations in peer-reviewed international journals. He is a recipient of the Yadupati Singhania Memorial Chair, an ASM Class of Fellow 2021, a Fellow of INAE 2021, a Fellow of NASc 2021, the Swarnajayanti Fellowship 2017–23, and the P.K. Kelkar Research Fellowship. He is co-inventor of three Indian patents (11 filed), the co-author of two books, and is a reviewer of 50+ journals. He is Editor in Chief of *Nanomaterials and Energy*, Associate Editor of *Journal of Thermal Spray Technology*, Principal Editor of *Journal of Materials Research* and Key Reader of *Metallurgical and Materials Transactions A*.

# Contributors

**Shipra Bajpai**
Department of Materials Science and Engineering
Indian Institute of Technology Kanpur
Kanpur, Uttar Pradesh, India

**Kantesh Balani**
Department of Materials Science and Engineering
Indian Institute of Technology Kanpur
Kanpur, Uttar Pradesh, India

**Alok Bhadauria**
Department of Materials Science and Engineering
Indian Institute of Technology Kanpur
Kanpur, Uttar Pradesh, India

**Shivani Gour**
Department of Materials Science and Engineering
Indian Institute of Technology Kanpur
Kanpur, Uttar Pradesh, India

**Rubia Hassan**
Department of Materials Science and Engineering
Indian Institute of Technology Kanpur
Kanpur, Uttar Pradesh, India

**Anup Kumar Keshri**
Department of Metallurgical and Materials Engineering
Indian Institute of Technology Patna
Patna, Bihar, India

**K. Vijay Kumar**
Department of Metallurgical and Materials Engineering
Indian Institute of Technology Patna
Patna, Bihar, India

**Priya Kushram**
School of Mechanical and Materials Engineering
Washington State University
Pullman, Washington, USA

**Moumita Mistri**
Department of Materials Science and Engineering
Indian Institute of Technology Kanpur
Kanpur, Uttar Pradesh, India

**Raghav Mittal**
Development Engineering
University of California
Berkeley, California, USA

**Divya Rana**
Department of Materials Science and Engineering
Indian Institute of Technology Kanpur
Kanpur, Uttar Pradesh, India

**Pooja Rani**
Department of Materials Science and Engineering
Indian Institute of Technology Kanpur
Kanpur, Uttar Pradesh, India

**Ariharan S**
Department of Coating Processes
Alexander Dubček University of Trenčín
Trenčín, Slovakia

**Roopal Singh**
Business Analyst
CARS24
Gurugram, Haryana, India

**Ritik Tandon**
Anod Plasma Spray Limited
Uptron Estate
Panki, Kanpur, Uttar Pradesh, India

**Ashutosh Tiwari**
Department of Applied Science and Humanities (Physics)
Rajkiya Engineering College
Banda, Uttar Pradesh, India

# Foreword

It is my pleasure to write the foreword of the book *Fundamentals of Thermal Spraying*, edited by Ariharan S, Rubia Hassan, Alok Bhadauria, Ashutosh Tiwari, Ritik Tandon, Anup Kumar Keshri and Kantesh Balani.

After reading this book, I can say that this is one among the most comprehensive presentations in this field, especially for undergraduate and postgraduate students as a perfect textbook. The book contains 11 chapters and covers all the fundamentals of thermal spraying, starting from powder characteristics to fabrication of the final part. The book starts its journey from the historical perspective of Schoop's patent, leads the reader to the need for surface coatings, takes effort in classifying the thermal spray techniques, and emphasizes myriad spraying parameters that affect the characteristics of the finally deposited coating. Various comparisons in terms of microstructure, performance of coatings, process parameters and so forth provide a good perspective of the differences between various thermal spraying techniques. The description of gun designs with plasma stability, gas parameters, powder feeding and injection, along with the science behind coating formation starting with single splat formation, is commendable. The details of powder characterization, powder production and powder feeders with its working principle highlight the rigor authors have taken to provide right understanding to readers. The aspects of Rayleigh criterion to heat transfer and splat layering and deposition to emanate with final microstructure is also very lucidly described. Authors contribute on the various aspects of coating reliability, providing assessment of adhesion tests and data analysis via Weibull distribution. The aspect of process diagnostics in providing thermal and kinetic history of the in-flight particles is also a key contribution of this book. The emerging aspect of machine learning is also a good addition. The extension of thermal spraying towards achieving freestanding and near-net shaped structures is also emphasized by the authors. Several selected case studies, presented by authors on biomaterials, corrosion resistance, thermal barrier coating and so forth are also noteworthy, as they provide a direct use of thermal spraying in various fields.

The book becomes more special because the authors have taken extra effort to present ideas with schematics and have provided solved numerical problems and questions for self-analysis.

In short, I will give my highest recommendation to this book as a textbook for undergraduate and postgraduate students. This book will also serve as a firsthand reference even for experienced researchers, developers and engineers working in the area of thermal spraying.

**Arvind Agarwal, PhD, FASM, FACerS, FAAAS**
Chair and Distinguished University Professor
Mechanical and Materials Engineering
Director, Advanced Materials Engineering Research Institute (AMERI)
College of Engineering and Computing
Florida International University
Miami, Florida, USA

# Preface

This textbook is a comprehensive version of a standard thermal spraying textbook, covering almost all the aspects of thermal spraying in brief and simple terms which makes it suitable for the curriculum at the undergraduate level. This compilation is produced with the intention of providing a background for thermal spray technology, which comprises a collection of coating techniques providing protection and/or improved performance to the substrate. Through this book, students will gain confidence for further reading beyond this text to gain expertise in the field.

Of the numerous books available on thermal spraying, most are designed for graduate level and as reference books providing access to a huge yet complicated volume of literature in the field. The current text aims at providing an understanding of the fundamentals of thermal spraying encompassing basic concepts from starting powder to final coated surface. The incorporation of intelligible schematics and inclusion of numerical problems throughout different sections will assist in comprehending the practical aspects of these concepts. In the process of covering all the major aspects of the field and simultaneously designing it to be simple and easily comprehensible, the authors confronted the need to keep the topics concise.

Eleven chapters are presented in this book. Chapter 1 is dedicated to different techniques for thermal spraying and starts with information about the evolution of thermal spraying processes. The history of thermal spraying began with Schoop's patent on flame spraying; later, Schoop also developed the electric arc–spraying process. A comparison of thermal spraying technology is made with other coating techniques, and thermal spraying provides an added benefit of depositing almost any material on practically any substrate to cover a wide range of applications. Further, the need for coating the surfaces of components in engineering applications is highlighted.

Chapter 2 is devoted to surface coatings, discussing the concept of surface and requirement of surface modification through different approaches of which coating is superior to the other techniques. It discusses the reasons for surface failure through wear, erosion, corrosion, high temperature oxidation and so on, and consecutively builds up the platform for utilizing different surface modifications to deal with various issues. Apart from serving as a protective shield against the surface damage of the substrate material, coatings can be applied to induce electrical insulation or electrical conductivity in the material and can also be utilized to impart an aesthetic look for decorative purposes. Of all the surface modification approaches, thermal spraying appears to be most advantageous owing to its versatility in terms of the materials deposited, as well as in terms of the properties of the coatings obtained.

Chapter 3 discusses in detail various thermal spraying techniques, highlighting pros and cons of each technique. The major classes of thermal spraying techniques, classified on the basis of heating source, include combustion spraying, electric arc spraying and plasma spraying. The cold-spraying process, which uses kinetic energy to accelerate the powder particles to supersonic velocities using convergent-divergent nozzles, has also been given a due position in this classification and is utilized in various important applications like medical and aerospace technologies.

Chapter 4 is dedicated to a discussion on spraying parameters. Each and every parameter involved in thermal spraying has an effect on the final deposited coating. In-flight condition of powder particles during spraying, expressed in terms of temperature and kinetic velocity, depends on powder characteristics like powder particle size and shape, powder feed rate, nature of carrier gas, enthalpy of plasma forming gases (for plasma spraying), distance between powder-spray gun and the substrate, and the preparation of substrate material prior to deposition. Presence of any undesirable impurity on the substrate could spoil all the efforts taken to optimize the sparing parameters and may compromise the integrity of the coating. Minimum wastage of powder and maximum deposition as coating needs to avoid entrapment of oxide/un-melted particles in the deposited coating.

Chapter 5 deals with the design of spray guns because it strongly influences spraying parameters like feed rate and particle velocity and temperature that determine the quality of the deposited coating. In cold spraying, an ultra-high velocity, required for deposition of particles as coating through plastic deformation, is attained using a special gun design with a converging and diverging type nozzle. On the other hand, spray guns for combustion-based techniques are relatively simple. Similarly, the gun design changes for other spraying techniques.

Chapter 6 discusses the powder characterization and powder synthesis. The feed material in thermal spray processes is usually in powder form, and the powder characteristics are reflected in the nature and quality of the coating. Thus, different powder morphology and powder size distribution induce different in-flight velocity and thermal characteristics. Different particle sizing techniques and associated equivalent diameters of particles are illustrated in this chapter. A spherical powder morphology helps in maintaining the flowability and regular feed rate. However, different synthesis techniques produce powders of different morphology and sizes, and modification may be required before feeding the powder in the thermal spray process. Modification of particle shape through techniques like spheroidization and particle size through mechanical reduction and consolidation are dealt with in the chapter.

Chapter 7 presents coating development through the formation of splats and layering of successive splats. Splat formation from incoming particles is dictated by the state of the particle (molten, partially molten or solid), heat transfer between the particle and the substrate, particle velocity and the preheating of the substrate. The mechanisms involved during its formation and successive deposition are discussed in the chapter. An interaction of incoming particles with the substrate in the making of splats and also between the successively depositing splats is associated with heat transfer, metallurgical bonding and mechanical integration. The final microstructure of the coating and its densification determine the functionality of the coating.

In Chapter 8, special attention is directed towards the testing of thermal spray coatings. The methods of coating characterization and preparation required prior to the standard test methods are discussed. Microstructural evolution, phase identification and phase quantification, estimation of basic physical, mechanical and thermal properties are also covered. The microstructure of the thermal spray coating, porosity and presence of any oxide or un-melted phase govern the properties and performance of the coating. Accordingly, characterization of microstructure and evaluation of performance through non-destructive or destructive testing is necessitated. This chapter discusses various approaches of porosity calculation in the microstructure, phase evaluation, mechanical property determination including hardness, fracture toughness and adhesive strength; thermal properties like thermal conductivity and thermal shock resistance; and wear properties under different conditions like fretting, abrasion, surface fatigue and erosion.

Chapter 9 provides insight to process diagnostics and online monitoring and control. Knowing that the kinetic velocity and temperature of particles are governing factors in thermal spraying, it becomes essential to measure and control these parameters for efficient deposition. In-flight particle parameters and their effect on microstructure, plume profiling, particle in-flight temperature, velocity distributions and their measurements using different diagnostic systems, and the limitations in the monitoring of in-flight particle in order to achieve constant repeatability, reproducibility and reliability is discussed in this chapter.

Chapter 10 engages the reader in development of nanostructured and near-net shapes of coatings. A wide variety of simple and complex configurations of coatings can be prepared through near-net shape processing. Bulk nanostructured coatings can also be engineered via thermal spraying. Due to high temperatures, almost all materials can be melted, which makes these processes suitable for spraying of near-net shape with specific thickness and geometries at different parts of the complex shapes. The development of a freestanding nanostructured near-net shape lies in the fact that nanostructure is retained without any significant alteration of grain size.

Chapter 11 covers some of few case studies on biomedical coatings, thermal barrier coatings, alternate chrome plating techniques, and corrosion-resistant and ultra-high-temperature coatings engineered through thermal spraying. These case studies facilitate the reader learning the need for thermal spraying in these important applications.

This book is a result of collective effort from all the authors, but the genesis of this project was realized after an enthusiastic involvement of Dr. Ariharan S, Dr. Rubia Hassan, Prof. Anup Kumar Keshri, Prof. Ashutosh Tiwari, Dr. Alok Bhadauria, Mr. Ritik Tandon, and Prof. Kantesh Balani. The idea was formed when Prof. Balani taught the course "MSE634: Fundamentals of Thermal Spraying" at IIT Kanpur in fall 2013 and then again in fall 2017 and felt the need to create a textbook. All the authors have strongly contributed to tailoring the writing to suit an undergraduate perspective by creating schematics, self-analysis questions and numerical problems where feasible. Authors AT and KB acknowledge *TARE* (TAR/2018/000310, Science and Engineering Research Board, Department of Science and Technology, Govt. of India). KB acknowledges the *Swarnajayanti Fellowship* (DST/SJF/ETA-02–2016–17, Department of Science and Technology, Govt. of India); *Indian Space Research Organization* (Vikram Sarabhai Space Centre, Space Technology Cell at IIT Kanpur); the Ministry of Education (erstwhile Ministry of Human Resources and Development), Govt. of India; and the *Yadupati Singhania Memorial Chair* of IIT Kanpur. AS acknowledges the *National Post-doctoral Fellowship*, and RH acknowledges the *Prime Minister Research Fellowship.* Authors extend a special gratitude to Prof. Arvind Agarwal for writing the foreword of this book. We acknowledge our host institutions including IIT Kanpur, IIT Patna, REC Banda, and Anod Plasma Spray Limited for facilitating research and encouragement in our respective departments.

**Ariharan S**
Department of Coating Processes
Alexander Dubček University of Trenčín, Slovakia

**Rubia Hassan**
Department of Materials Science and Engineering
Indian Institute of Technology Kanpur, India

**Alok Bhadauria**
Department of Materials Science and Engineering
Indian Institute of Technology Kanpur, India

**Ashutosh Tiwari**
Department of Materials Applied Science and Humanities (Physics)
Rajkiya Engineering College Banda, Uttar Pradesh, India

**Ritik Tandon**
Anod Plasma Spray Limited, Kanpur, India

**Anup Kumar Keshri**
Department of Metallurgical and Materials Engineering
Indian Institute of Technology Patna, India

**Kantesh Balani**
Department of Materials Science and Engineering
Indian Institute of Technology Kanpur, India

# Abbreviations

| | |
|---|---|
| AIP-PVD | Arc ion plating physical vapor deposition |
| $Al_2O_3$ | Alumina |
| APS | Air plasma spraying |
| Ar | Argon |
| ASPS | Axial suspension plasma spraying |
| BeO | Beryllium oxide |
| BET | Brunauer-Emmett-Teller |
| CA | Contact angle |
| CAH | Contact angle hysteresis |
| C/C | Carbon/carbon |
| CCD | Central composite design |
| $CeO_2$ | Cerium oxide |
| CFD | Computational fluid dynamics |
| CNT | Carbon nanotube |
| COC | Ceramic-on-ceramic |
| COP | Ceramic-on-polyethylene |
| CS | Critical speed |
| CTE | Coefficient of thermal expansion |
| CVD | Chemical vapor deposition |
| DC-TRAP | DC-triple plasma atomization process |
| D-gun | Detonation gun |
| Dj | Diamond jet |
| DOE | Design of experiments |
| EB-PVD | Electron beam physical vapor deposition |
| EDS | Energy dispersive spectroscopy |
| EIGA | Electrode induction gas atomization |
| EMI | Electromagnetic interference |
| EPMA | Electron probe micro-analyzer |
| FWHM | Full width half maximum |
| GAP | Gas atomization process |
| $H_2$ | Hydrogen |
| HAp | Hydroxyapatite |
| HCF | Hip contact forces |
| HCPEB | High-current pulsed electron beam |
| He | Helium |
| HIP | Hot isostatic pressing |
| HPCS | High-pressure cold spraying |
| HPPS | High-pressure plasma spraying |
| HVAF | High-velocity air fuel |
| HVOF | High-velocity oxy-fuel |
| Hyperpoc | Hydrogen pressure reducing powder coating |

| | |
|---|---|
| ICDD | International crystal diffraction data |
| ICIC | Inflammatory cell-induced corrosion |
| IGA | Inert gas atomization |
| IPP | In-flight particle pyrometer |
| IPS | Inert gas plasma spraying |
| LCS | Low-carbon steel |
| LDH | Layered double hydroxides |
| LFA | Low-friction arthroplasty |
| LPCS | Low-pressure cold spraying |
| LPPS | Low-pressure plasma system |
| LVDT | Linear variable differential transformer |
| MAC | Mechanically assisted corrosion |
| ML | Machine learning |
| MMC | Metal matrix composite |
| MOM | Metal-on-metal |
| MOP | Metal-on-polyethylene |
| MSR | Mechanical size reduction |
| $N_2$ | Nitrogen |
| N.A. | Numerical aperture |
| NC | Nanostructured coating |
| NDT | Non-destructive testing |
| NIR | Near infrared |
| OA | Osteoarthritis |
| OCP | Open circuit potential |
| ODS | Oxide-dispersion strengthened |
| OR | Outside range |
| PCA | Process control agent |
| PLD | Pulsed laser deposition |
| PREP | Plasma rotating electrode process |
| PSD | Particle size distribution |
| PVD | Physical vapor deposition |
| PWT | Plasma wind tunnel |
| RA | Rheumatoid arthritis |
| REO | Rare earth oxides |
| REP | Rotating electrode process |
| RFI | Radiofrequency interference |
| RF-IPAP | RF-induction plasma atomization process |
| RRR | Repeatability, reproducibility and reliability |
| SCP | Spray conversion processing |
| SDC | Spray and deposit control |
| SEM | Scanning electron microscope |
| SHS | Self-propagating high-temperature synthesis |
| SMAT | Surface mechanical attrition treatment |
| SOFC | Solid oxide fuel cells |
| SPPS | Suspension/solution precursor plasma spraying |

| | |
|---|---|
| SPS | Spark plasma sintering, suspension plasma spray |
| SS | Stainless steel |
| SVM | Support vector machine |
| T | Temperature |
| TBC | Thermal barrier coating |
| TEM | Transmission electron microscopy |
| THA | Total hip arthroplasty |
| THR | Total hip replacement |
| TKA | Total knee arthroplasty |
| TMAH | Tetramethylammonium hydroxide |
| TPS | Thermal protection system |
| TSC | Thermal-sprayed coating |
| TSR | Thermal shock resistance |
| UHMWPE | Ultra-high-molecular-weight polyethylene |
| UHTC | Ultra-high-temperature ceramics |
| VIGA | Vacuum inert gas atomization |
| VOC | Volatile organic compound |
| VPS | Vacuum plasma spraying |
| WC | Tungsten carbide |
| WCA | Water contact angle |
| XRD | X-ray diffraction |
| YSZ | Yttria-stabilized zirconia |
| $ZrO_2$ | Zirconia |

# Symbols

| | |
|---|---|
| $A$ | Cross-sectional area, projected area, empirical constant |
| $a_c$ | Diagonal length of the indent |
| $A_e$ | Exit area |
| $A_i$ | Area of nozzle exit |
| $A_t$ | Nozzle throat area |
| $\alpha$ | Linear expansion coefficient, thermal diffusivity |
| $\alpha_l$ | Mean thermal expansion coefficient |
| $\alpha_v$ | Fractional change in volume with temperature |
| $B$ | Biot number |
| $c$ | Half crack length, velocity of light |
| $C_d$ | Drag coefficient |
| $CI$ | Confidence interval |
| $C_i$ | Volume fraction |
| $c_i$ | Velocity of the liquid drop |
| $C_p$ | Specific heat capacity |
| $d$ | Sliding distance |
| $D$ | Theoretical density, flattened particle diameter |
| $d_0$ | Arithmetic mean |
| $d_a$ | Projected area diameter |
| $d_s$ | Surface diameter |
| $d_v$ | Volume diameter |
| $d_{32}$ | Sauter diameter |
| $D_p$ | Particle diameter |
| $\delta$ | Resolution of the microscope, offset |
| $E$ | Young's modulus |
| $E_x$ | Electric field |
| $E(\lambda)$ | Thermal flux |
| $e$ | Electron charge |
| $e_T$ | Dilatational strain |
| $\eta$ | Fluid viscosity, arithmetic mean of experimental dataset, Poisson's ratio, transformation zone factor |
| $F$ | Feret diameter, applied load |
| $F_N$ | Applied load |
| $F_T$ | Structure factor |
| $g$ | Acceleration due to gravity |
| $\gamma$ | Specific heat ratio, surface tension |
| $H$ | Substrate thickness, hardness |
| $h$ | Coefficient of heat transfer, coating thickness, droplet height, phase transformation zone |
| $H_v$ | Vickers hardness |
| $I$ | Solidification rate |

| | |
|---|---|
| $I_c$ | Index of crystallinity, crack length from the corner of the indentation |
| $I_i^P$ | Diffracted intensity |
| $I_o$ | Incident beam intensity |
| $j_I$ | Spherical Bessel function |
| $K$ | Sommerfeld number |
| $k$ | Wave number, thermal conductivity |
| $k_B$ | Boltzmann constant |
| $K_{Ic}$ | Fracture toughness |
| $L$ | Length at temperature $T$, thickness of the specimen |
| $L_c$ | Critical load |
| $L_f$ | Enthalpy of fusion |
| $LP$ | Lorentz polarization factor |
| $\lambda$ | Wavelength of light, scaling parameter |
| $M$ | Mach number, Martin diameter |
| $m$ | Rate of mass flow of gas, Weibull modulus |
| $m_e$ | Mass of electron |
| $M_g$ | Molar mass of the gas |
| $M_{hkl}$ | Multiplicity factor |
| $M_o$ | Total mass of the particles |
| $m_p^o$ | Mass flow rate |
| $n$ | Total number of measurements |
| $n_c$ | Critical speed |
| $n_e$ | Number density of electrons |
| $n_x$ | Ranking of a sample failing at strength $x$ |
| $P$ | Load |
| $P$ | Packing factor |
| $P_a$ | Atmospheric pressure |
| $p_e$ | Partial pressure of the plasma |
| $Pe$ | Exit pressure |
| $Pi$ | Inlet gas pressure, density of the $i$th phase |
| $P_m$ | Hardness of the softer counterpart |
| $P^P$ | Packing factor of the pure phase |
| $P_t$ | Throat pressure |
| $P(x)$ | Probability of the element under test to support a load $x$ |
| $\phi_c$ | Work function |
| $Q$ | Heat energy, heat flux |
| $Q_0$ | Specific energy of ignition |
| $q_l$ | Feed rate |
| $\rho$ | Density |
| $\rho_g$ | Gas density |
| $\rho_m$ | Fluid density |
| $\rho_p$ | Particle density |
| $R$ | Gas constant, radius of the ball mill container, radius of the goniometer |
| $R_a$ | Roughness |
| $RA$ | Relative accuracy |

$R_{et}$ Particle Reynolds number
$R_{th}$ Thermal contact resistance
$r/r_m$ Radius of the particle, radius of the ball, roughness, base radius
$\sigma$ Electron conductivity in plasma
$\sigma_d$ Standard deviation
$\sigma_f$ Fracture stress
$S,e$ Spalt thickness
$s$ Solid layer thickness, separation distance between the slits
$t$ Thickness of the coating
$T_c$ Cathode temperature
$T_s$ Substrate temperature
$T_t$ Melting point of the splat
$\Delta T$ Temperature change
$T_{gi}$ Initial gas temperature
$T_p$ Particle temperature
$\theta$ Water contact angle
$\Theta$ Solidification parameter
$u$ Fluid viscosity, coefficient of friction
$u_g$ Gas velocity
$u_{gi}$ Initial gas velocity
$u_p$ Velocity of particle
$V$ Volume of the particle
$V_a$ Unit cell volume
$V_f^t$ Volume fraction of tetragonal $ZrO_2$
$V_i$ Effective diffraction volume
$v$ Relative velocity of the torch (w.r.t. substrate)
$v$ Sonic velocity, fluid viscosity, frequency, Poisson's ratio
$v_l$ Injection velocity of the slurry
$v_i$ Specific volume of gas
$v_t$ Terminal velocity
$W$ Work of adhesion
$w$ Angular velocity
$W_e$ Weber number
$W_i$ Weight fraction
$X_a$ Amorphous fraction
$\overline{x}$ Mean
$X_i$ Weight fraction of *i*th component
$X_{ic}$ Refined weight fraction
$X_m$ Mass absorption coefficient
$x_o$ Characteristic stress
$X_{sc}$ Refined weight fraction of the internal standard
$X_s$ Weight fraction of the original internal standard
$x_u$ Minimum stress required for failure
$\xi$ Degree of flattening

# Introduction

*Fundamentals of Thermal Spraying* discusses the thermal spraying concept, and its uses are explained elaborately. Starting from the historical development of thermal spraying, the book is very comprehensive and is pitched more at the informed reader. Herein, self-explaining schematics highlight the conceptual notes and depict the concepts deeply for better understanding for graduates of programs including Materials Science and Engineering, Metallurgical Engineering, Mechanical Engineering, Production and Manufacturing Engineering, and Design. It also aims at the staff of companies that are specifically working on surface engineering with thermal spraying. The chapters include schematics, keeping undergraduate students in mind for its adoption as a textbook. Each aspect, from starting powder, spraying parameters, diagnostics and coating deposition, is provided for in-depth understanding. The major parts of the instrumentation, the spraying gun, which is the fundamental aspect of different thermal spraying conditions, are discussed as well. Solved examples, numerical problems and descriptive questions are included for self-assessment for the purpose of self-learning. Important case studies from domains including aerospace, biomedical, manufacturing, wettability and so on are highlighted with practical application of thermally sprayed coatings. Thus, the book can inspire and motivate readers to become involved in the research and development of thermal spray coatings.

# 1 Development of Thermal Spray Techniques and Applications

*Raghav Mittal, Ariharan S, Rubia Hassan, Alok Bhadauria and Kantesh Balani*

## CONTENTS



DOI: 10.1201/9781003321965-1

Surface engineering, a broad sub-discipline of Materials Science, encompasses several methods and techniques to modify the properties of a solid surface. The motivation to alter surface properties is to enhance their interaction with the environment, be it via force, heat, electricity, magnetic fields, biological interaction and so forth. Among others, thermal spraying is an important surface engineering technique with widespread industrial applications. This chapter covers the origins of thermal spraying, its evolution, the emergence of new spraying techniques and some of its leading applications across industries.

## 1.1 SCHOOP'S PATENT AND GUN DESIGN

The idea of spraying metallic particles to obtain homogeneous and continuous deposits on a substrate is accredited to the Swiss inventor Max Ulrich Schoop. In 1909, he filed two patents, one in Switzerland and another in Germany, explaining the design for spraying process (metal), which delivers metallic coating with highly dense structure [1]. Chronologically, flame spraying is the *first* spraying technique developed by Schoop. The patents were described in a process with an oxyacetylene welding torch. He claimed that the invention was based on an observation made while seeing his children using Flobert guns for sport shooting. The lead bullets shot on their garden wall had formed, more or less, a connected coating. This prospect of being able to alter the properties of a surface excited Schoop to conduct experiments and verify its viability. Particularly, the modified process was described with a thin wire of lead and tin as the feed. Later, the powdered materials were utilized in the advanced torches of hot jet flow, where it heated while being accelerated towards the substrate. In 1908, Schoop also patented an *electric arc–spraying* process to allow most of the metals and electrically conductive materials to be sprayed. The proper control on the electric arc–spraying process could be able to coat zinc on steel or stainless steel and lead to successful development of the metallizing industry for the prevention of the corrosion of structural components. After extensive research, Schoop explained the underlying mechanism as follows:

> "The fundamental principle of the present invention consists in imparting a high velocity to solid metal or metal oxide particles projected towards a non-porous surface not coated with adhesive, and not heated to the melting point of the particles, an intimate connection between the particles and the receptive surface being obtained by means of the violent impact."

Further, Schoop suggested that the action could be governed via accumulation of the external heat [2]. Some of the ways of achieving this included heating of:

- An agent with pressure carrying particles (vapor, compressed air or gas);
- The surface of the substrate (to be coated);
- Powder particles (before being discharged into the gun or during their flight).

For example, the powder particles added in stream of gases, which are ignited just as they exit the discharge nozzle. In this manner, the enthalpy content of the gases would be transferred to the particles. As a result, these particles become more ductile and hence susceptible to deformation upon impact. Along similar lines, Schoop argued that the mixture containing air and metal could be heated with the help of a tube (generally electrically heated) or a flame that can be arranged at the exit from the delivery nozzle. For a flame, however, the difficulty would arise in ensuring homogeneous heating of the particles. Thus, Schoop's vision is commended for acknowledging that to deposit, it is not necessary for all particles to be in an entirely liquid state before impinging the surface. If subjected to adequate pressure, the particles (except those electrically non-conductive) are rendered a plasticity which is quite sufficient to form a soldering or fusion of particles among each other, which results in the formation of a uniform and continuous metallic layer. To execute the process described, he devised the apparatus illustrated in Figure 1.1. For inspiration, in those times it was natural for him to turn to appliances, such as the sand blasting apparatus used for surface treatment. The only resemblance Schoop's design had, however, was regarding the method of working and not in regard to the action. The effects produced by the two were completely different.

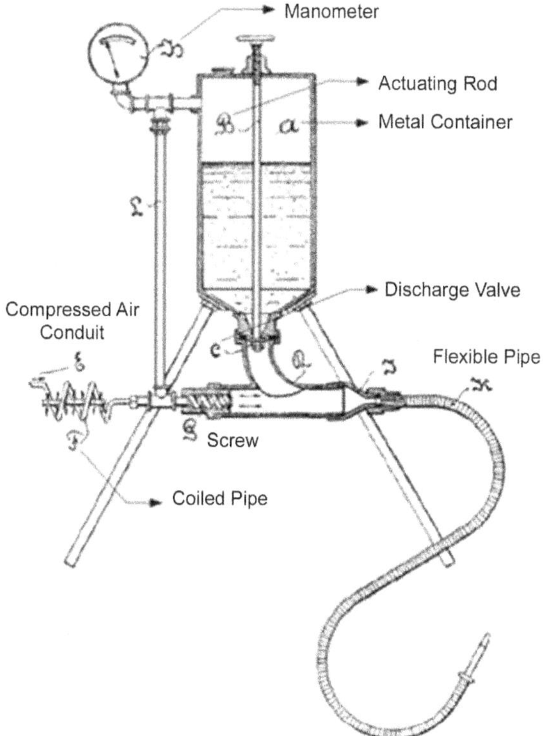


**FIGURE 1.1** Schoop's apparatus for projecting powdered solid particles at very high pressures.

*Source*: Courtesy of JTST Historical Patent #22 United Kingdom patent office [3].

In engineering parts and components, the surface bears the brunt of all major interactions during use. It is the surface that first negotiates with the external stimuli; the bulk's role comes subsequently. This makes it highly vulnerable to a list of surface phenomena which could cause potential damage or degradation of the component. Functionalized coatings may be deposited in order to prolong the life of components and augment desired properties. Some of the essential motivations for coating structural parts include:

1. Improved functionality, for example, biocompatibility in titanium-based hip implants;
2. Improved component life by increasing wear, erosion and corrosion resistance;
3. The solution to previously insurmountable engineering problems;
4. Enhanced part life by salvaging worn-out parts;
5. Conservation of scarce material resources and reduction of component costs;
6. Reduction of power consumption and effluent output.

## 1.2 EVOLUTION OF THERMAL SPRAY TECHNIQUES

Thermal spray processes are explained as heating the particles to a fully molten or partially molten state followed by accelerating them toward a prepared surface with the help of pressurized gases or liquid atomization jets [3]. The high pressure from gases provides momentum to the particles, causing them to distort, flatten and subsequently fragment into smaller drops and ligaments. Upon impact on the substrate, these drops form splats which cool down to form lamellar structures. Successive impact of particles results in layers of many such lamellae which form the coating. The rate of cooling is in the order of $10^6$ $Ks^{-1}$ for metallic particles, though they generally exceed this magnitude. The nature of the bond could be either mechanical (topographical interlocking) or metallurgical (atomic bonding/diffusion).

## 1.3 DEVELOPMENT OF NEW PROCESSES

Today there exist several variants of thermal spray techniques which utilize either a high degree of superheating or very high particle velocities to deposit lamellar coatings. Densities as high as 99% can be achieved through these processes. The significant expansion of the thermal spray technology did not occur after the first development of the technique until 1945, more than three decades later. However, powder, electric arc and plasma spraying were developed and subsequently introduced in that development [4]. Improvements in thermal spray processes have been made with the basic operating principles developed by Schoop and other scientists. After World War II, powders are fed directly into the combustion flames of the system to generate high-velocity jets. The high-velocity jets will alter the characteristics of the coated materials. Other than the development in the thermal spray processing, additionally, tailoring of starting feedstock materials have improved the process. Figure 1.2 summarizes the development history

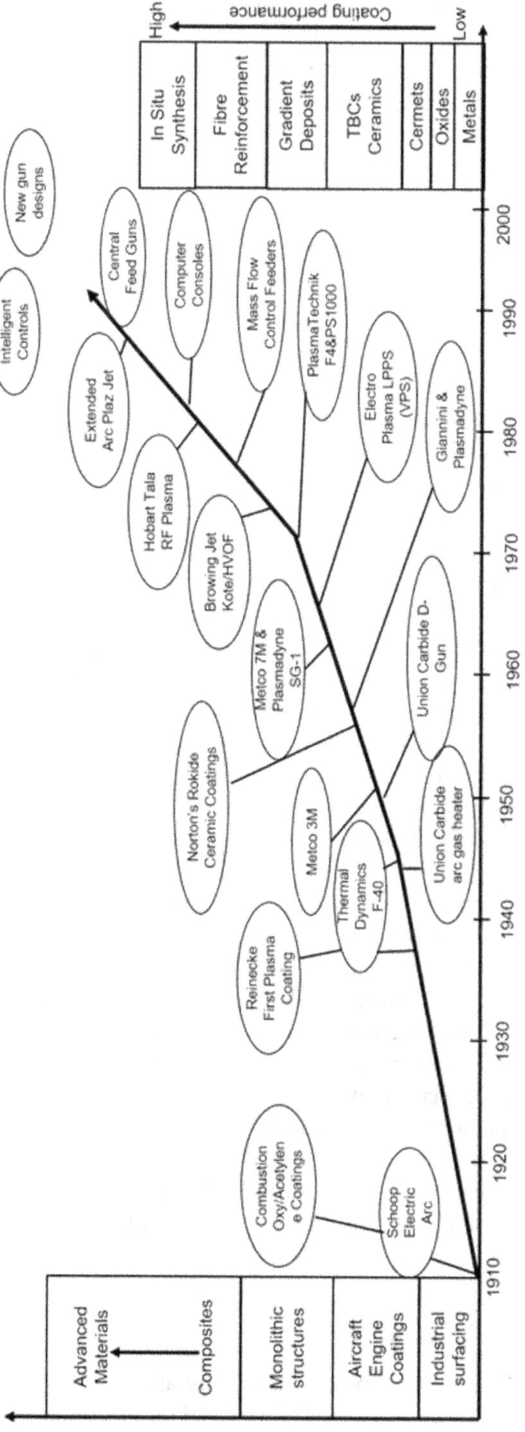


**FIGURE 1.2** Timeline thermal spray developments, equipment and processes.

*Source:* Adapted and redrawn from [5].

of the thermal spray process, applications and materials, highlighting significant milestones in the area.

The selection and process parameters can be tweaked to realize porous coatings as well, which can be classified based on their source of heating as follows:

1. Flame spray: conventional and powder
2. Electric arc wire spray
3. Plasma spray
4. High-velocity oxy-fuel spray
5. Detonation gun spray.

### 1.3.1 Flame Spray

The evolution of combustion processes was driven by development in the scientific and technical innovations and the needs of the market. Historically three types were used: (1) flame spraying, (2) high-velocity oxy-fuel flame and (3) detonation spray. Flame spraying predominantly uses oxyacetylene torches that can achieve a premixed maximum combustion temperature of about 3,000 °C with flame velocities below 100 m/s. Spray gases common in use include acetylene, propane or hydrogen gas, along with oxygen. Generally, the sprays are fed into the spraying gun in the axial direction. The combustion spray guns are designed to perform feeding, acceleration, heating and providing direction to the flow of the material. In the powder flame spray, the spraying materials in powder form will be fed into the vicinity of flame. The powder particles are injected radially in the spray torch. Since most of the materials are available in powder form, the powder flame spray offers versatility in the aspect of utilization.

In the wire flame spray, the wire is axially fed into the spray torch so that it melts in the gaseous oxy-fuel flame. The only difference between the powder and wire flame–spray processes are the feeding materials. The feeding materials should be in the form of wire in wire flame spray. The flame temperatures and characteristics are determined by the ratio of oxygen to fuel gas and their pressures. The ratio determines whether the flame is oxidizing, neutral or carburizing. The molten droplets atomize upon interaction with the compressed air that also directs them towards the substrate. The shortcoming of the conventional combustion spraying process is that the maximum gas temperature is provided by the adiabatic flame temperature of the combustion gas mixture. So, it is limited for spraying materials with high melting temperatures, such as ceramics. Further, the environment of the combustion process can lead to chemical reactions between the spraying materials and the combustion gases. Still, flame spraying is a commonly used coating technique because it is economical.

### 1.3.2 Electric Arc Wire Spray

The electric arc (wire-arc) spray process uses wire as a feeder material. Heating is obtained by creating an arc between the wire and the electrode. In this process, an arc is generated when two oppositely charged metallic wires are brought into contact. The resultant heat is sufficient to melt the feed, which is subsequently atomized and projected by compressed air onto a substrate ready to be coated. Exposure of metals in the harsh environment during the electric arc spraying was better controlled compared to that of flame spraying.

The coating composition can be tailored by the use of dissimilar metals or cored wires. In general, electric arc wire spraying exhibits higher bond strength than flame spraying. Electric arc wire spraying is advantageous over flame spraying in terms of providing a higher bond strength, deposition rate and deposition efficiency.

### 1.3.3 Plasma Spray

Interest in thermal spraying applications grew with the plasma spraying process introduced for the first time in 1939 by Reinecke. Prior to the development of plasma spraying, a confined arc gas heater was established. Primarily, it was developed to assist in cutting and joining as similar to the welding technology. An earlier plasma spray coating process was first demonstrated with the powders injected as the feed into a high-temperature plasma arc. The plasma arc could melt powder particles and accelerate the molten particles toward a substrate simultaneously. One of the advantages of plasma spraying was significantly higher process temperatures than that of the combustion spray. Also, there is no dependence of the feed material on the heat source, whereas feed material is completely dependent in the wire arc spray process. A high-frequency electric arc initiated between cathode (made of tungsten) and anode (water-cooled copper) ionizes the gas flowing in between, creating a high-pressure plasma. It is sustained using DC power. The gases considered for plasma generation include Ar, He, $H_2$ and $N_2$. At temperatures as high as 15,000 °C, the particles melt and are directed onto the workpiece.

Plasma processes have high heating potential and particle speed, which improve the coating quality. In the early 1980s, subsequent growth occurred in the applications of high-technology thermal spray. In the late 1950s, much of the thermal spray equipment technology saw its parallel and separate advancement in flame/combustion, electric arc and plasma spray. Successively, the detonation gun (D-Gun) process was introduced from the advancement in the arc gas heater technology for powder-spray applications. In the early 1970s, the plasma spray emerged as one of the most extensively used advanced technologies among thermal spray coatings. Thus, the developments in the plasma spray equipment as well as materials were aimed at for further advancement. Initially, the plasma spraying guns were designed with the maximum power of 40 kW, but the advancement in the equipment design improved the maximum to 250 kW. Also, continuous utilization of the system was enabled with effective water-cooling design and electrodes along with the use of high arc voltage. Simultaneously, plasma gas flow rates in the plasma guns have increased, and the speed of gas exit has promptly improved from subsonic to supersonic values. Successive increases in the particle speeds due to high gas exit from the plasma gun increased the density and bond strength of the deposited coating. Moreover, the development rate of plasma spraying device is not so much. So, the research and development in thermal spray now focus on the process control, computer-controlled supports, real-time sensors for the process control, and automated or robot handling systems.

### 1.3.4 High-Velocity Oxy-Fuel Spray

Combustion spraying, with the exception of D-Gun, has caused an improvement with the development of high-velocity oxy fuel (HVOF) spraying system. The HVOF spray has enhanced the combustion spray jet by improving the particle temperatures. Also, the high speeds of molten particles are achieved through confinement of combusting gases and particles. The resulting high particle velocity and temperatures enhance the coating density and bond strengths. HVOF spraying is one of the coating processes that challenge the D-Gun and plasma spray coating processes in the coating technology. It is also reported that high particle speeds reduce the overheating of the particles. Ultimately, it leads to prevention of the oxidation and decarburization (in the case of carbides) which are mostly seen in the plasma-sprayed coatings.

Fuel gas, including propane, propylene, methyl-acetylene propadiene propane (MAPP), hydrogen or liquid kerosene, is burned in a chamber with the help of oxygen. Powder particles are introduced in the products of this combustion and projected towards the substrate with high-velocity compressed air. As the gas expands through a nozzle, it attains supersonic velocity which helps ensure dense coatings. The deposition mechanism is more dependent on plastic deformation of the particle rather than complete melting.

### 1.3.5 Detonation Gun Spray

Initially design of the D-Gun proved that it is capable of producing dense, reliable and adherent coatings with high bond strength among the thermal spray technologies. Till date, no process has emerged over the D-Gun process to produce efficient coatings. However, the recently developed HVOF coatings are challenging at this juncture. Before the beginning of the D-Gun process, aero-engine did not mention the type of thermal spray technique involved for the coating. It is just because of relatively high porosity and poor adherence of the coatings with the aero-engine components. But the plasma spray had experienced a similar application in many high-temperature materials with its high materials flexibility to process.

A mixture of oxygen and fuel gas (generally acetylene) is ignited through high-frequency cyclic (~10 times per second) sparks to cause detonations. The detonation waves so produced heat the particles and accelerate the powders to supersonic velocities as a result of their high temperature and high pressure. As the shock wave expands in all directions, the barrel is purged with nitrogen gas to avoid choking of the powder feed from the back end. Based on the application of thermal spray, the modified plasma arc heaters were utilized in the aerospace industry to improve coatings characteristics.

## 1.4 CLASSIFICATION OF THERMAL SPRAY PROCESSES

Thermal spray deposition techniques are classified (Figure 1.3) into four types: flame spraying, plasma arc spraying, electric arc spraying and kinetic spraying. Many of these major spraying process shall be subdivided based on the environment chosen, type of feed and so forth. Each of these processes includes many more subcategories with their own characteristics. A coating is characterized by its microstructure (porosity and inclusions), homogeneity, thickness and bond strength. A suitable coating process needs to be designated for particular materials and specific application with special characteristics.

## 1.5 COMPARISON WITH OTHER COATING TECHNIQUES

Surface coatings can be realized via many routes, all of which have their own merits and demerits. Based on the coating's thickness, structural homogeneity, density, adhesion strength, other physical/chemical/mechanical properties and costs, we select an appropriate deposition technique. What distinguishes thermal

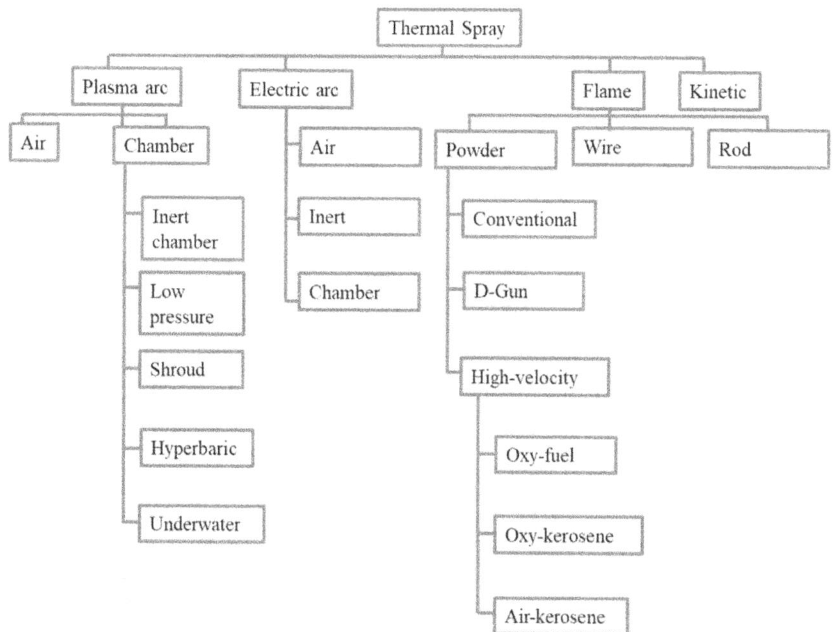


**FIGURE 1.3** Hierarchy of thermal spray techniques.

*Source*: Adapted and redrawn from [6].

spraying from other coating approaches is that it does not deposit material in the form of individual atoms, molecules or ions; thermal spraying processes are non-atomistic. What is deposited instead are large size (about a few microns) molten/semi-molten liquid droplets or powder particles which are accelerated with compressed air. The material can be fed in a variety of forms: wire, powder or rods. Other than plasma, the heating sources are also different, as they include electric arc or oxy-fuel combustion. Here we discuss alternative processes, their working principles, main characteristics of the microstructure and their applications.

### 1.5.1 Electro/Electroless Plating

*Electro-deposition*: Under the application of an external voltage, metal ions from the solution are deposited on the cathode surface. Thus, the part to be plated acts as a cathode (site of reduction of metal ions) and the source metal as an anode (site of oxidation/dissolution). The electrodes remain immersed in an electrolyte which contains at least one dissolved salt or other ions permitting the flow of electricity. Ions are continuously replenished in the electrolyte by the dissolution of atoms/ions from anode. The utility of the process lies in the enhancement of chemical (corrosion resistance), physical (appearance, color, thickness, conductivity) and mechanical performance (tensile strength, hardness, wear resistance, etc.).

*Electroless deposition*: The process is based on chemical reduction (by a reducing agent) of ions in aqueous solution which form a coating without electrical energy. It is catalytic, wherein the substrate is such that it allows specific deposition to initiate and continue without spontaneous decomposition of electrolyte (Figure 1.4). Areas of application of such electroless coatings include temperature

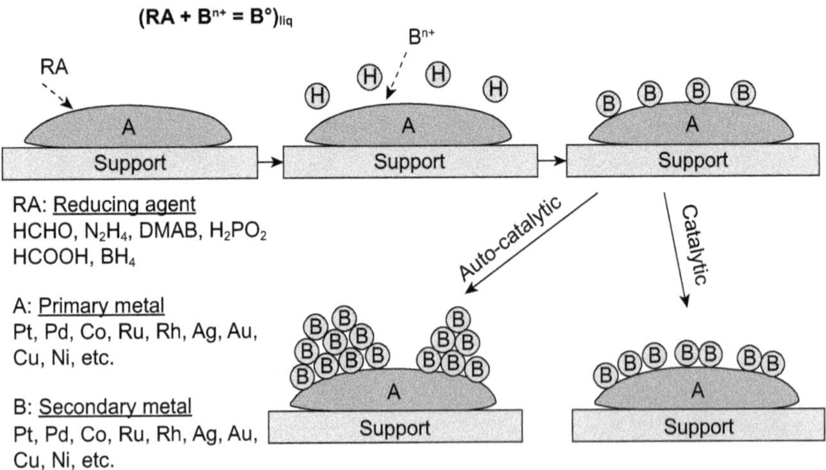


**FIGURE 1.4** Schematic of the stages involved in electroless deposition.

*Source*: Adapted and redrawn from [7].

sensors, valves for fluid handling, printed circuits and resistors, magnetic tapes, gears crankshafts, hydraulic cylinders and so on.

*Immersion deposition*: The spontaneous nature of the reaction (driven only by electrode potential) brings about the displacement of metal ions from the anode to the cathode. Therefore, the process is carried out in the absence of an external power source. Disadvantages of the deposit properties include that it is difficult to control and may be porous and poorly adherent.

### 1.5.2 Chemical Vapor Deposition

Chemical vapor deposition (CVD) involves the formation of non-volatile solid film by reaction of vapor phase chemical or reactants containing required species (Figure 1.5). Gaseous precursors—such as fluorides, chlorides, iodides, bromides, ammonia complexes, phosphorus and hydrocarbons—are thermally decomposed and transported via forced convection to the reaction region (i.e., the substrate). Once the reactant reaches the substrate, the following steps take place to ensure deposition [8]:

1. Diffusion of reactant to surface
2. Absorption of reactant to surface
3. Chemical reaction
4. Desorption of gas by-products
5. Out-diffusion of by-product gas.

The process is usually carried out at relatively lower pressures (i.e., between 13 and 100 kPa) to avoid unwanted reactions and develop uniform coating thickness

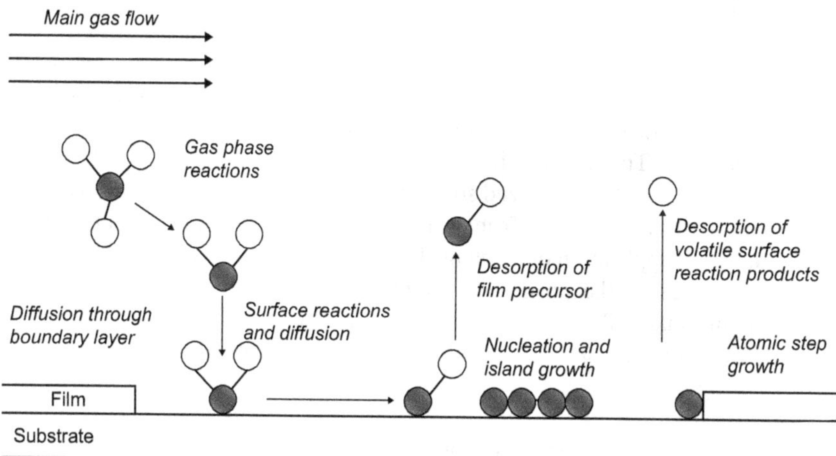


**FIGURE 1.5** Schematic of steps involved in the CVD process.

*Source*: Adapted and redrawn from [9].

on the substrate. A major advantage of the process is that it is omnidirectional, making it easy for the mixtures of gases to easily go through blind holes to support access to the most intricate features, even also in internal regions of the porous components.

CVD is a very versatile process to provide powders, fibers, thin films, and monolithic parts in addition to producing coatings. The CVD process is applicable to produce several varieties of metallic and non-metallic materials (e.g., carbon, silicon); also it produces oxides, nitrides, borides, sulfides, silicides, carbides and intermetallics. The parameters for controlling the microstructure are temperature of substrate and chamber and the degree of supersaturation. The elevated temperature of substrate allows diffusion between substrate and coating resulting in better adhesion due to diffusion bonding.

### 1.5.3 Physical Vapor Deposition

The physical vapor deposition (PVD) process comprises evaporation, sputtering and ion plating (Figure 1.6). This process is the line-of-sight technique, where processes are generally carried out in a vacuum at temperatures between 150 °C and 500 °C. Here as well, the heating enhances diffusion and thus improves bonding, while the low pressures decrease the deposition rates [8]. PVD includes:

1. *Evaporation*: In this process, solids are vaporized by different means (resistive, laser, electron beam, arc discharge, etc.) and condensed to the substrate after almost collision-less line of sight transport. Ceramics, metals and alloys can be deposited by tailoring the process and arranging reaction with gases during the dwell time of vapor particles. Limitations include non-uniformity of thickness because of the dependence on the angles and radius of the substrate surface which is defined by the cosine law. This hurdle is overcome by ensuring of multiple sources or by subjecting substrate to oscillatory movement. The coating is characterized by relatively weak adhesion since the particles impact the substrate with relatively low kinetic energies (0.1–0.5 eV).
2. *Sputtering*: This process involves ejecting atoms from a "target" with the aid of a highly energized sputtering gas (~100s of eV, up to 1000 eV). The sputtered/chipped-off material flux is directed towards a substrate under the application of a voltage between target (cathode) and substrate (anode). Since the positive terminal is on the substrate side, nonconducting materials such as ceramics cannot be deposited by DC alone. Radiofrequency or reactive DC sputtering are invoked to deposit such coatings.

In reactive DC sputtering, the sputtered-off metal atoms are made to react with the gases to make the desired ceramic materials (e.g., aluminum reacts with oxygen and forms $Al_2O_3$). This reaction is further improved with a radiofrequency coil, disposed where reactive gases are injected near substrate. Furthermore, the

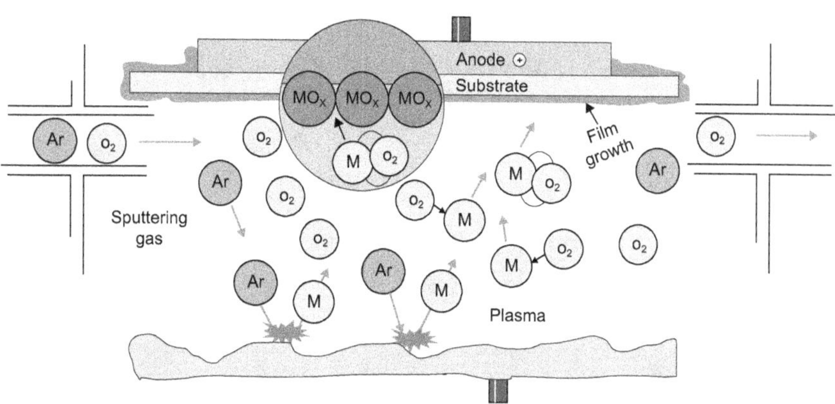


**FIGURE 1.6** Schematic showing the steps involved in the PVD process.

*Source*: Adapted and redrawn from [10].

pressure is required to be 0.1–1 Pa to increase the impingement rate of atoms (i.e., their collision frequencies) to improve the reaction kinetics.

When a magnetic field is used to concentrate high-energy ions on the target to eject atoms, it is known as magnetron sputtering. The field increases the electron density and also that of the sputtered flux of ions. It is widely used to improve the corrosion resistance of steel and Al, Ti and Mg alloys.

### 1.5.4 Pulsed Laser Deposition

The pulsed laser deposition (PLD) technique uses a laser beam with high power (~$10^8$ W/cm$^2$) to melt, vaporize and ionize the material from the target surface [11] (Figure 1.7). The beam is focusing periodically onto the substrate with laser pulses of a few tens of nanoseconds and frequencies of a few tens of Hz.

The interaction with the laser pulse produces an intense plasma plume, the length of which extends to the substrate. The evaporated ions collect at the substrate and condense to form a thin film coating. This technique can be applied to produce insulating and superconducting circuit components for improving the biocompatibility and wear properties for medical applications.

A comparative characteristic of various coating techniques, presented in Table 1.1, emphasizes the versatility that is offered by selecting appropriate deposition techniques. Comparison in terms of equipment and operation cost, process environment, thermal damage, surface finish and choice of coating material highlights the need to identify important variables before choosing the appropriate criteria and selecting the deposition technique. Being able to deposit almost any materials, as well as the ease of depositing thicker coatings (> 50 µm), makes thermal spraying an attractive technique for scaling up and industrial reclamation of large structural parts.

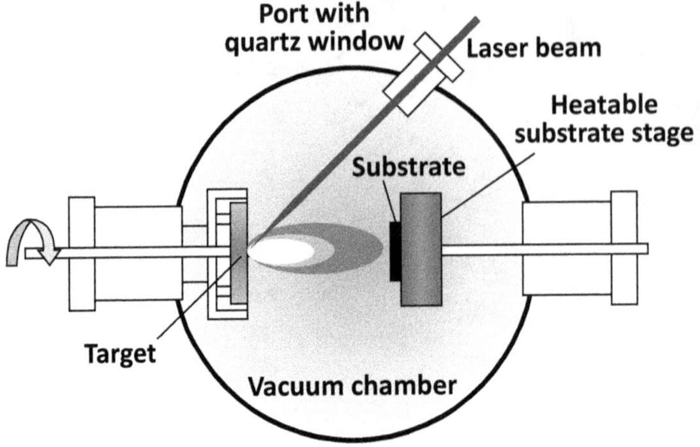


**FIGURE 1.7** Schematic of pulsed laser deposition.

*Source*: Reprinted with permission from [11].

**TABLE 1.1**
**Various Physical Characteristics of Different Coating Techniques [1, 6]**

| Characteristics | Electro/ electroless Plating | Chemical Vapor Deposition | Physical Vapor Deposition | Thermal Spray |
|---|---|---|---|---|
| **Equipment cost** | Low | Moderate | Moderate to high | Moderate to high |
| **Operating cost** | Low | Low to moderate | Moderate to high | Low to high |
| **Process environment** | Aqueous solutions | Atmospheric to medium vacuum | Hard vacuum | Atmospheric to soft vacuum |
| **Coating geometry** | Omnidirectional | Omnidirectional | Line of sight | Line of sight |
| **Coating thickness** | Moderate to thick (10 μm-mm) | Thin to thick (10 μm-mm) | Very thin to moderate | Line of sight thick (50 μm-cm) |
| **Substrate temperature** | Low | Moderate to high | Low | Low to moderate |
| **Adherence** | Moderate mechanical bond to very good chemical bond | Very good chemical bond to excellent diffusion bond | Moderate mechanical bond to good chemical bond | Good mechanical bond |
| **Surface finish** | Moderately coarse to glossy | Smooth to glossy | Smooth to high gloss | Coarse to smooth |
| **Coating materials** | Metals | Metals, ceramics, polymers | Metals, ceramics, polymers | Metals, cermets, ceramics, polymers |

## 1.6 APPLICATIONS OF THERMAL SPRAY

Thermal spray offers to deposit almost all type of materials with almost all type of substrate; this is why it covers a wide variety of applications (Figure 1.8). Today, its application has become more critical for improving the performance and functionality of a myriad of substrates in the core engineering sector. From household items like frying pans to giant structures like bridges, biological implants for arthroplasty to paper manufacturing rolls, turbine engine blades to the same engine's combustion chamber, valves to heat exchangers, impeller shafts to drilling equipment, the engineering techniques to deposit coatings have become indispensable. Owing to the reliability and credibility of this technology, manufacturers rely heavily on them to provide good corrosion, wear, erosion and thermal protection; refurbishing of damaged/worn out components; improved electrical properties; biocompatibility; and even electrochemical activity.

Recent developments in thermal spraying have achieved ingenious feats such as production of freestanding structures, ceramic/metal matrix composites, superconducting oxides and so on. Some of the other common applications of thermal spraying are discussed in detail below.

### 1.6.1 Wear/Erosion Resistance

Any interaction/relative motion between two surfaces or bounding surfaces of solids causes unwanted removal of material leading to dimensional loss (mostly lateral) [13, 14]. As a result, significant financial losses are incurred within most

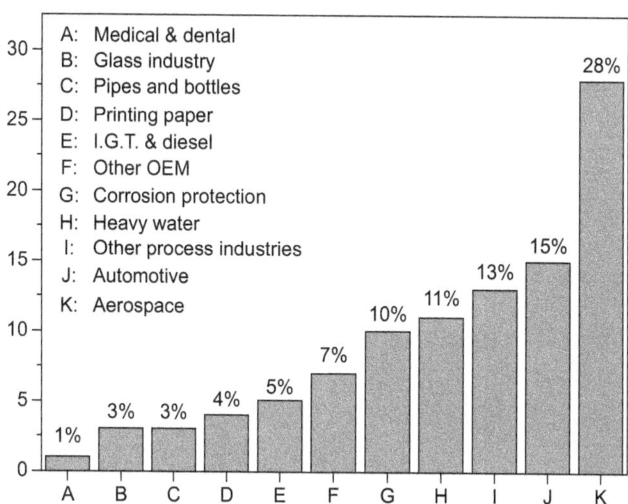


**FIGURE 1.8** Bar graph presenting the thermal spray technology applications in Europe in 2001.

*Source*: Adapted and redrawn from [12].

industry segments. Wear-resistant coatings not only provide an increase in performance efficiency but also prolong the life of components. Additionally, we also benefit from the enhanced harness and increased ductility which provides the ability to avoid micro-cracking or increased surface fatigue. To find an appropriate material for such a coating, we must understand the different types of wear mechanisms, which include abrasive, fretting, erosive, adhesive and so on. Some of the most common examples of wear and abrasion resistant coatings include:

- Tungsten carbide: an excellent replacement for hard chrome plating in construction machinery;
- Non-stick materials coated on frying pans: a hard coating with low-friction coefficients;
- Aluminum oxide: resists wear by fibers, and also resists erosion in high temperatures.

### 1.6.2 Oxidation/Corrosion Resistance

Corrosion resistant coatings prevent or delay surface deterioration, which is triggered by reaction to severe environmental conditions and chemical/oxidation action. They can be distinguished in the following categories [15, 16]:

*Cathodic*: Such coatings comprise a layer of a material which is electrochemically more active than the metal surface, which is vulnerable to corrosion. Owing to their higher "active voltage" than the substrate metal, cathodic coatings are preferentially consumed, playing a sacrificial role.

*Barrier*: Coatings in this category consist of several layers of corrosion resistance materials, which help to improve the corrosion resistance (act as a barrier) applied on a corrosion vulnerable surface. Barrier coatings can be made more effective by decreasing its porosity levels and using extra sealers for enhanced protection.

*Corrosion and Wear*: Cermets combine the hardness and wear/abrasion resistance of ceramics (say carbides) with the corrosion resistance of nickel or chrome binders, tackling both together. There are also several other ceramics materials (e.g., $Al_2O_3$–$TiO_2$ and $Cr_2O_3$); these are hard and also act as corrosion resistance materials.

### 1.6.3 Thermal Barrier Coatings

Thermal barrier coatings (TBCs) are refractory-oxide ceramic coatings [17, 18] which provide thermal insulation to metallic components operating at elevated temperatures (Figure 1.9). Their success has resulted in a steep increase in the operating temperatures of the metallic turbine blades beyond the suggested limits. Associated with their incorporation has been a dramatic improvement in the efficiency and performance of gas turbine engines.

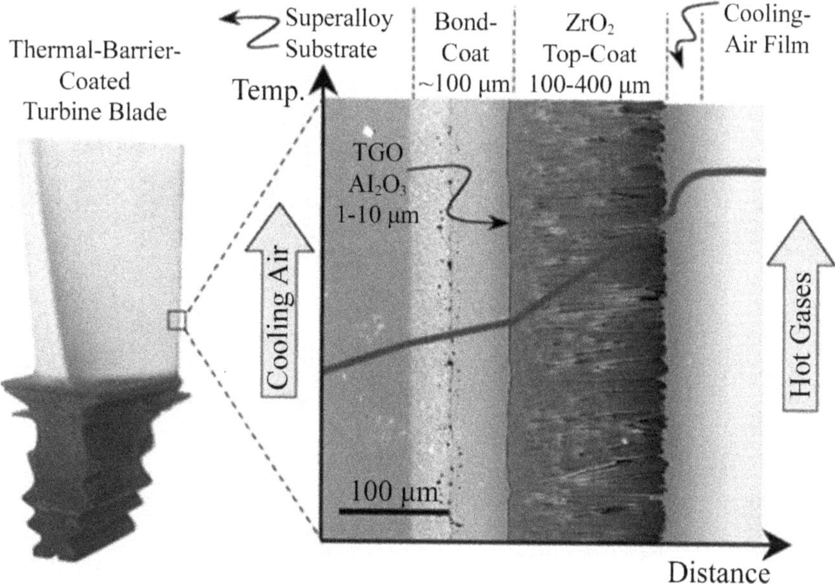


**FIGURE 1.9** Layers included in the TBC of a turbine blade.

*Source*: Reprinted with permission from [19].

Materials selection for turbine blades demands low thermal conductivity, high coefficient of thermal expansion and good erosion and oxidation resistance.

Some of the commonly used coating materials include MCrAlY (M = Fe, Co, Ni) alloy, zirconia ($ZrO_2$), yttria-stabilized zirconia (YSZ), alumina ($Al_2O_3$) and so on. As illustrated in Figure 1.9, TBCs these days are vertically segmented to accommodate the mismatch in expansion coefficients which generate stress during the heating/cooling cycles. Other alternative materials are also being explored—for example, pyrochlore structure-based ternary metallic oxides, which unlike zirconia shows no phase transformation at high temperature.

### 1.6.4 Biomaterial Coatings

Based on the extent to which they interact with biological systems, the coatings are broadly classified into bioactive, biocompatible or bio-inert [20]. Bioactive coatings try to emulate the characteristics of the body part they augment/repair/replace (therapeutic). For example, hydroxyapatite (HAp), or tricalcium phosphate, is similar to bone material in the sense that it facilitates new bone growth attached to a medical implant (Figure 1.10). Biocompatible materials are generally composed of titanium alloys because of their surface characteristics, such as surface texture (porosity), steric hindrance, binding sites and hydrophobicity (wetting). They are optimized to elicit the desired cellular response.

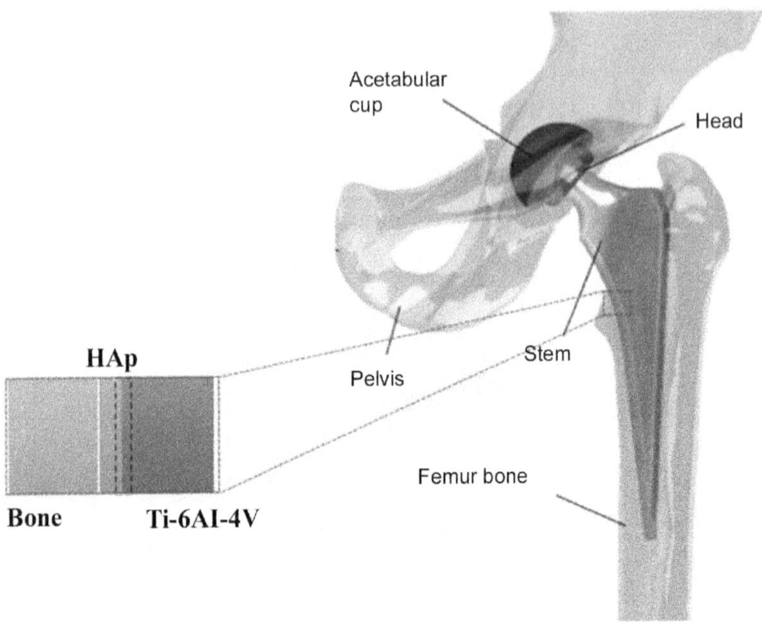


**FIGURE 1.10** Apatite coating on hip implant to improve biocompatibility and hip implant placed at the acetabulofemoral joint as a replacement.

*Source*: Open access from [21].

In order to enhance the limited stability of HAp and its thermally decomposed products, novel plasma-sprayed coatings of $CaO–P_2O_5–TiO_2–ZrO_2$ are being used nowadays. These stoichiometric compositions improve mechanical integrity and chemical stability as compared to that of the pure calcium phosphates. For example, $CaTiZr_3(PO_4)_6$ displays at least an order of magnitude lower solubility than that of hydroxyapatite in simulated body fluids. Furthermore, it shows better adhesion to the substrate of Ti6Al4V. The main applications are in orthopedics (bone implants) and orthodontics (dental implants).

### 1.6.5 Electrical Conductive Coatings and Heat Transfer Improvement

Materials such as silver, copper, aluminum, molybdenum, tin alloys and bronze alloys are used to make electrical contacts. Their conductivity, however, depends on both the material and the spray technique adopted; it is generally between 40% and 90% of the bulk. They can be applied on a wide range of materials including ceramics and polymer composites. Crucial areas of their application include heating elements, contact points, slip rings, flexible circuits, in-situ thermocouples, commutator segments and so on. Similarly, copper or aluminum is used for high-thermal conductivity coatings. The ceramic beryllium oxide (BeO) is deployed in areas where electrical insulation is necessary but thermal conductivity is desired.

### 1.6.6 Electromagnetic Shielding

Electromagnetic interference (EMI) and radiofrequency interference (RFI) are two major problems that can severely affect other electronic components or distort low-level signals. Application of metals like aluminum and zinc on gasket areas of electronics enclosures (e.g., doors, plates, cabinets), and filters can be employed for EMI and RFI shielding. These not only prevent device malfunctions but also provide attenuation and good point-to-point resistivity. Some common products which are coated with EMI and RFI coatings include medical devices (e.g., X-ray equipment), marine electronic equipment, computer motherboards, process control components and so forth.

### 1.6.7 Electrically Resistive or Insulating Coatings

Dielectrics, oxygen sensors, high-temperature strain gauges, and heating coil insulation have insulating surfaces (mostly oxides)—for example, alumina ($Al_2O_3$) and alumina + titania ($Al_2O_3$ + $TiO_2$) coatings commonly used for ultra-high resistance and regulating conductivity. Engine components, crimper wheels, cable channels, corona rolls, circuit segments and rolling contact bearing shells are some of the most common parts where they are applied [22].

### 1.6.8 Electrochemically Active Coatings

An important application of spray coatings is found in solid oxide fuel cells (SOFC) [23]. SOFCs generate electricity (with negligible footprint) by oxidation of fuel across electrolytes which are impermeable to gas but conduct ions. Their functional layers (i.e., anode, cathode, and electrolyte) are made of ceramic materials. Since high conductivity is critical for enhanced performance, the cells are operated at elevated temperatures. For this, the electrodes must be porous and composite to provide selective permeability to gases, while maximizing the triple boundary area where electrolyte, fuel, and anode are present. Typically, YSZ coatings are plasma sprayed onto metallic substrates to obtain such functionality.

### 1.6.9 Restoration and Repair

Equipment and components in every industry are subjected to wear, erosion and in some cases even corrosion. Prolonged interaction (physical or chemical) at the interface (mostly between surfaces) causes a decline in their performance and loss of dimensional integrity. Thermal spray techniques help salvage these parts at a cost which is much less than the alternative of replacement. Common coating materials used for repair include the following [24]:

- Metals and alloys: Mo, Al, brass, bronze, Zn, SS, Hastelloy, nickel alloys, steel;

- For added wear, abrasion and erosion resistance: tungsten carbide, Ni-Co-Br alloys, stellites (Co-Cr alloys), ceramics;
- For added corrosion resistance: Hastelloy, ceramics, Monel (Ni-Cu), SS, chrome alloys.

### 1.6.10 Abradable/Abrasive Coatings

A major application of abradable/abrasive coatings is found in gas turbine engines. Here, the engine casing and the tips of rotating turbine blades are designed in such a manner that there is precise clearance control between them. To seal the gases within a particular chamber and prevent them from bypassing the blades, the blade tips are coated with abrasion-resistant materials while the internal wall of the casing is coated with a relatively soft abradable material. This way, the turbine blades cut grooves in the soft coating and block the flow of gases. This technique has demonstrated an increase in the engine performance. These coatings are typically nickel/graphite, nickel/bentonite, aluminum/polyester, or other hard/soft material combinations. To reduce the wear of blade tips, abrasion-resistant materials such as ceramics and carbides may be applied as coatings.

## 1.7 SUMMARY

In summary, thermal spraying techniques offer an economically effective way to improve the surface properties of a solid for any intended application. As discussed, these techniques find application including wear resistance, corrosion resistance, thermal barrier properties, biological compatibility and electrical insulation, and even facilitating electrochemical activity. Today, there exists a range of applications where the technique is deployed to also restore old components by making up for the wear and tear losses during erosive/tribological/corrosive interactions. This chapter has reviewed the emergence, the evolution and the current scenario associated with thermal spraying techniques. Subsequent chapters discuss the different techniques in detail, covering the issues and merits associated with each.

**Questions for Self-Analysis**

1. What are some of the leading motivations for coating structural parts and components?
2. What is thermal spraying? How is it different from other coating techniques?
3. How are thermal spray processes classified?
4. What materials can be thermally sprayed?
5. What are some common applications of thermally sprayed coatings? Specify industries where thermal spraying techniques have been adopted.
6. What advantages does thermal spraying have over other coating technologies?
7. How are thermally sprayed coatings created or built up?

8. What are some of the key issues associated with high-temperature processes during thermal spraying techniques?
9. How thick are typical thermal spray coatings deposited?

## REFERENCES


[1] Siegmann, S., & Abert, C. (2013). 100 years of thermal spray: About the inventor Max Ulrich Schoop. *Surface and Coatings Technology*, 220, 3–13. http://doi.org/10.1016/j.surfcoat.2012.10.034.

[2] Schoop, M.U. (2001). Early thermal spray application—JTST historical patent #21. *Journal of Thermal Spray Technology*, 10. http://doi.org/10.1361/105996301770349484.

[3] Schoop, M.U. (2001). Early thermal spray application—JTST historical patent #22. *Journal of Thermal Spray Technology*, 10, 37–39.

[4] Fauchais, P.L., Heberlein, J.V., & Boulos, M.I. (2014). *Thermal spray fundamentals: From powder to part.* Cham, Switzerland: Springer Nature Switzerland AG.

[5] David, J.R. (Ed.). (2004). *Handbook of thermal spray technology.* Materials Park, OH: ASM International, pp. 13–14.

[6] David, J.R. (Ed.). (2004). *Handbook of thermal spray technology.* Materials Park, OH: ASM International, pp. 44–45.

[7] Tate, G., Kenvin, A., Diao, W., & Monnier, J.R. (2019). Preparation of Pt-containing bimetallic and trimetallic catalysts using continuous electroless deposition methods. *Catalysis Today*, 334, 113–121.

[8] http://www-inst.eecs.berkeley.edu/~ee143/fa10/lectures/Lec_13.pdf.

[9] Pedersen, H., & Elliott, S.D. (2014). Studying chemical vapor deposition processes with theoretical chemistry. *Theoretical Chemistry Accounts*, 133, 1476.

[10] www.alcatechnology.com/en/blog/magnetron-sputtering/

[11] Marianna, B., Agatino, D.P., Yurdakal, S., & Palmisano, L. (2019). Chapter 2—Preparation of catalysts and photocatalysts used for similar processes. In: Marcì, G., & Palmisano, L. (eds.), *Heterogeneous photocatalysis.* Amsterdam, Netherlands: Elsevier B.V., pp. 25–56.

[12] Ducos, M., & Durand, J.P. (2001). Thermal coatings in Europe: A business perspective. In: Berndt, C.C., Khor, K.A., & Lugscheider, E.F. (eds.), *Proceedings of international thermal spray conference.* Singapore: ASM International, Materials Park, OH, pp. 1267–1271.

[13] Dorfman, M.R. (2005). Thermal spray coatings. In *Handbook of environmental degradation of materials.* Amsterdam, Netherlands: Elsevier, pp. 405–422.

[14] www.thermalspray.com/applications/wear-and-abrasion-resistance/

[15] www.thermalspray.com/applications/corrosion-protection/

[16] www.asbindustries.com/chrome-oxide-coatings

[17] Wellman, R.G., & Nicholls, J.R. (2007). A review of the erosion of thermal barrier coatings. *Journal of Physics D: Applied Physics*, 40(16), R293.

[18] Heimann, R.B. (2006). Thermal spraying of biomaterials. *Surface and Coatings Technology*, 201(5), 2012–2019.

[19] Liu, Q., Huang, S., & He, A. (2019). Composite ceramics thermal barrier coatings of yttria stabilized zirconia for aero-engines. *Journal of Materials Science & Technology*, 35(12), 2814–2823.

[20] http://apsmaterials.com/biomedical/titanium-coating/

[21] Nagentrau, M., Mohd Tobi, A.L., Jamain, S., & Otsuka, Y. (2019). Contact slip prediction in HAp coated artificial hip implant using finite element analysis. *Mechanical Engineering Journal*, 6(3), 18–00562.

[22] Legrand, M., Batailly, A., & Pierre, C. (September 26, 2011). Numerical investigation of abradable coating removal in aircraft engines through plastic constitutive law. *Journal of Computational and Nonlinear Dynamics*, 7(1), 011010, January 2012.

[23] www.engineeringnews.co.za/article/sofc-and-thermal-spray-general-2017-08-29

[24] www.thermalsprayusa.com/applications/dimensional-restoration/

# 2 Surface Coatings

*Shipra Bajpai, K. Vijay Kumar, Anup Kumar Keshri and Kantesh Balani*

## CONTENTS



With the growing necessity of improving industrial performance, there has been continuous effort to increase component functionality. Also, the machinery and its components must be useful up to a high standard and should be reliable for a long time. During its service conditions, machinery or a component interacts with other parts and components in various gaseous or liquid environments. Due to these interactions, the component may become degraded. Thus, in order to protect the component in such an environment, several surface modification and coating deposition techniques are developed.

During the service period, the surface of a component may undergo various interactions; for example, stress cycles that lead to its damage like wear, fatigue, oxidation and corrosion. In order to protect the surface from damage, surface treatments are a must to protect the component. Ti and its alloy-based biological

DOI: 10.1201/9781003321965-2

implants also get damaged as they encounter bodily fluids. Therefore, they are also surface treated. Surface treatment processes include surface hardening (e.g., shot peening, flame hardening) and surface coating techniques (e.g., chemical vapor deposition, physical vapor deposition, thermal spray technique). Among all the techniques, thermal spray techniques are found to be most advantageous because of their versatility in terms of the materials deposited and in terms of the properties of the coatings obtained. It has been observed that plasma-sprayed coatings, compared with physical vapor deposited, help in the reduction of thermal conductivity that is the essential requirement of thermal barrier coatings.

Surface modification techniques not only improve the performance of materials within specific environment but also enhance the life of materials. Surface modification can be classified as mechanical hardening, thermal treatment, chemical cure or surface coatings. Of the many types of surface modification, surface coatings tower most of the other techniques. Surface coatings are chosen depending on the application. In addition to protecting the substrate, coatings also enhance the performance of the substrate. For example, in the case of optical coatings, the substrate is coated in order to protect it from wear and erosion; thus, the optical properties of the substrate are not affected. Furthermore, coatings are not only used to protect the substrate against wear, corrosion and so forth, but they are also used for some decorative purposes, implant prostheses and electrical isolation/conduction. Depending on the technological needs of society in the future, more complex coating processes may be developed.

The main focus of this chapter is to make the reader familiar with the surface failure processes like wear, erosion, corrosion, and high-temperature oxidation and to introduce various cures for these failures, especially the surface coatings. Different surface coating technologies for protection against these processes depending upon the particular problem will also be discussed in detail.

## 2.1 SURFACE

In many dictionaries, a *surface* is defined as the upper layer or exterior face of a material. But in a more detailed way, surface can be termed as the interface between a solid object and its surroundings. Surface can be any two-dimensional geometrical shape, like a sphere or cylinder, that separates the object from the environment. It is the very first part of any component or material that is perceived to undergo degradation (visually) by an observer, as any physical or chemical phenomena starts from this portion of the materials. Hence, it becomes vital to make any surface sturdy and resilient to undergo any unintended deformation during operation.

## 2.2 WHY MATERIALS FAIL

During the processing of a component, various stresses act on the component due to processes like polishing, grinding and so on. In these processes, unwanted material loss occurs in significant amounts. Just as they do during the service period of the components, they have to undergo various stresses and environments that may degrade the surface. Nonetheless, when the machinery and/or components are intended to operate in an aqueous or humid environment, their surfaces become corrupted by continuous corrosion in the same media. In the upcoming sections, the two materials deteriorating factors and their several types are briefly described.

### 2.2.1 Wear of Components

There have been multiple definitions for the wear of materials, including removal of material from a surface, movement of the material from one surface to another, or transfer of the surface material from one place to another on a single component. Simply, *wear* can be defined as the progressive loss of material due to the sliding motion between two materials or components. Wear is caused by the interaction of the solid particles in motion or by the mechanical action of the fluid in motion with respect to the solid. It is an undesirable phenomenon, as it leads to increased friction and thus failing of the material. Lubrication regime, nature of load, surface hardness and roughness, and contaminants in the lubricating oil are some factors that decide the extent of wear during the service period of a component. Wear processes are classified in Figure 2.1 [1].

#### 2.2.1.1 Adhesive Wear

When two material surfaces in contact are moving relatively under the applied load, the point of contact between the surfaces becomes damaged and these surfaces become welded locally due to interatomic forces. When these bonds are stronger than the material, the bonded material shears or plastically deforms and work hardens. In this case, the material transfers from one surface to another surface, resulting in adhesive wear. Figure 2.2 shows the mechanism of adhesive wear, where the impacting particle gets locally welded to the substrate and removes the material from the surface.

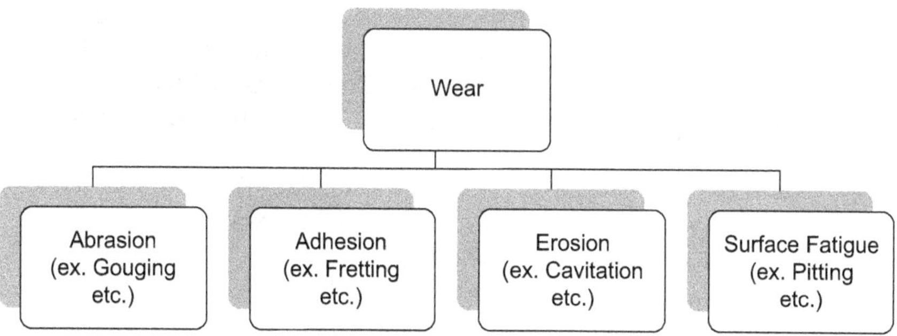


**FIGURE 2.1** Classification of wear categories and wear modes.

*Source*: Adapted from [1].

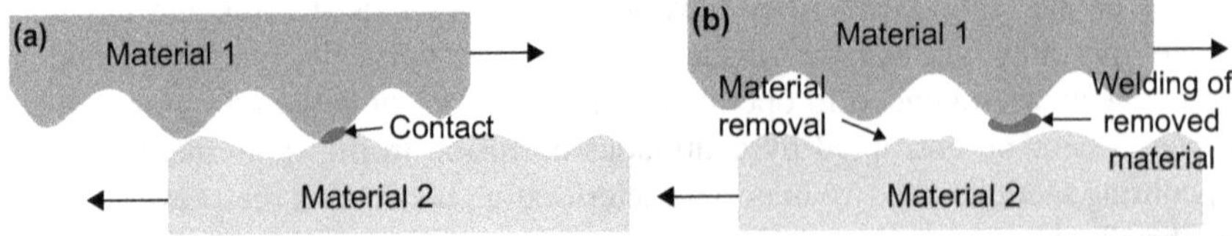


**FIGURE 2.2** Schematic representation of adhesive wear: (a) two surfaces coming in contact and (b) removal of soft material and its adhesion to the hard material.

*Source*: Adapted and redrawn from [2].

Adhesive wear is typically generated under plastic deformation. When the plastic contact occurs between two similar types of materials, in terms of hardness and/or elastic modulus, there originates an interface with adhesive bonding. At this interface, the removal of material from one surface occurs as a result of strict adhesion at the interfacial interaction. The debris of one surface becomes bonded or welded to another component or surface, and the new system follows a "stick and slip" motion and may further cause abrasive wear to the other substrate or component. For example, in knee implants, a minute portion of ultra-high molecular weight polyethylene becomes attached to the titanium implant during its continuous operation.

#### 2.2.1.2 Abrasive Wear

Abrasive wear is produced when hard particles or protuberances are forced against a solid surface. In this case, material removal takes place due to a shearing or cutting action. The term *harder* means that the particle producing abrasion should be harder from the surface on which wear damage is taking place. One specific requirement for abrasive wear is that the impacting particles should have some sharp, angular edges so as to produce the shearing or cutting action.

During its working life, one material or component progressively scratches and grooves its softer counterpart during a relative motion between them. There are basically two forms of abrasive wear: two-body wear and three-body wear. Two-body wear involves the removal of material from a softer surface due to continuous rubbing of another, harder material on its surface and the debris is completely removed from the pair in motion. This means that the debris is not involved further in the motion between the two rubbing bodies. In contrast, in three-body wear, the debris also takes part in the rubbing between the two rubbing surfaces and acts as abrasive material for increasing the wear. It is assumed that among all the type of wear, more than 50% of the wear-based failure are due to abrasive wear. The damaging of crankshaft journals in a reciprocating compressor is a well-known example of abrasive wear. Hard dirt particles break down in the lubricating medium and then cut down the crankshaft at its softer regions.

In Figure 2.3, two surfaces having sharp edges are moving towards each other. When two of the edges come in contact, the harder material cuts the softer one, resulting in the abrasive wear.

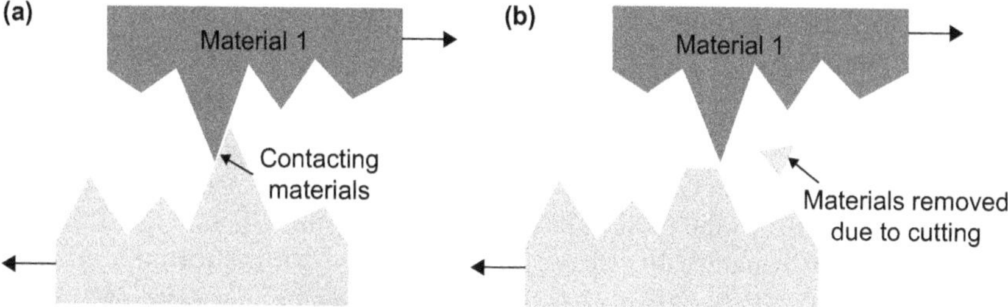


**FIGURE 2.3** Schematic representation of abrasive wear: (a) two moving materials in contact and (b) abrasion of material due to cutting.

*Source*: Adapted and redrawn from [3].

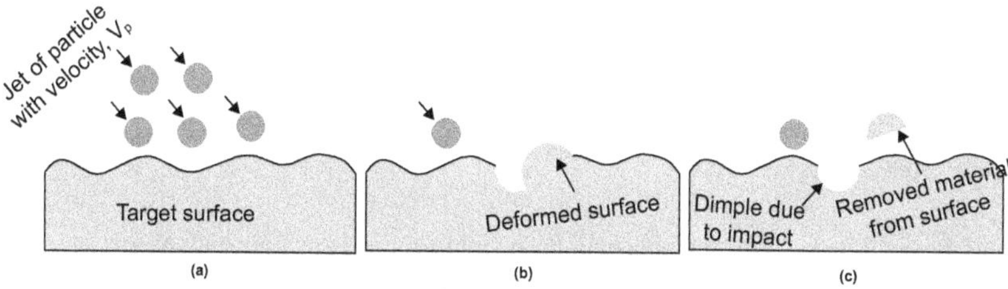


**FIGURE 2.4** Schematic representation for erosion mechanism. (a) Solid particles impacting the target surface, (b) deformation on the surface due to impact and (c) dimple formation as well as material removal from the surface after impact.

*Source*: Adapted and redrawn from [4].

### 2.2.1.3 Erosion

Erosion is a process where the surface becomes damaged due to solid particles present in the moving fluids. The fluids may be any medium, such as air, water, oil or lubricants. These solid particles attack the surface, and erosion takes place due to wear. In other words, during erosive wear, hard abrasive particles are directed towards a solid surface. Velocity or inertial force of the impinging particles act as the dominating factor for scratching the component. The wear volume or the material removal during erosive wear mainly depends upon the velocity of the impinging abrasive particles and the total mass of the particles attacking over the working surface. An example of erosion is in gas or water pipes where solid particles present in the high-velocity fluid stream erode the internal surface of the pipe. The mixture of gas and solid flowing through a rubber hose is another good example of erosive wear.

The schematic of erosive wear is shown in Figure 2.4. When solid particles repeatedly impact the surface, craters and platelets will form; these craters will grow with subsequent particle impact, and platelets are easily removed into the flow.

#### 2.2.1.4 Surface Fatigue

Surface fatigue is the type of wear where a surface becomes damaged due to cyclically applied stresses. In many cases, although there is sufficient lubrication between the moving parts, significant wear is noticed for them. Here, the continuous and repeated contact of the debris between the surfaces produces very high local stresses between the moving parts. This contact may be of an elastic nature or an elasto-plastic nature. When this debris rolls or slides between the parts for a large number of cycles, then wear of the surfaces come into existence by the fracture produced by fatigue. The wear of material in the case of fatigue wear is governed by crack nucleation, crack advancement and fracture mechanics. Since the cycle is repeated a large number of times, the worn surface of the material possesses an extreme amount of strain as compared to the bare and undamaged part of the surface.

Main types of surface fatigue wear are:

1. Pitting wear
2. Impact wear
3. Spalling
4. Brinelling.

Presence of asperities at the surface are the main reason for wear. Surface roughness decides the degree to which these asperities are present on the surface. Wear can be lowered by polishing the surface. In this case the asperities are lowered and hence the chances of wear are mitigated. Titanium and other bioimplants are the examples, showing suffering from the fatigue wear.

Titanium and its alloys are most widely used implant materials for hip joints replacement. Due to a mixed lubrication regime in the artificial joints, wear always occurs in this region. During motion of the hip joint implants, macroscopic particles are generated which get trapped between the tissues of these joints, leading to the unwanted reactions and wear in the body. Thus, surface treatments of the implants are required to reduce the wear in the biological materials.

### 2.2.2 Corrosion and High-Temperature Oxidation

Corrosion is the gradual degradation of a material due to chemical environment or stresses depending upon the service conditions. Corrosion is an undesirable phenomenon where (mostly) metal objects are exposed to various environmental agents, such as air, water or temperature. Corrosion results in the most stable form of a metal, such as oxides, hydroxides and carbides, and degrades the material's performance including strength and appearance. For example, iron and its alloys possess an attractive tensile strength and good rigidity and elasticity. But once the iron corrodes, it becomes brittle and flaky and loses its sound characteristics. Sometimes, oxidation is also considered as a type of corrosion where materials are exposed to high temperature, such as in re-entry vehicles. Here they undergo oxidation while entering or exiting the earth's atmosphere and form oxides, which leads to the failure of the component. There are various factors that affect the rate of corrosion of a material, including exposure of materials to oxygen-rich gases (e.g., $CO_2$, $SO_2$ and $SO_3$), a humid environment or high temperatures, or the presence of salt-like impurities.

Materials which are implanted in vivo encounter bodily fluids like blood and interstitial fluids. These fluids contain enough chloride ions to corrode implants. The presence of amino acids and protein in the blood results in accelerated corrosion of the implantation and change in the pH value. Titanium has already proven to be highly stable and corrosion resistant, however there are some reports which show the growth of titanium ions in the interface between tissue and implant, signifying the release of metal ions and corrosion in the implant.

## 2.3 SURFACE TREATMENTS

Previous sections discussed the mechanisms causing surfaces to become damaged and how this leads to the degradation of a material in a service environment. These damages are required to be minimized during their operation, and the components should perform reliably and be durable. In the upcoming sections, surface modification and surface protection techniques or processes will be discussed. During surface treatment, either the metals or materials surfaces are strained via plastic deformation, or their surfaces are covered with hard nitride or carbide films. In other processes, the materials surfaces are covered with various coatings (depending on the application of the material), as in case of chemical vapor deposition or thermal spraying. The type of surface treatment to be employed on a sample depends upon the nature of the application it is going to conduct.

The surface of the implant material is highly critical in triggering biological responses. During the processing of titanium alloy–based implants, oxidation and contamination of the surface occurs, which forms a layer on the surface resulting in degraded properties of the implant. Thus, the surface becomes unsuitable for biomedical applications. Good wear and corrosion-resistant, titanium-based implants also require specific surface properties. Therefore, surface treatments are required.

Surface treatments are classified into surface hardening and surface coating [5] (Figure 2.5), as detailed in the following sections.

### 2.3.1 Surface Hardening

Surface hardening means to harden the surface of a material but keeping the core soft so as to achieve a good combination of properties that can withstand high stresses and fatigue. Surface hardening can be divided into three major groups: mechanical, thermal and thermochemical treatments.

#### 2.3.1.1 Mechanical Treatments

Mechanical treatment of a surface is done to incorporate roughness on the surface, improve adhesion and/or to remove contaminations from the surface. During mechanical treatment, in-plane residual stress is generated just below

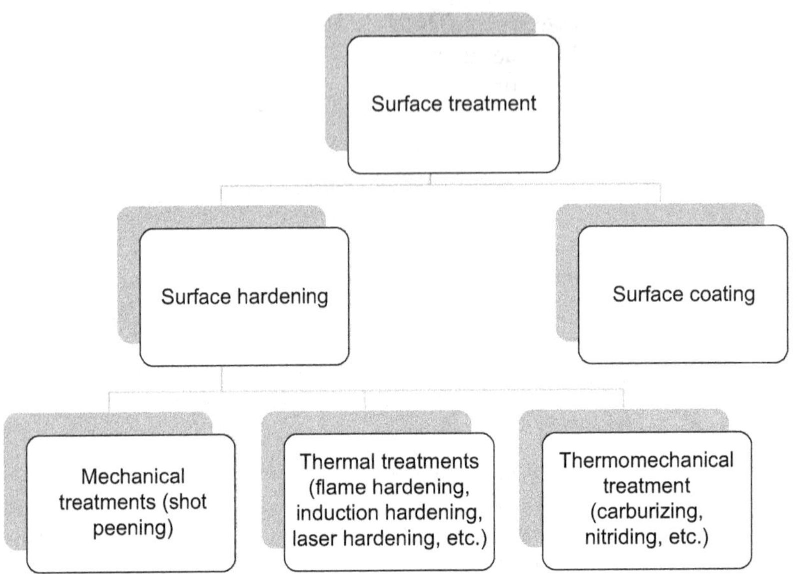


**FIGURE 2.5** Classification of surface treatment processes.

the surface of the engineering component. The primary aim of this method is to induce grain nanocrystallization of the surface and generate a large amount of plastic strain, making the component resilient towards wear and corrosion. Shot peening and surface mechanical attrition treatment are mature examples of mechanical treatment.

*Shot peening* is the mechanical operation done to introduce surface-compressive residual stresses. It includes impacting a surface with shot (rounded metallic, glass or ceramic particles) with sufficient force as to cause plastic deformation at the surface. The size of the shot varies from 0.2 mm to 1 mm, and they impact the surface at a very high speed of ~100 m/s. Also, the direction of the impingement of the shot is kept nearly perpendicular to the component surface.

Azar et al. [6] have worked on increasing the fatigue and corrosion resistance of 316L stainless steel (SS) in Ringer's solution using shot peening. The peening process was carried out with 1–2 mm spherical balls for a period of 5 to 25 minutes. The shot-peened surface showed an increase in surface hardness and fatigue resistance as compared to the bare specimen. After shot peening, the surface roughness improved and showed an elevation in the strain hardness in the near surface region. Also, the residual stress showed an improved value. The treated samples also showed significant curb against corrosion, and all the treated samples showed lower corrosion rates with respect to the parent sample. This was attributed to the development of a stable passive layer on the surface after the treatment, which inhibited the infiltration of harmful corrosive ions into the specimen. In this way, shot peening improved the durability of the SS substrate, making them fit for long-term use as bio-implants.

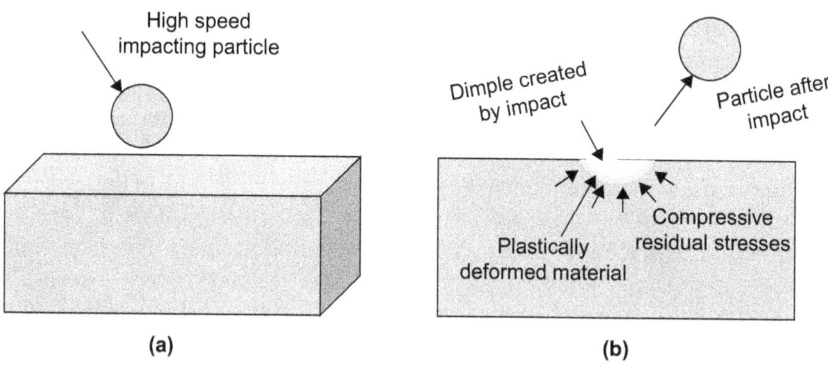


**FIGURE 2.6** Schematic representation of shot peening process: (a) impact of high-speed particles on surface which creates plastically deformed zone, compressive residual stresses and (b) dimple on the surface.

*Source*: Adapted and redrawn from [7].

The basic principle of *surface mechanical attrition treatment* (SMAT) is the same as shot peening: spherical shots are made to impact over the sample surface. But the major difference lies in the size of balls and the velocity of impact. The ball size is kept from 2 mm to 20 mm, and the velocity of the balls is approximately 5–20 m/s. Also, the balls are impinged over the sample from random directions, not perpendicularly (as in case of shot peening).

In Figure 2.6, shot impacts the surface with higher velocities, creating dimples at the surface. Presence of this dimpling leads to generation of compressive stresses at the surface. Thus, the surface becomes hardened and has a nanostructured microstructure.

In the last few decades, SMAT has been a good alternative for the shot peening purpose, although they have similar working principles. SMAT of low-carbon steel (LCS) was performed by Singh et al. [8] with different ball sizes varying from 4 mm to 8 mm in diameter. Post treatment, the corrosion behavior of LCS was studied. SMAT treatment decreased the grain size of the LCS from ~20 nm to ~11 nm. This decrease in grain size was due to the presence of highly densified mechanical twins within the surface microstructure of LCS after impacting the surface. The open circuit potential surged, and the corrosion current density decreased after the SMA treatment. Grain refinement after the surface treatment, blockage to the corrosive electrolyte ion and increase in the breakdown potential of the passive layer together resulted in an increase in the corrosion resistance of the sample after SMAT processing.

#### 2.3.1.2 Thermal Treatments

When the surface of a component is subjected to high temperature and its chemistry remains unchanged, then the processes are termed as thermal treatments for hardening. Thermal treatment is a procedure of heating and cooling metals following certain predetermined methods, which manipulates the surface of the material to enhance its preferred properties. Thermal treatments may be classified depending upon the thermal medium utilized for the hardening process. Flame, induction, laser and electron beam hardening are well-known thermal treatment methods.

*Flame hardening* is one among the simplest form of hardening. In this process, heating of surfaces of large workpiece and complex cross sections is done by an oxy-acetylene torch followed by rapid water-cooling jet/spray. This hardening treatment is generally followed by stress-relieving treatment, where the part is heated to 180–200 °C in a furnace or oil bath. This stress-relieving process does not reduce the hardness of the part.

Flame hardening is usually carried out for iron-based alloys, where (in the simplest form) steel is heated to its hardening temperature and then is cooled in air, water, or oil media. Acetylene and propane are the main gases used for generating flame during flame hardening. These gases are mixed with air in a certain predefined ratio to flame under pressure. A burner points this flame towards the component to increase its temperature up to 800 °C. This is the preferred process for larger parts and lower volumes, such as large gears.

Generally, flame hardening processes are classified as follows:

1. Stationary (both burner and workpiece are stationary);
2. Progressive (burner is combined with the water spray);
3. Spinning (work piece rotates while burner remains stationary);
4. Progressive spinning (burner moves over a rotating workpiece).

Figure 2.7 shows the schematic representation of the progressive flame hardening. In this process, the burner moves over the surface and heats it. At the same time, a water jet attached to the burner cools the surface and hardens it.

Recently, flame hardening has been worked out to modify the surface of low-carbon steel. Thamilarasan et al. [10] used a high-pressure oxy-acetylene gas mixture to increase the temperature of a material up to its recrystallization

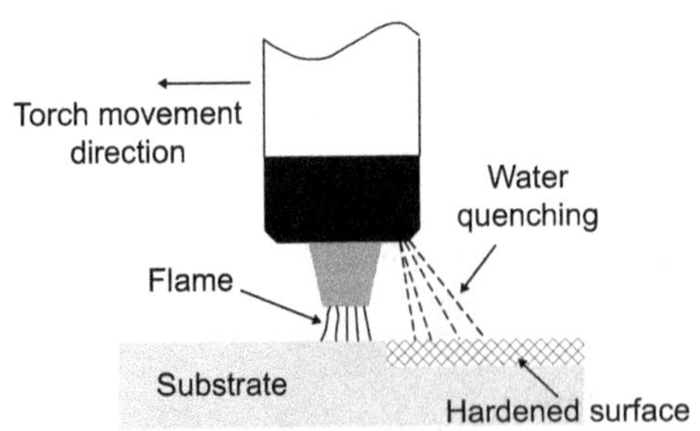


**FIGURE 2.7** Schematic representation of the progressive flame hardening process.

*Source*: Adapted and redrawn from [9].

temperature and then quenched it in a water medium. The recrystallized structure showed an improvement in the micro-hardness of the material. The flame temperature, torch cap diameter and cooling rate of the heated material are the vital factors affecting the hardness of the material after the hardening process.

In the *induction hardening* process, a component is heated by electromagnetic induction. A conductor coil carries an alternating current of high frequencies. Then the sample is placed in the magnetic field of the coil. The sample surface is heated up by the eddy current generated by the electromagnetic field. As a result of this, induction heating of the component takes place and generates a temperature up to 850 °C. High frequency of the alternating current gives a narrow depth for the temperature rise. The hardened depth due to induction hardening is lower than that of flame hardening. Due to skin effect, the heat generated remains only at the surface of the component; thus, only the near surface region gets heated. During induction hardening, the surface is heated for a very short duration and is then cooled with a jet of water. After quenching, a martensitic structure forms, and thus the surface is hardened. Induction hardening treatment can be used as a local surface treatment.

Figure 2.8 shows the induction hardening process where the component is heated by electromagnetic induction. After heating the component is submerged in water and quenched, and the surface becomes hardened. Generally, Cu coil is used for the induction of current.

Induction hardening has been used for hardening railway axles and increasing their remaining life span. The fatigue resistance of a railway axle with a crack 4 mm deep was induction hardened to generate a residual stress varying from 400 MPa to zero [12]. The remaining life of the damaged axle showed a threefold improved life consistency due to the instituted compressive residual stress, as

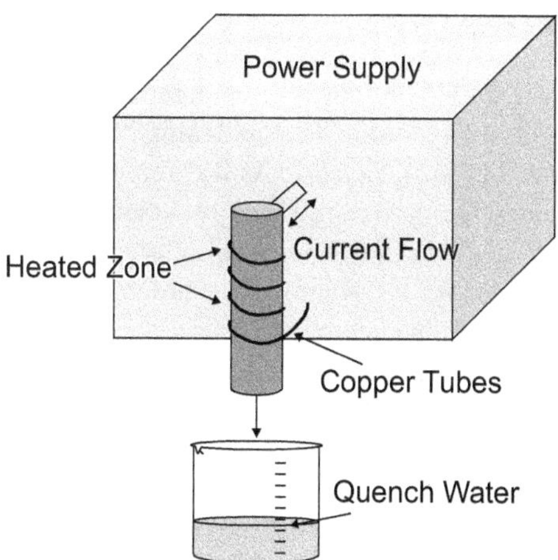


**FIGURE 2.8** Diagrammatic representation of the induction hardening process.

*Source*: Adapted and redrawn from [11].

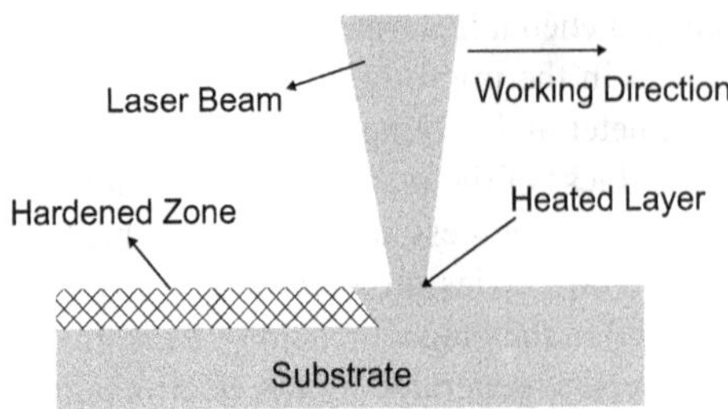


**FIGURE 2.9** Schematic representation of the laser hardening process.

*Source*: Adapted and redrawn from [13].

compared to the axle without any residual stress. Also, the induction hardening helped in restricting the further extension of the crack in the axle and improved the life span of the railway axle.

In the *laser hardening* process, laser beams are used to heat a surface and then quickly let it cool in the surrounding air. This process is used for large components or those with a complicated shape. This process uses a high-powered diode that transfers the heat energy at specified local regions of component surface by roaming across its surface. The laser generated heats the surface in an instant and brings the temperature as high as the austenitizing temperature of the component. The rapid cooling post-heating results in the formation of a martensitic structure in a component and hardens this material. Because a laser beams has a very high intensity, it can melt a surface. Thus, to avoid melting of a material, a lens is used in between. Laser hardening increases hardness and wear resistance, which leads to the reduction of abrasive wear.

Khoury et al. [14] incorporated hardening into bearing steel using laser exposure using a high-power diode laser, executed at a range of 910–1030 nm and at a maximum power of 2.5 kW. The surface temperature increased to 1050 °C with a laser beam width of nearly 7 mm. The hardening of the bearing was in relation to the laser feed rate. The surface hardness was reported to increase, which subsequently resulted in an increase in the wear resistance of the material. The wear resistance increased twofold after the hardening process. In other cases, such as for AISI 420 steel blades for a turbine, laser hardening is a new way to replace flame hardening. The rapid quenching after laser annealing produces a fine martensitic structure in a random dispersal over the surface, improving the surface hardness and corrosion resistance.

*Electron beam hardening* is used for components that typically distort during the induction hardening process. In this process, the component is generally kept in a vacuum and an electron beam is focused on it to heat the surface. With time, the power input is decreased to avoid the melting of the workpiece. This method is usually carried out for hardening a thin surface layer or only the outermost part of any surface. The inner properties of a component remain unchanged, as thickness near to ~0.1 mm is usually hardened using the electron beam hardening.

Since the whole process is carried out in vacuum, the surface that is heated due to the electron beam undergoes self-quenching. The ratio of area affected by the electron beam to the unaltered surface area is so small that the unaffected regions causes cooling of the heated area fast enough that the material gets hardened simultaneously.

Electron beam hardening has been successfully employed for strengthening the low-alloy structural carbon steel. The microhardness of the steel showed a twofold to threefold increase in its value on the beam exposed area, with negligible variation in the roughness. The regions with a lower cooling rate faced a relatively low hardness value as compared to those with a higher cooling rate. They also illustrated that the chemical composition of the alloy, before and after hardening, showed not much variation. Only some percentage of minor elements was decreased after the electron beam exposure due to continuous cooling and heating of the material.

### 2.3.1.3 Thermochemical Treatment

In thermochemical treatment, temperature is applied for heating and some chemical reactions also take place. Examples include carburizing, nitriding and carbonitriding.

*Carburizing* is the most widely used surface hardening process for steel. In this process, carbon is allowed to diffuse from the steel surface by heating above its transformation temperature (generally 900–930 °C) and holding the steel in the presence of carbonaceous material. During holding, carbon is diffused into the surface of the material and thus increases the hardness of the surface. Carburizing process is divided into three categories:

1. Pack carburizing
2. Liquid carburizing
3. Gas carburizing.

During carburizing, the parts are heated to a high temperature, therefore it may lead to the grain coarsening, as well as carbide formation. In order to refine the microstructure and to break the carbide network, some post carburizing treatments are given to the component.

In the *nitriding* process, nitrogen is diffused into the material to increase the hardness of the surface. This process is typically useful for the alloys which form stable nitrides. There are various methods by which nitriding can be done:

1. Gas nitriding
2. Salt bath nitriding
3. Plasma nitriding.

In the *carbonitriding* process, carbon and nitrogen diffuse interstitially into the metal. This diffusion takes place in the temperature range of 700–900 °C. It creates barriers to slip, and increases hardness and modulus near the surface of the metal.

Carbonitriding is typically useful for enhancing the wear resistance of the mild, plain carbon and/or low alloy steels. The process is carried out at lower temperatures in the presence of gas mixture consisting of carburizing gas and ammonia. Carbonitriding produces thin, high hardness case that provides high strength and wear resistance while resulting an increased resistance to high temperature softening.

### 2.3.2 Surface Coating

The first method of providing protection to the surface is to alter the surface properties by some treatments such as hardening of a surface. Another method includes the protection of surface by coating with some resistant materials. The best feature of applying coating over a material is that the parent material is cut off from the outer environment and the various environmental agents cannot reach the material to degrade its properties. Also, all the surface damaging factors first attack the coating before damaging the actual component. Once the coating is damaged, it is quite easy to redeposit the coating on the component, hence the damage cost of any industrial component can be mitigated after applying the coating over its engineering component. Several other features of coatings include the following:

1. Improvement of component performance, as coatings enable high-temperature contact of a material via using thermal barrier coating;
2. Reduction in wear of the components due to erosion, corrosion and so on, thus increased durability of the component;
3. Improvement of component life, because the worn parts can be rebuilt to the original shape and size;
4. The need for replacement of entire component is circumvented;
5. Reduction in component cost due to improvement in the performance after applying a coating.

Surface coatings can be applied by various methods depending upon the application required. Figure 2.10 shows several surface coating techniques.

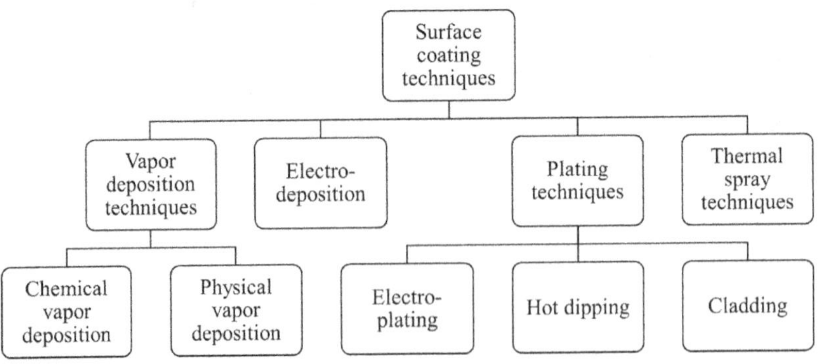


**FIGURE 2.10** Surface coating techniques.

#### 2.3.2.1 Electrodeposition

Electrodeposition is a process that uses an electric current to dissolve the metal cations so that they can form a coating on the surface of the material. The substrate is considered as a cathode and the metal to be deposited as the coating is considered as an anode. Both the electrodes are immersed in the electrolyte that permits the flow of electricity due to the presence of salts and ions.

The metal to be coated is put into a salt-containing solution in an inert container. The salt necessarily includes the coating metal ion. The workpiece is the cathode and the anode is a rod or sheet made of the coating material. Once the voltage is introduced between the two terminals, then the coating takes place on the cathode material because of electron exchange from anode to cathode via the salt solution. The anode can also be made of inert material when a suitable salt solution is used with enough of the coating metal ion.

#### 2.3.2.2 Physical Vapor Deposition

Physical vapor deposition (PVD) is used to deposit the thin layer of material, which is typically in the range of few nanometers to micrometers thick. During PVD, the target material is evaporated in a vacuum to form atoms or ions, which are transported and deposited on the substrate surface owing to condensation and reactions of metal ions to the substrate. Formation of multi-component, highly dense coatings during deposition at low substrate temperatures are important characteristics of PVD process.

During PVD, certain basic methods such as sputtering or evaporation produce vapor, in various terms, such as atoms or ions. These atoms or ions are of the material that is to be coated on the substrate and are directed using a target. Being directed from the target, they are deposited on the substrate in the form of a coating. One advantage of this material is that the coating process can be carried out in relatively low temperatures in the form of thin films. PVD uses a clean and dry vacuum environment, where the ejected atoms or ions are deposited over the entire area at the same time, not in some localized areas.

PVD has also been used in coating of nanocrystalline ceramic powders. Adhavan and colleagues [15] have deposited nanocrystalline ceria-zirconia compound on AISI 304 using the electron beam PVD method. They further studied the oxidation behavior of the coating during exposure to high temperature. Their study implied that the powders with a ceria percentage equal to or greater than 50% exhibited three to four times higher oxidation resistance at high temperature as compared to the bare or uncoated samples. In this way, not only did the performance of the steel improve, but the durability also increased because the parent material was never directly exposed to the high temperature.

#### 2.3.2.3 Chemical Vapor Deposition

Chemical vapor deposition (CVD) involves the deposition of the coatings on substrate owing to chemical reactions between the chemicals' gaseous phases and substrate. The main stream involves precursor gases that chemically react/

decompose and deposit the non-volatile coating on the substrate, and the volatile by-products return to the main gas stream. During CVD, high-standard, well-performing coatings are obtained in a vacuum chamber, between an organometallic or halide compound and other gases to produce well-adhered, solid thin films on desired samples, mainly for the semiconductor industry. CVD is generally used for fabricating composite thin films and also producing different nanomaterials. While PVD is a straightforward colliding type of material, CVD is a multidirectional, deposition-type coating.

Since long, the glass increase showed their specific interest towards the coating of nitrides of silicon and titanium [16]. Oxides of silicon, aluminum, zinc and other transition materials are also coated using CVD, showing an increase in the electrical and optical properties of the glass. Various chemical sources (depending upon the type of coating to be produced) and reactions are hired to coat these materials. The various factors which differ for each coating are termed to be the type of precursor used, the deposition rates attained during the coating, cost and toxicity of the precursor. The different coatings mentioned above perform different functionalities. For example, infrared reflectivity is obtained by titanium nitride coating, hence used as an efficient solar control glass; silicon or its oxide have high reflectivity, which is quite vital for any mirror coating showing high reflectivity to visible light, whereas thin amorphous aluminum oxide film effectively blocks the infiltration of sodium out of soda-lime glass.

#### 2.3.2.4 Thermal Spray Technique

Thermal spray is a coating technique where the stream of particles (metallic or non-metallic) are deposited on the substrate in a molten or semi-molten state. The particles to be deposited are heated with the help of electric arc and combustion processes. When the molten or semi-molten particles impact the surface, they get deformed and form bonds with the substrate surface. Upon impact, the particle stream flattens and forms splats. These splats are the foundation of any thermal coating. Several layers of these splats form the coating. The feed materials can either be in powder or wire form. Powder or wire feeding can be done either axially or radially. The exception of this definition is cold-spray process, a process in which the particle is neither heated nor molten prior to the formation of the coating.

Thermal spraying process involves heating the particle using either combustion or electric arc, impacting them towards the substrate at a high velocity using some carrier gas and then finally plastically deforming them at the substrate, where they form a layer called as coating. During flight, the atomization of the molten and semi-molten particles take place. Figure 2.11 displays the typical appearance of a thermal spray coating.

The substrate surface is made rough prior to the coating to ensure better adhesion of the coating on its surface, otherwise the coating may get delaminated from the substrate. Figure shows the presence of some porosity, molten/semi-molten droplets, oxides, un-melted particles and lamellae. These are some irresistible flaws of thermal spray coating that are needed to be mitigated as per the application of the coating.

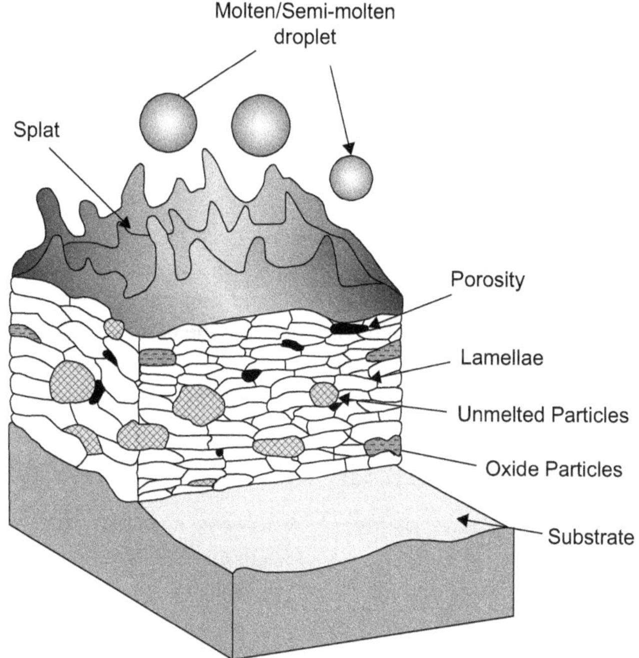


**FIGURE 2.11** Schematic of thermally sprayed deposit.

*Source*: Adapted and redrawn from [5].

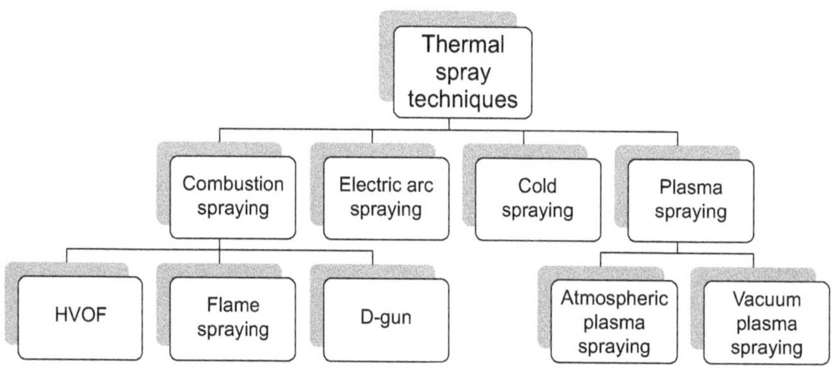


**FIGURE 2.12** Different thermal spray techniques.

Thermal spray techniques are further classified as shown in Figure 2.12.

Combustion spraying involves the generation of heat by the combustion of fuel. The heat generated melts the particle and deposits it on the substrate. Electric arc spraying utilizes an electric arc for the generation of the heat. During the cold-spraying process, as the name suggests, high temperatures are not required.

In this case, high velocities of the particles are utilized to deposit the particle on the substrate, but the velocities are lower than the HVOF process. In the plasma spraying process, the plasma gas is struck between arcs, where it gets ionized and generates heat that melts the particle and deposits them on the substrate. Plasma spraying is also done by two methods: first in the presence of air, or atmospheric plasma spraying, and second in the presence of a vacuum, which is vacuum plasma spraying.

Most of the spraying techniques work in the open atmosphere in the presence of air that may lead to the formation of oxides in the coating. This oxide content increases with the increase in temperature of thermal spray. Vacuum plasma spraying, high-velocity oxy-fuel and cold spray are the processes where the oxidation content is reduced due to the absence of air, high velocity of the particle and low thermal spraying temperature, respectively.

## 2.4 COMPARATIVE STUDY OF ALL COATING TECHNIQUES

A brief introduction of all the coating methods is now complete. Next, a comparative study of all the coating techniques is carried out in this section to describe the processes suitable for particular applications. Table 2.1 lists some of the general characteristics of various coating techniques.

It can be observed from Table 2.1 that thermal spray techniques are used when thick coatings are required, whereas the CVD and PVD techniques are used to deposit thin coatings. Thermally sprayed coatings are resistant to wear and corrosion, and that is the main advantage of the thermal spray technique over other deposition techniques. Thermal spray techniques are versatile with respect to the deposited material; for instance, in the case of thermal spray coatings, both metals and ceramics can be deposited, whereas other techniques are generally limited to one type of material.

The best example for any type of coating in our everyday life are the implant materials, such as hydroxyapatite (ceramic), titanium (metal) or ultra-high-molecular-weight polyethylene (UHMWPE; polymer) orthopedic coating

**TABLE 2.1**
**General Characteristics for Various Coating Techniques**

| Coating Technique | Coating Thickness | Coating Material | Characteristics |
|---|---|---|---|
| **PVD** | <5 μm | Metals and ceramics | Wear resistant |
| **CVD** | 3–50 μm | Metals, ceramics and polymers | Wear resistant |
| **Electro/electroless plating** | 10–100 μm | Metals | Corrosion resistant |
| **Thermal spray** | 0.05–3 mm | Ceramic and metallic coating | Wear resistant and corrosion resistant |

*Source:* Adapted and modified from [5].

implants. Hydroxyapatites (HAs) have a composition like that of the mineral phase of human bone and teeth, and that is why it is the best-known biocompatible material. Due to its good biocompatibility, it is widely used as implant material for the artificial hip joint. Titanium and its alloys are also considered as good choices for implants, but they require some surface modification for successful use as bioimplants. Table 2.2 lists some of the methods for surface modifications of Ti-based and Ti alloy–based biological implants.

Surface modification in case of Ti-based metallic implants has been found as the most promising means of obtaining increased biocompatibility and tissue growth. Several deposition methods have already been tried to deposit the HA-based coatings for biomedical implants; among them, plasma-sprayed coatings are widely used with thin coatings, improved blood compatibility and increased wear and corrosion resistance [18–20]. Plasma-sprayed HA-based coatings on titanium and titanium alloy have already been used as artificial and dental implants. Therefore, plasma-spray deposition of HA-based coatings on Ti and its alloys is discussed further.

Joshi et al. [19] deposited the HA-based coating on the Ti6Al4V alloy by using the plasma-spray process with primary gas Ar and secondary gas $H_2$. The cross-sectional micrograph of coating revealed a good bonding of coating with the substrate along with a thickness of 75–100 μm. Topographical analysis of the coating confirms the presence of porosity at the surface, which often is desirable for good

**TABLE 2.2**
**Overview of Different Methods for Surface Modifications of Titanium and Its Alloy Based Biological Implants**

| Method for Surface Modification | Characteristics of Layer | Objectives |
|---|---|---|
| Mechanical treatments like machining, grinding, etc. | Subtraction of the surface leading to smooth and rough surface | Clean surface and rough, improved adhesion bonding |
| Chemical methods like sol gel, alkaline treatment, CVD, etc. | Thin layers of thickness ~10 nm to 5 μm of calcium phosphate, $TiO_2$, silica, TiN, etc. | Improved biocompatibility, bioactivity or bone integration, surface-specific patterns, improved wear resistance, removed surface oxide and contaminations, etc. |
| Physical methods like PVD, ion implantation, etc. | TiN, TiC and diamond-based coatings with thickness of ~1 μm | Improved wear and corrosion resistance with the coatings |
| Thermal spray | Coatings of HA, Ti, etc. with thickness of ~30–200 μm | Improved bioactivity of the implants |

*Source:* Adapted and modified from [17].

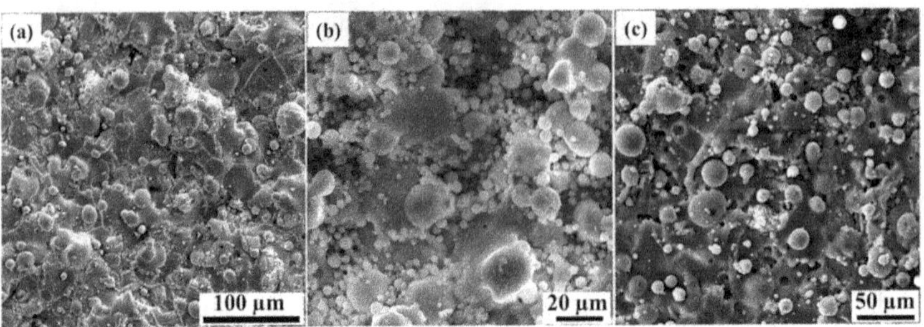


**FIGURE 2.13** SEM micrographs revealing the morphology of HA coatings deposited on Ti substrate preheated at (a) 20 °C, (b) 200 °C and (c) 400 °C.

*Source*: Reprinted with permission from [20].

biocompatibility. Fomin and his group [20] has also plasma-sprayed HA-based coatings on titanium substrate with induction preheating in the temperature range of 400–600 °C. Morphological analysis of the coating surface by scanning electron microscopy (SEM) reveals the significant effect of preheating temperature on the grain size, which increases from 5.7 ± 3.3 μm for preheating temperature of 20 °C to 12.4 ± 5.3 μm when preheated at temperature of 400 °C, as shown in Figure 2.13.

Another important application of the surface modification techniques is the TBCs. In this case, the surface is modified by coating different layers of materials to reduce the thermal conductivity. TBC systems are thermally insulating coatings used to insulate turbine blades from experiencing high temperatures. Complementing the thermal barrier coatings with external film cooling and internal component air cooling, temperature reductions of up to 165 °C are possible. A typical TBC is made up of two main layers: (1) a bond coat (aluminide or overlay MCrAlY) to resist oxide diffusion and (2) a ceramic top layer (typically 7YSZ) to provide thermal insulation. The ceramic top layer is usually deposited via plasma spraying or electron beam physical vapor deposition (EBPVD) [21].

Ghasemi et al. [22] has deposited the nanostructured yttria stabilized zirconia (YSZ)-based TBCs on the Ni-based superalloys and investigated the adhesion strength as well as thermal insulation of these coatings. The microstructural analysis of the coatings carried out by SEM revealed the bimodal microstructure which consists of both the nanosized particles and columnar grains, as shown in Figure 2.14. The presence of a bimodal microstructure has resulted in crack termination; thus a higher adhesion strength of ~38 MPa was obtained for plasma-sprayed nano-YSZ coatings.

A comparative study of TBCs is performed [21] where the low-density plasma-sprayed coatings have resulted in better dispersion of heat flux due to inherent phonon scattering and thus better thermal insulation. In contrast to PVD coatings, the columnar grains and lack of large splat boundaries restricted the heat flow and resulted in poor thermal insulation when compared to plasma-sprayed coatings.

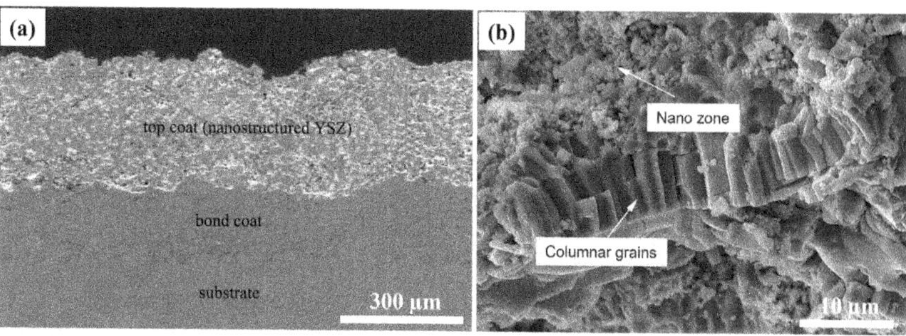


**FIGURE 2.14** SEM images of nano-YSZ coating deposited on Ni superalloy-based substrate with NiCrAlY bond coat. (a) Cross-sectional micrograph; (b) higher magnification micrograph of coating showing the presence of both nanosized particles and columnar grains.

*Source*: Reprinted with permission from [22].

From the two practical cases of the coating processes in our day-to-day life, as well as industrial applications, we can say that the thermal spraying process has certain advantages over other coating processes. Some of the unique characteristics of thermal spray are as follows:

1. A coating can be deposited of almost any type of materials, be it metal, ceramics, or polymers.
2. The substrate on which a coating is to be deposited may vary from metal to ceramic or to polymer type.
3. The materials to be converted into quality coating can be initially in any form, such as powder, wire or rod.
4. A mixed variety of materials or wide area of mixture can also be deposited on a substrate with ease, such as ceramic added to other ceramics, metals or polymers.
5. A uniform coating is produced in a very short time frame.
6. A high deposition rate, in $mm.s^{-1}$, is obtained with not a very high experimental cost.
7. The whole coating process can take place in any working environment (e.g., air, at reduced pressure, in a vacuum).
8. Thick, freestanding coatings can be obtained in a net-shape fashion, which can be used for different applications or for microstructural verification.

## 2.5 SUMMARY

A *surface* is the boundary that connects a material with its surroundings. During a service period, the surface of a component may undergo various interactions with the environment, as well as other components, and it may undergo various stress cycles that lead to its damage, like wear, fatigue, oxidation and corrosion. In order to protect the surface of the component from damage, surface treatments are a must. Ti and its alloy-based biological implants also are damaged as they encounter bodily fluids. Therefore, they are also surface treated. Surface treatment processes include surface hardening (e.g., shot peening, flame hardening) and surface coating techniques (e.g., CVD, PVD, thermal spray technique). Among all the techniques, thermal spray techniques are found to most advantageous because of their versatility in terms of the materials deposited and in terms of the properties of the coatings obtained. Thermal spray is a one-step process where a wide variety of materials can be coated over a great range of substrates. Among all the thermal spraying processes, plasma-sprayed HA-based Ti implants are used widely in the biomedical implants. It has been observed that plasma-sprayed coatings, compared with that of PVD, help in the reduction of thermal conductivity that is the essential requirement of the TBCs. In the next chapters, various thermal spraying processes are discussed along with their advantages, disadvantages and applications.

### Questions for Self-Analysis

1. What causes machine components to degrade? What can be done to keep it from deteriorating?
2. What is a surface, and how are you going to change it?
3. What causes materials to fail?
4. What is wear? What causes wear? What impact does wear have on materials? What factors influence the amount of wear?
5. What are the different ways to wear? Also, explain them.
6. What are the different types of fatigue wear on the surface?
7. What are the advantages of titanium and Ti alloys used in hip joint implants?
8. What is corrosion, and how does it happen? What is the difference between corrosion and wear?
9. What factors influence the rate at which a material causes corrosion?
10. What are the benefits of mechanical treatments? Mention some examples.
11. Why is thermal treatment of materials required? Explain various kinds of thermal treatments.
12. What are different thermochemical treatments? Explain them.
13. What are the features of surface coating? Explain the various types of coatings.
14. According to your knowledge, what is the best surface coating technique?
15. What is thermal barrier coating? Why do we use it in gas turbine engines?

## REFERENCES

[1] R. Tarodiya, A. Levy, Surface erosion due to particle-surface interactions—A review, *Powder Technol.* 387 (2021) 527–559. https://doi.org/10.1016/J.POWTEC.2021.04.055.

[2] M.C. Tanzi, S. Farè, G. Candiani, Sterilization and degradation, in: *Found. Biomater. Eng.*, Elsevier, Amsterdam, Netherlands, 2019: pp. 289–328. https://doi.org/10.1016/b978-0-08-101034-1.00005-0.

[3] M. Salot, Study on wear resistance of Al-Si alloy using a 3-body dry abrasive wear testing machine, in: *NCIME*, Indus Univ., Ahmedabad, 2016: pp. 1–6. www.researchgate.net/publication/332780432_STUDY_ON_WEAR_RESISTANCE_OF_Al-Si_ALLOY_USING_A_3-_BODY_DRY_ABRASIVE_WEAR_TESTING_MACHINE (accessed March 23, 2022).

[4] R. Tarodiya, A. Levy, Surface erosion due to particle-surface interactions—A review, *Powder Technol.* 387 (2021) 527–559. https://doi.org/10.1016/j.powtec.2021.04.055.

[5] P.L. Fauchais, J.V.R. Heberlein, M.I. Boulos, Overview of thermal spray, therm. *Spray Fundam* (2014) 17–72. https://doi.org/10.1007/978-0-387-68991-3_2.

[6] V. Azar, B. Hashemi, M. Rezaee Yazdi, The effect of shot peening on fatigue and corrosion behavior of 316L stainless steel in Ringer's solution. *Surf. Coatings Technol.* 204 (2010) 3546–3551. https://doi.org/10.1016/j.surfcoat.2010.04.015.

[7] H. Kumar, S. Singh, P. Kumar, Modified shot peening processes—a review, *Int. J. Eng. Sci. Emerg. Technol.* 5 (2013) 12–19.

[8] S. Singh, K.K. Pandey, S.K. Bose, A.K. Keshri, Role of surface nanocrystallization on corrosion properties of low carbon steel during surface mechanical attrition treatment, *Surf. Coatings Technol.* 396 (2020) 125964. https://doi.org/10.1016/j.surfcoat.2020.125964.

[9] D.K. Dwivedi, Surface engineering by changing the surface metallurgy, in: *Surf. Eng.*, Springer, New Delhi, 2018: pp. 73–90. https://doi.org/10.1007/978-81-322-3779-2_4.

[10] J. Thamilarasan, N. Karunagaran, P. Nanthakumar, Optimization of oxy-acetylene flame hardening parameters to analysis the surface structure of low carbon steel, in: *Mater. Today Proc.*, Elsevier, Amsterdam, Netherlands, 2020: pp. 4169–4173. https://doi.org/10.1016/j.matpr.2021.02.680.

[11] A.M. Guthrie, K.C. Archer, Induction surface hardening, *Heat Treat. Met.* 2 (1975) 15–21. https://doi.org/10.1007/978-94-007-2739-7_834.

[12] Y. Hu, Q. Qin, S. Wu, X. Zhao, W. Wang, Fatigue resistance and remaining life assessment of induction-hardened S38C steel railway axles, *Int. J. Fatigue.* 144 (2021) 106068. https://doi.org/10.1016/J.IJFATIGUE.2020.106068.

[13] R.S. Lakhkar, Y.C. Shin, M.J.M. Krane, Predictive modeling of multi-track laser hardening of AISI 4140 steel, *Mater. Sci. Eng. A.* 480 (2008) 209–217. https://doi.org/10.1016/j.msea.2007.07.054.

[14] M. El-Khoury, M. Seifert, S. Bretschneider, M. Zawischa, T. Steege, S. Alamri, A. Fabián Lasagni, T. Kunze, Hybrid processing of bearing steel by combining Direct Laser Interference Patterning and laser hardening for wear resistance applications, *Mater. Lett.* 303 (2021) 130284. https://doi.org/10.1016/J.MATLET.2021.130284.

[15] R. Aadhavan, P. Bera, C. Anandan, S. Kannan, K. Suresh Babu, Phase evolution of EBPVD coated ceria-zirconia nanostructure and its impact on high temperature oxidation of AISI 304, *Corros. Sci.* 129 (2017) 115–125. https://doi.org/10.1016/j.corsci.2017.09.026.

[16] R. Gordon, Chemical vapor deposition of coatings on glass, *J. Non. Cryst. Solids.* 218 (1997) 81–91. https://doi.org/10.1016/S0022-3093(97)00198-1.

[17] X. Liu, P.K. Chu, C. Ding, Surface modification of titanium, titanium alloys, and related materials for biomedical applications, *Mater. Sci. Eng. R Reports.* 47 (2004) 49–121. https://doi.org/10.1016/j.mser.2004.11.001.

[18] X. Zheng, M. Huang, C. Ding, Bond strength of plasma-sprayed hydroxyapatite/Ti composite coatings, *Biomaterials.* 21 (2000) 841–849. https://doi.org/10.1016/S0142-9612(99)00255-0.

[19] S.V. Joshi, M.P. Srivastava, A. Pal, S. Pal, Plasma spraying of biologically derived hydroxyapatite on implantable materials, *J. Mater. Sci. Mater. Med.* 4 (1993) 251–255. https://doi.org/10.1007/BF00122276.

[20] A. Fomin, M. Fomina, V. Koshuro, I. Rodionov, A. Zakharevich, A. Skaptsov, Structure and mechanical properties of hydroxyapatite coatings produced on titanium using plasma spraying with induction preheating, *Ceram. Int.* 43 (2017) 11189–11196. https://doi.org/10.1016/J.CERAMINT.2017.05.168.

[21] A. Feuerstein, J. Knapp, T. Taylor, A. Ashary, A. Bolcavage, N. Hitchman, Technical and economical aspects of current thermal barrier coating systems for gas turbine engines by thermal spray and EBPVD: A review, *J. Therm. Spray Technol.* 17 (2008) 199–213. https://doi.org/10.1007/s11666-007-9148-y.

[22] R. Ghasemi, H. Vakilifard, Plasma-sprayed nanostructured YSZ thermal barrier coatings: Thermal insulation capability and adhesion strength, *Ceram. Int.* 43 (2017) 8556–8563. https://doi.org/10.1016/j.ceramint.2017.03.074.

# 3 Classification of Thermal Spray Techniques

*Alok Bhadauria, Divya Rana and Kantesh Balani*

**CONTENTS**



DOI: 10.1201/9781003321965-3

Thermal spraying process is an extensively used technique for deposition. There is probably no limitation available for the combination of substrate and coating materials. So, thermal spraying coatings have a broad area of applications, both for repair and manufactured new parts. Today, there are many thermal spraying techniques, and many new techniques have been emerging. Every spraying technique has some advantages and disadvantages, and use of these techniques is dependent on the shape and size of the parts being treated, the atmospheric/service conditions, accessibility of thermal spraying equipment, and human knowledge. Thermal spraying is growing scientifically and is connected to many fields (e.g., materials science, tribology, medicine). Recently, a hybrid between vapor deposition and plasma spraying called hypersonic plasma particle deposition is developed, in which vapor precursors are injected with high-velocity plasma to make nanostructured coatings. The new coating materials like nanostructured and quasicrystalline have also been investigated apart from new thermal spraying techniques. Powder mixed from micrometer-sized and nanometer-sized particles has been inspected as well. As the thermal spraying is used to preserve components from wear, from corrosion at high temperature, and to restore damaged surfaces, nowadays these coatings are also used in biomedical applications. In this chapter, several thermal spraying techniques are discussed with their advantages, disadvantages and applications.

Thermal spraying involves a class of different coating techniques in which the feedstock in different forms is heated up to or above the melting point of precursor particles to form a coating. A stream of molten particles is generated by heating it electrically (arc or plasma) or chemically (combustion flame). Coatings are produced when particles undergo plastic deformation to impact on a surface, in the molten or solid state and with sufficient momentum and/or heat [1, 2].

Thermal spraying provides several ranges of coating thickness (several μm to several mm), which depends on the available feedstock and process. Coatings via thermal spraying can cover a large area with a high deposition rate (1–25 kg/h) compared with other conventional coating techniques. A large variety of materials can be thermally sprayed (e.g., metals, alloys, ceramics, polymers, composites) [2]. The materials are injected into the spraying gun in different forms, wire/powder, which due to heat are in the fully molten or partially molten state to be accelerated by a gas stream towards the substrate surface to deposit the coating. Combustion and electric arc discharge are two basic sources of energy for thermal spraying processes. The gas stream can gain energy input from other sources; for example, by focusing a laser beam or a halogen lamp, photon heating occurs, which may create a plasma volume that can be used for thermal spraying to produce reliable coatings [3]. The details of several thermal spraying techniques are discussed in this chapter with their principle advantages, disadvantages and applications.

## 3.1 THERMAL SPRAYING PROCESS

Thermal spraying is governed basically by two parameters: temperature and velocity. High temperature melts the particle to a state of plasticity, and velocity provides the required momentum to the molten/semi-molten particles, which gets accelerated and impacts the substrate to form a coating [2, 3]. Figure 3.1 shows a schematic of thermal spraying. The feedstock materials for coating are melted by some heat source. The molten material is then driven by the process gases and sprayed onto the substrate, where it solidifies and forms a layer.

### 3.1.1 The Structure of Thermally Sprayed Coating

Molten metal or partially melted particles impact on the substrate resulting in flattening, cooling and solidification of the particles and subsequent coating formation. Generally, this type of structure is called lamellar structure, which is a layered and sandwich-type structure made by deposition of successive layers. The splats (flattened particles) are disk shaped and the structure is not isotropic, which means the properties vary with the direction. The structure of thermal-sprayed coatings is heterogeneous due to the difference in cooling conditions of individual particles on impacting the substrate. The substrate materials and temperature also influence the cooling conditions. So, it is quite challenging to anticipate the structure and performance of the coatings via thermal spraying.

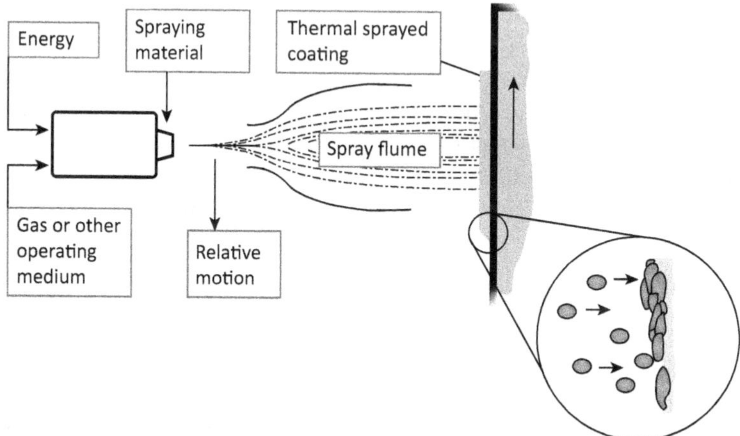


**FIGURE 3.1** General schematic of highlighting the thermal spraying process.

*Source*: Adapted and redrawn from [3].

During the coating process, some particles remain unmeted and they influence to create voids (say by bouncing off after impact), called porosity. All thermal spraying coatings involve some amount of porosity, which is very common to these coatings and depends on the particle velocity (impact energy), particle temperature, size and spraying angle. Particle velocity is inversely proportional to porosity. Generally, ceramic coatings are characterized via open porosity in the sample, so to improve the corrosion properties, sealing or impregnating treatment is applied. The major benefit of the thermal spray technique over other deposition processes is that coatings are deposited on the place of exploitation of certain a part (portability, on-spot repairs; e.g., bridges, offshore object repair), and also coating applications are not limited by the structure of the component for the deposition of coating. So a relatively thick coating (~10 mm) can be deposited at low deposition costs. Some more advantages of thermal spraying are as follows:

- The cost of repairing the machine components (damaged due to wear and tear of the surface) is much cheaper than replacing them with new ones.
- Thermal spraying is very versatile, in that any materials can be thermal sprayed (materials ranging from soft to hard, ceramics, superalloys, all metals and plastics). Cold thermal spraying is used to coat at a low temperature such that even an apple can be coated.
- Depending on the spraying system and the materials, coating thickness ranges between 0.001 inch to more than 1 inch thick, and the spraying rate ranges from 3 to 60 lb/hr.

Apart from the advantages of thermal spraying, there are also some disadvantages, like low coefficient of efficiency of the applied coating material and also the noise, fumes and dust produced during the spraying processes, which can affect human health. Also, a surface having a complex shape is difficult to reach, and coating can be applied only to that portion of the substrate which the gun or torch can reach easily. Thermally sprayed coatings possess a lamellar, sandwich-like structure and have different properties with the directions. As today, various thermal spraying methods and different coating materials are possible, so thermal spraying has covered many applications in different industries like automotive, power generation, agriculture, chemical, glass, textile, steel and so on. Thermal spraying processes can be classified into three major categories, taking into consideration the energy heating source: (i) combustion spraying, (ii) electric arc spraying, and (iii) plasma spraying. Figure 3.2 describes the classification of thermal spraying coating processes. One more recent addition to the thermal spraying processes is *cold spray*, which utilizes kinetic energy to energize and accelerates the powder particles to a supersonic range of velocities using convergent-divergent nozzles [1, 2, 4]. Table 3.1 shows a general comparative review of different thermal spraying techniques. It provides important information, however very generalized data for each process. All the data shown in the table related with general interpretation of data from many sources [5, 6]. Also, the temperature during the coating processes is very

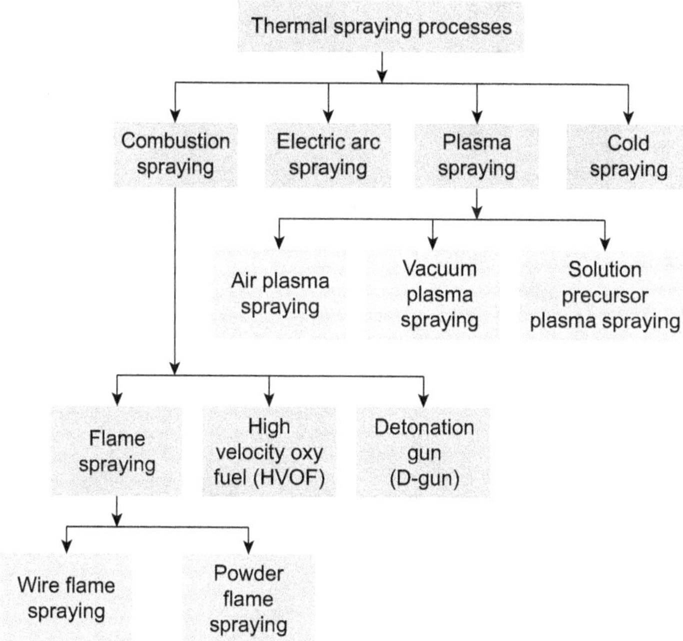


**FIGURE 3.2** Classification of thermal spraying processes.

*Source*: Adapted and redrawn from [3].

crucial for the thermomechanical properties of the substrate. Figure 3.3 represents the graph between the temperature involved in the coating processes to the coating thickness, which is obtained in the several respective processes. The coating thickness varies from several μm to mm (100 μm to 1 mm).

## 3.2 COMBUSTION SPRAYING

The exploration of combustion spraying was driven over many years by development in both scientific and technical aspects providing novelties and according to market requirements. For example, in 1983, Browning developed HVOF spraying, developed tungsten carbide–cobalt (WC–Co) cermet coating exhibiting improved properties [3]. Combustion spray guns are tools used for proper feeding of the spray material. There are three basic types: (1) flame spraying, (2) HVOF spraying and (3) detonation gun [7, 8, 9].

### 3.2.1 Flame Spraying

Flame spraying is one of the conventional and most popular thermal spraying methods. In the beginning of the 19th century, Swiss engineer Schoop developed this process (Schoop and Guenther, 1917). Initially, it was used for metals

**TABLE 3.1**
**Comparison of Different Coating Properties Deposited via Thermal Spraying Methods**

| Thermal Spraying Process | Flame Temperature (°C) | Particle Impact Velocity (m/s) | Adhesive Strength (MPa) | Porosity (%) | Oxide Content (%) | Relative Process Cost (low = 1, high = 5) | Maximum Spray Rate (kg/h) | Power (kW) | Typical Coating Thickness (mm) |
|---|---|---|---|---|---|---|---|---|---|
| Flame | 3000 | 40 | 8 | 10–15 | 10–15 | 1 | 2–6 | 25–100 | 0.1–15 |
| Detonation gun | 4000 | 800–1200 | >70 | 1–2 | 1–5 | 4 | 0.5–2 | 100–270 | 0.05–0.3 |
| HVOF | 3000 | 600–800 | >70 | 1–2 | 1–5 | 3 | 2–8 | 100–270 | 0.1–2 |
| Electric arc | 4000 | 100 | 12 | 10 | 10–20 | 2 | 10–25 | 4–6 | 0.1–15 |
| Air plasma | 12000 | 200–400 | 10–70 | 1–5 | 1–3 | 4 | 2–10 | 30–80 | 0.1–1 |
| Vacuum plasma | 12000 | 400–600 | >70 | <0.5 | 0 | 5 | 2–6 | 50–100 | 0.1–1 |
| LPCS | 200–650 | 300–500 | 5–30 | <0.5 | 0 | 1 | 0.5–3 | 3.3 | 0.2–2 |
| HPCS | 500–1000 | 400–800 | 10–40 | <0.5 | 0 | 4 | 4–12 | 17–47 | 0.3–4 |

*Abbreviations*: HPCS = high-pressure cold spraying; HVOF = high-velocity oxy-fuel; LPCS = low-pressure cold spraying.

*Source:* Adapted from [6].

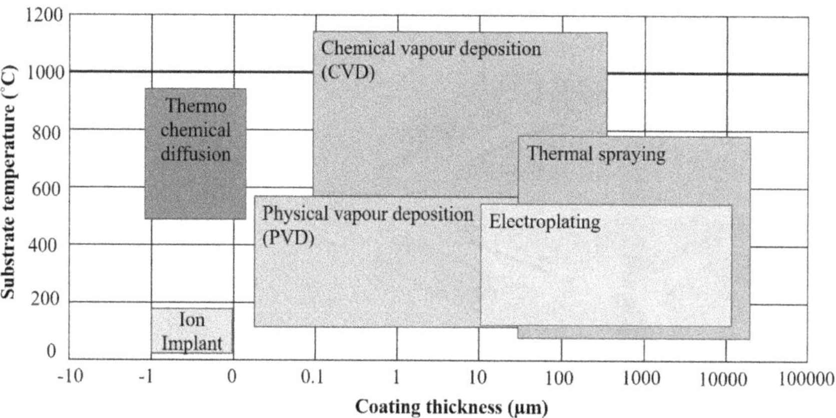


**FIGURE 3.3** Qualitative comparison of coating processes.

*Source*: Adapted and redrawn from [7].

with a low melting point (e.g., lead, tin) and then extended to refractory materials [9]. Because of low initial capital investment and simplicity of operation, it has maintained its utility in today's modern thermal spraying world. A wide range of materials is possible to be sprayed in powder, wire or rod form, ranging from polymers to ceramics [8]. Thermal spraying uses heat energy derived from fuel combustion with oxygen to melt the spraying particles which are made to impact on the substrate to form a coating. Stoichiometry shows full conversion of C into $CO_2$ and the H atom into $H_2O$, according to the following reaction:

$$2C_2H_2 + 5O_2 \rightarrow 4CO_2 + 2H_2O \tag{3.1}$$

Flames can be oxidizing and reducing according to the application; for metal spraying, the flames are set to be reducing to minimize oxidation. Coatings produced through flame spraying exhibit a porosity content of >10% and low adherence (<30 MPa), but still the process is versatile, portable and involves low capital investment.

#### 3.2.1.1 Wire Flame Spraying

##### OPERATION PRINCIPLE

Figure 3.4 shows the schematic of the wire flame–spraying technique. The spraying material is in the wire form injected into flame to melt the wire tip. This process relies on the heat energy produced by the chemical reaction between combustible fuel gas and oxygen. A stream of carrier gas causes atomization of the molten tip of the wire and drives the droplets towards the substrate, where they will solidify after impact at a particular cooling rate. The produced heat during the process develops a stream of gas with excess temperature of around 3000 °C and maintains the balance between acetylene and oxygen. In wire flame spraying, the atomization

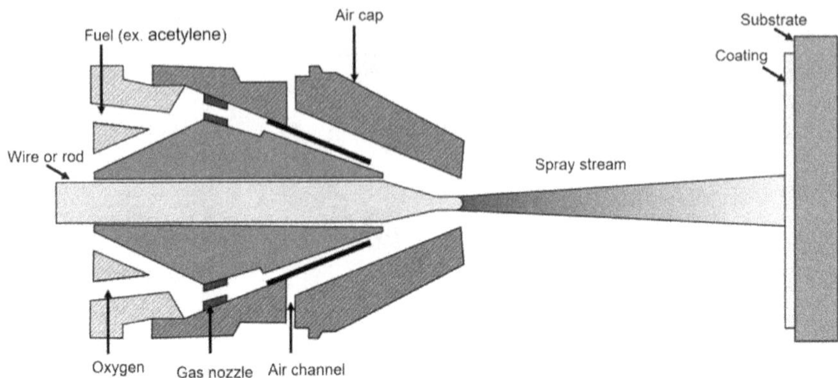


**FIGURE 3.4** Schematic of wire flame spraying.

*Source*: Adapted and redrawn from [7].

process produces finer droplets resulting in finer coatings. Apart from the other techniques, wire flame spraying is usually used for the resistance of the corrosion of materials applications [10, 11]. It is desirable to optimize process parameters so as to get better coating properties (e.g., porosity, adhesion). The main process parameters include feedstock material, service temperature (< 3200 °C), fuel (propane, hydrogen and acetylene) and spray gas speeds below 100 $ms^{-1}$.

On the major upper edge of the wire flame–spraying process, a fine molten metal spray is generated to produce denser coatings due to the continuous melting of the wire. Some other advantages are as follows:

- The wire spray technique is capable of a high deposition rate of coating on the substrates. This coating can be used for complex-shaped features where more durability of the coating is required.
- The cost of the wire form is less than the powder form because the processing of nanopowder includes a high cost.
- Wire flame coating is comparatively simpler in the operation because wire is fed into it and a roller will provide the path in extrusion. Also, the reproducibility is easier in comparison to other coatings.
- Gases are involved to provide the velocity and direction to the coating. The arrangement of the components is portable in wire flame spraying [10, 11].

The wire flame–spraying process has major disadvantages of low bond strength and shear strength between the coating/substrate. Apart from this, some other disadvantages include the following:

- It is limited to spraying materials as supplied in wire form.
- High thickness of the coating is difficult to deposit by wire flame spraying.

The coatings deposited via wire flame spraying are used for protection against atmospheric corrosion, abrasion resistant coatings and nonferrous coatings (copper, bronze, etc.). It is also used as electromagnetic shielding.

#### 3.2.1.2 Powder Flame Spraying

##### OPERATION PRINCIPLE

In the powder flame–spraying process, the materials used for coating are in powder form. Coating material is fed into the gas flame where it melts due to the heat produced during combustion (Figure 3.5). A carrier gas accelerates, and the molten particles are transported to the substrate where they solidify to form a coating. The fuel gases generally used are acetylene and hydrogen. The chemical reaction involved in this method lies between the oxygen and combustion of fuel to generate the source of heat. Because of this high heat source, a stream of gas has a temperature (>3000 °C) and maintains a balance of both gases [3, 12].

The feeding of the material into the flame-spray torches is done either by carrier gas or by gravity. Carrier gas devices use a gas to transfer the powder sample from the external powder feeder to the spray gun via a hose. Gravity devices use bottles that are directly mounted on the top of the torch to feed the powder. Powder flame spraying is employed to process alloy powders which requires post treatment before the desirable application. The parameters which define the process involve the feedstock, fuel used, flame temperature attained and the in-flight velocities of the particle. Flame temperatures attained are as high as 2800–3000 °C and the particle velocity as high as 50 $ms^{-1}$. This coating technique falls under the category of low-velocity flame spraying.

In cases where it is difficult to form a wire out of a material, and when wire flame and other methods are unsuccessful to advocate the coating on some

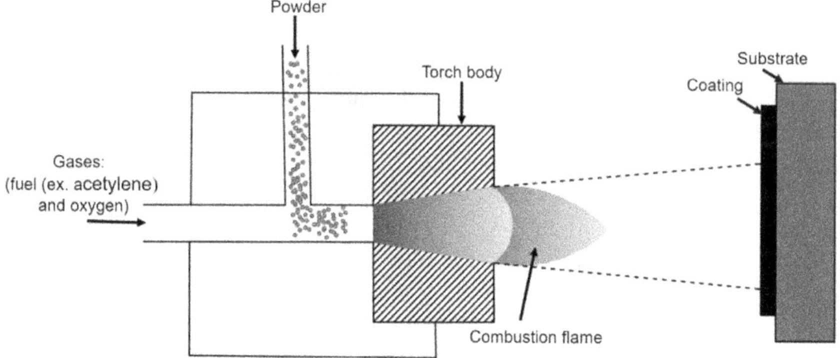


**FIGURE 3.5** Schematic representation of the powder flame–spraying process.

*Source*: Adapted and redrawn from [5].

special alloys, powder flame spraying is used. Some other advantages include the following:

- The setup of powder flame spraying is portable due to the easy mechanism involved in the process.
- The powder form of any material is easily available in the market. Hence, powder flame spraying has a large choice of materials.
- Reproducibility of the coating quality is easier to control.

Owing to its simplicity and flexibility, powder flame spraying is among the oldest and widely used spray technique. Nevertheless, there are some inherent disadvantages, as follows:

- The working temperature of powder flame spraying is around 3000 °C. This temperature is low enough to melt the coarse particles of ceramic materials but is high enough to melt the fine metallic particles.
- Higher heat is transmitted to the substrate leading to distortion of the substrate.

Powder flame spraying is used for maintenance and restoration purposes, deposition of composite coatings, spraying of thermoplastics and depositing wear-resistant coatings.

## 3.3 DETONATION GUN

Poorman et al. first developed detonation spraying in 1955, and it was developed as the detonation gun (D-gun) process by Union Carbide in 1955. Later, in 1960, the Paton Institute (Kiev, Ukraine) also developed the detonation gun [5].

### OPERATION PRINCIPLE

The detonation gun (D-gun) is the first thermal spray technique utilizing high velocity. It consists of a long barrel with a water-cooling arrangement, and it is closed on one side and open on the other side. The barrel is 1 inch in diameter and associated with some valves for powder and gases. A schematic of the technique is shown in Figure 3.6. The fuel gas (most commonly acetylene) and oxygen mixture are fed into the gun barrel along with the powder coating material (particle size <100 μm). For the ignition purpose a spark plug is used; it ignites the fuel gas, oxygen and powder mixture, resulting in detonation that melts and further accelerates the particles with a speed of 600–750 m/s. Nitrogen gas is used for purging after every detonation. Every detonation resulted in deposition of a splat (disk) of coating, which is a few micrometers thick and close to 1 inch in diameter. The coating is fabricated by many overlapping splats [4, 13].

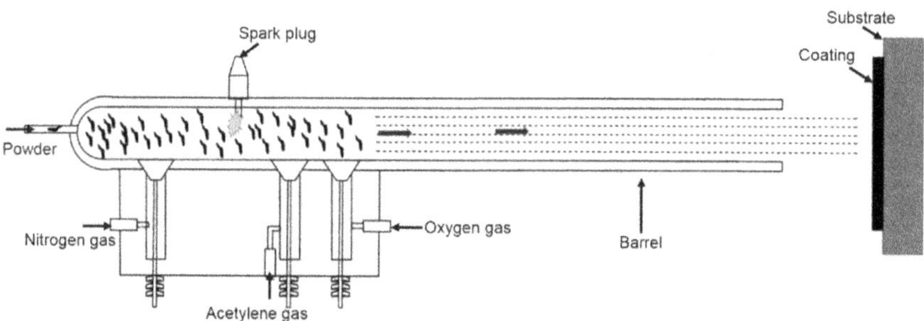


**FIGURE 3.6** Schematic representation of D-gun spraying.

*Source*: Adapted and redrawn from [5].

The main characteristic parameters for D-gun are as follows:

Barrel length and diameter (1800 and 22 mm), fuel (acetylene) and oxygen mixture, carrier gas (nitrogen), frequency of firing (2–4 Hz), coating thickness (5–20 μm) and power consumption/hour (2 kW) [14].

### 3.3.1 Advantages of D-Gun

The D-gun spraying process produces coatings with a good adhesive strength, improved corrosion resistance, high density (low porosity) and almost no oxidation. There are several more advantages of the D-gun spraying process, as follows:

- The D-gun spraying technique uses a wide range of materials, including cemented carbide with a metal matrix composition and all types of ceramics.
- The range of coating thickness is significant (0.05–2 mm).
- These types of coating represent good, densified coating (<1% porosity).
- The D-gun spraying technique is good for effective amorphous coating due to its high particle velocity (500–1000 m/s) and low processing temperature [14, 15].

### 3.3.2 Disadvantages of D-Gun

The major disadvantage of D-gun is that these coatings are strong in compression and weak in tension. Due to this reason, the coating cannot be applied on malleable or expanding components. The other disadvantages are as follows:

- The D-gun spraying process is performed at specifically designed locations due to its significant large size and loud process. So, this system is set up in a soundproof room [15].
- D-gun spraying creates a mechanically bonded coating which is less strong than metallurgically bonded coating.

D-gun sprayed coating is used to protect the coating against mechanical and wear damage. For example the coating made by the D-gun process shows higher hardness and densification (50 HV and ~1.2 ± 1.0% porosity) than supersonic plasma spraying, which resulted in good wear performance at room (~25 °C) and elevated temperatures (500 °C) [16]. Applications of the D-gun process include machinery components (shafts, bushings, seals, bearings) and aviation (e.g., rotor/stator blades, guide rails) [4].

## 3.4 HIGH-VELOCITY OXY-FUEL

The high-velocity oxy fuel (HVOF) torch was developed at the beginning of the early 1980s. The working principles of HVOF and D-gun are more or less the same; the only difference is the method of combustion of fuel in oxygen. In the case of HVOF, the burning is continuous, whereas for D-gun it is repetitive. The initial work on HVOF was carried out at Browning Engineering (USA) and by Thayer School of Engineering (USA). Nowadays, HVOF has become a standard technique for spraying carbides and alloys [3, 5, 16].

### OPERATION PRINCIPLE

HVOF is a thermal spraying system which utilizes various fuel gases (hydrogen, propylene, propane and kerosene) combined with oxygen. In this process, a mixture of fuel and oxygen is injected into the combustion chamber and mixed with a sprayed powder sample. Combustions of fuel/gas generate high pressure and high temperature conditions in the combustion chamber, causing the gases to flow at supersonic speed via the nozzle (diamond shaped pressure waves), as shown in Figure 3.7. Each diamond formation in carrier gas signifies crossover to the next Mach speed (i.e. speed of sound in air). Typical carrier gas velocities reach 500–2000 m/s). The precursor particles are changed into the partial/fully molten state when particles travel through the nozzle and into the chamber. The hot gas stream and powder mixture is accelerated towards the substrate to be coated. The temperature of the flame is between ~2500 and 3200 °C depending on the kind of fuel used, the

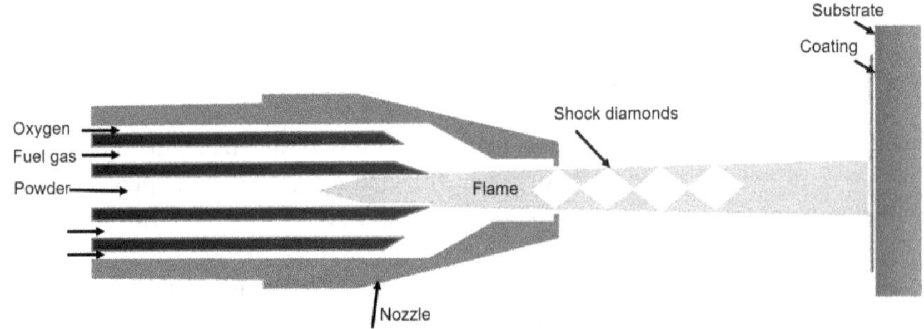


**FIGURE 3.7** Schematic representation of HVOF using gas for fuel.

*Source*: Adapted and redrawn from [7].

oxygen/fuel ratio and the pressure of the gas. Factors including flame temperature, the melting point of material, particle dwell time and thermal conductivity are all responsible for full particle melting or partial melting [5]. The HVOF process depends on several parameters, such as standoff distance, air/oxygen pressure, fuel/air flow rate, powder feed rate and so on.

### 3.4.1 Advantages of HVOF

There are many benefits of the HVOF spraying process. For example, these coatings can be applied on materials including tungsten carbide, which provides coatings with excellent wear resistance (~10 times more than hard chrome plating). Some other advantages of HVOF coatings are as follows:

- The low oxide content (0.25–0.9 wt%) in HVOF-sprayed coatings is due to there being less in-flight exposure time [5, 17].
- There is higher density (lower porosity) due to higher particle impact velocities (400–1000 m/s).
- In HVOF, the protection from corrosion improves owing to less through-thickness porosity (<1%).
- $Al_2O_3$ and $ZrO_2$ coating deposited via HVOF shows 15 and 20 times improvement in wear resistance than that of atmospheric plasma spraying (APS), respectively [18].

### 3.4.2 Disadvantages of HVOF

The main shortcoming of the HVOF technique is that it is quite impossible to do coatings on the internal surfaces of cylindrical components because this process requires the line of sight of the surface, and stand-off distance is ~200–250 mm. Some other demerits are as follows:

- In HVOF, powder particle sizes are limited to 5–60 μm, as this process requires narrow particle size distributions and qualified, well-experienced personnel to ensure better and safe operation and achieve quality coating.
- HVOF spraying equipment needs a larger investment than that of other thermal spraying techniques (e.g., flame and electric arc spraying).
- HVOF requires automated manipulation of the spray gun rather than manual operation.

HVOF-sprayed coatings are used for replacing the chrome plating on navy aircraft hydraulic components and the landing gear [18]. HVOF also provides corrosion resistance and erosion wear resistance in the hydroelectric power industry [5, 18]. These sprayed coatings are useful to metallic interconnects which are used in "solid oxide fuel cell" stacks to resist extremely high temperature oxidation of chromium species. The HVOF coatings have wide applications in industries such as aviation, chemical, power, mining, automotive, petrochemicals and so on.

## 3.5 ELECTRIC ARC SPRAYING

The electric arc–spraying process, with a novelty of using twin wire, was originally developed and patented by Max Schoop in 1914. Later, he continued his research on thermal spraying including plasma spraying [5].

### OPERATION PRINCIPLE

Electric arc spraying is a thermal spraying technique utilizing two types of metallic wires as a material for coating. These wires, being electrically charged, are fed by a wire feeder to bring them to the nozzle at an angle of 30°. The opposite charges on these wires produce an arc to melt the wire, and then jet-compressed air at high pressure atomizes the molten metal and accelerates the molten particles onto the workpiece (substrate; see Figure 3.8). The power source for arc producing is a constant voltage, direct current welding apparatus. The main process parameters in electric arc spraying are voltage, current, type of materials, air pressure and wire feed speed.

For example, Chmielewski et al. [19] used the following parameters for the spraying process: voltage (34 V), current (250 A), wire materials (aluminum and steel), wire diameter (1.6 mm), air pressure (0.3 MPa) and wire feed speed (2.6 mm/min).

### 3.5.1 Advantages of Electric Arc Spraying

Electric arc spraying is a relatively cost-effective method and one of the most comprehensive thermal spraying methods. Also, the process is a simple push-button operation. It is easily capable of offering engineering coatings which change the surface properties of sprayed objects. Some other benefits of the coatings are as follows:

- The relatively high temperature (~4000 °C) of the arc melts the wires faster, and deposited particles have a large heat content and higher fluidity than in the flame-spraying technique.

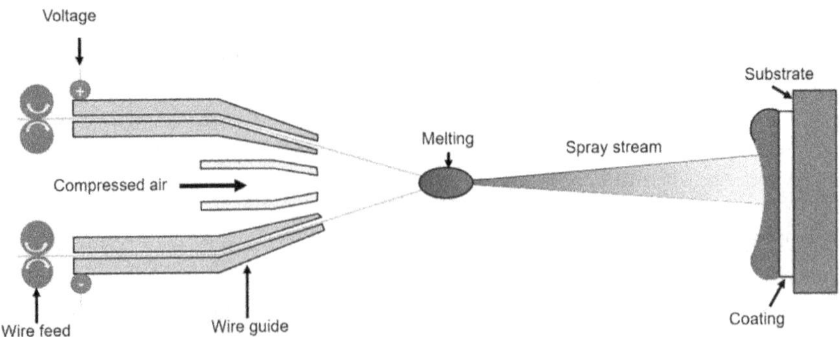


**FIGURE 3.8** Schematic representation of electric arc spraying.

*Source*: Adapted and redrawn from [7].

- The deposition rates are higher (15 kg/h or more), which is 3–5 times that of other thermal spraying processes.
- The power input is 5–10 kW, which is very low as compared to 50 kW for HVOF and plasma spraying. As this process does not use high pressure combustion gases, it poses lower risks compared to other thermal spraying systems.

### 3.5.2 Disadvantages of Electric Arc Spraying

Apart from the many advantages, there are several disadvantages of the electric arc–spraying process:

- Only electrically conductive materials can be sprayed. It is not possible to spray the cermet or any ceramics; however, hard particles can be sprayed via cored wires.
- Electrical arc spraying can produce huge amounts of fume and dust, so the operator needs to wear protective clothing and use a breathing device.
- Arc spraying has a low particle velocity (100 m/s), which results in low bond strength and high porosity in the sprayed samples [3].

Electric arc spraying is used to improve the corrosion resistance of steel structure on ships and steel coating for engine block bores. Other applications include bridges, boilers, rolls, crankshafts, gearbox, shielding of X-ray rooms, slab marking and so on [1, 5].

## 3.6 PLASMA SPRAYING

Dr. Schoop of Zurich invented and developed the plasma spraying process in 1911. Plasma spraying is used to fabricate dense coatings by the association of high particle velocities and temperature. Plasma is a fourth state of matter such that high temperatures lead to the ionization of the gas, making it electrically conductive to the extent that electrical and magnetic fields govern its behavior. Plasma spray allows spraying of metals and ceramics onto a wide range of materials [1, 3, 5].

### 3.6.1 Air Plasma Spraying

#### OPERATION PRINCIPLE

The plasma torch incorporates a cathode of thoriated tungsten and a copper anode. An arc of very high frequency is struck between the two electrodes. The gas which generates plasma usually contains a primary gas such as argon (Ar) and a secondary gas such as hydrogen ($H_2$) or helium (He) to improve the heat transfer characteristics. This mixture of plasma gases flows around the electrodes, and a high voltage discharge causes an arc to form between the electrodes. This heating from the arc causes the ionization and dissociation of the gases to form

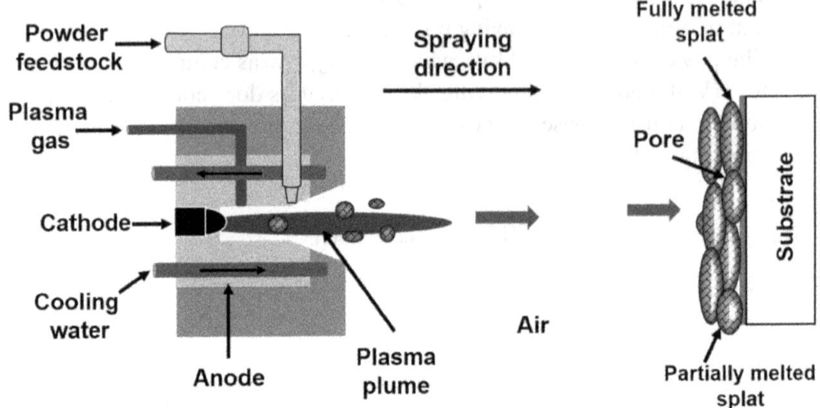


**FIGURE 3.9** Schematic of air plasma spraying.

*Source*: Adapted and redrawn from [8].

a plasma. Figure 3.9 shows the schematic of air plasma spraying (APS). As the plasma flows around the constricted nozzle-shaped anode, a plasma plume of several centimeters of length is developed with the temperatures of about 15,000 °C just at the exit of the nozzle to about 6000 °C as it moves away from the exit [20]. The powder-spray material is injected radially into plasma plumes, where they melt and are accelerated towards the substrate by the gases, where they rapidly cool and solidify to form a coating. Many complex parameters are involved in the process for the precise control and optimization of the process. These include plasma gases (Ar, He, $N_2$, $H_2$), plasma temperature (6000–15,000 °C), particle velocity up to ~500 m/s and good flow properties of the powders, which is in turn associated with their morphology. A constant powder feeding rate is required for good-quality coating.

### 3.6.1.1 Advantages of APS

The APS method produces a highly dense coating by the combined effect of high temperature, relatively high particle velocities and an inert spraying medium. The other benefits of this method are as follows:

- Versatile thermal spraying process and materials with large melting point; for example, refractory materials like W, $Al_2O_3$ can be sprayed.
- Air plasma spraying can produce dense coatings (>90%).
- Materials with varying composition such as functionally graded material can be sprayed.
- Deposition rates are quite high, up to 4 kg/hr.

#### 3.6.1.2 Disadvantages of APS

In spite of many advantages of this process, there are some drawbacks:

- In this process, the problem is that coating internal surfaces like bores of smaller diameters also requires automated gun manipulators.
- Generation of by-products, such as amorphous $Ca_3(PO_4)_2$ and bioactive $Ca_3(PO_4)_2$, leads to mechanical and adhesive instability of the coating.
- Plasma guns experience fast deterioration of the internal gun electrodes and other internal parts, which leads to frequent replacement of these electrodes and other parts.

APS is highly cost-effective and a very versatile process since many material types including metals, ceramics and composites can be deposited as a coating. It has found its application in thermal and electrical insulation, to prepare coatings for wear/abrasion resistance and TBCs. It is preferably suitable for deposition of the coatings on the larger-sized components.

### 3.6.2 Vacuum Plasma Spraying

#### OPERATION PRINCIPLE

Vacuum plasma spraying (VPS) is a thermal spray technique in which fine powder particles (10–100 μm) are heated to a fully molten/partially molten state and accelerated towards the substrate, where they solidify and flatten to form individual splat. With continuous deposition, more molten droplets are deposited, which forms a coating. Figure 3.10 shows the schematic of VPS, where the chamber atmosphere is kept under vacuum (< 900 mbar). A very high intensity arc is generated between the anode (copper) and cathode (tungsten). As the arc current increases, high-temperature plasma (~12,000 °C) is created, which results in a higher degree of ionization. Argon is used as a primary gas for generating plasma; to improve the heat conductivity of plasma, gases such as $N_2$, He, and $H_2$ can be mixed with argon [21].

Powder feedstock is injected into the hot plasma plume, where it melts owing to the extremely high temperature of plasma (~12,000 °C) and is directed towards the substrate. The spray distance is usually of the order of 100–400 mm. The molten particles flatten and solidify rapidly on the substrate, forming individual splats. As for the perfect plasma system, all the molten particles should reach the substrate, so with time particles continue to deposit, forming a coating consisting of many splats layered one over the other [4, 5]. These splats are linked via chemical and mechanical bonding. The outcome of the process is affected by the type and temperature of both substrate and coating material. The purity of the coating is also affected by the kind of gas used and the rate at which powder is accelerated towards the substrate.

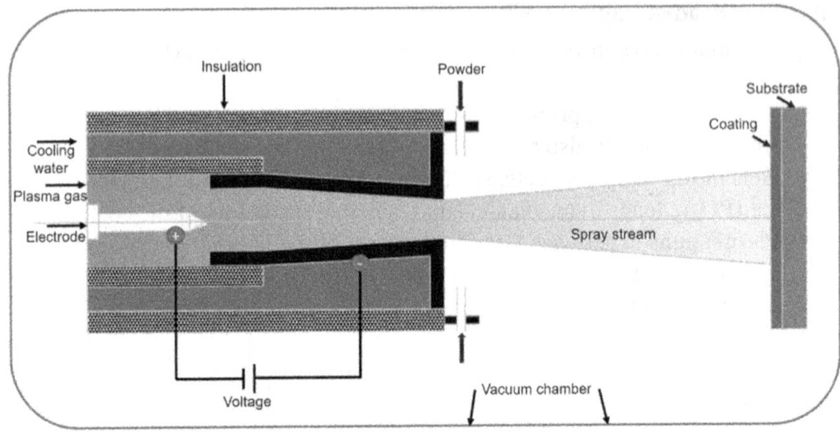


**FIGURE 3.10** Schematic of vacuum plasma spraying.

*Source*: Adapted and redrawn from [7].

#### 3.6.2.1 Advantages of VPS

VPS is carried out at low pressures, enabling the increase in the particle velocity that suppresses the formation of any reaction products. Quality of the coating is improved as well as its adhesive strength along with other advantages, including the following:

- This process can manufacture near-net shaped structures of materials such as ceramic metal matrix–reinforced composites and functionally graded materials.
- The coating developed in VPS is oxide-free coating and easy to characterize; less residual stress develops during the coating.
- Because of plasma and the sophisticated arrangement in vacuum plasma spraying, it can be used for longer spray jets. The efficiency of deposition is high in comparison to other techniques.
- Better control of coating thickness is possible in vacuum plasma spraying.

#### 3.6.2.2 Disadvantages of VPS

The main disadvantage with this technique is the high substrate surface temperature, but this is compensated for by the good quality of the resultant coating. Some other drawbacks are as follows:

- The major problem with VPS is to hold the substrate in the chamber, due to the size limit of the chamber, so a large substrate cannot be kept.
- It is costly compared with the other techniques because of the vacuum generation setup. In fact, this technique is very sophisticated and needs proper precautions and maintenance.

- Coating made by this process can vary in properties than initial feedstock owing to the formation of meta-stable phases during plasma spraying.

This spraying process is extensively used to deposit coatings of highly reactive materials and a high melting point. Since the operation is performed in a controlled atmosphere (i.e., a vacuum), the oxide-free coatings can be produced. VPS is also used to produce coatings for biomedical applications.

### 3.6.3 Suspension/Solution Precursor Plasma Spraying Technique

APS is a conventional plasma spraying method which was explained in section 3.6.1. In APS, solid powder particles in the range of 10–100 μm particle size are fed into the plasma jet via carrier gas. The particles impact on the substrate, resulting in a coating formation with micrometer-sized features. The use of powder feedstocks of this size makes it difficult to use fine-sized microstructures. Therefore, to overcome the limitations of conventional plasma spray techniques, the suspension/solution precursor plasma spraying (SPPS) technique has emerged as a new technique that is used to permit coatings with comparatively improved features via submicron/nanometer–sized particles [22].

#### OPERATION PRINCIPLE

Figure 3.11 shows the schematic of the SPPS technique. In SPPS, the liquid phase, which contains dispersed fine particles, is fed into the plasma plume via liquid feedstock system as a stream, or stream of droplets. The steps involved in SPPS

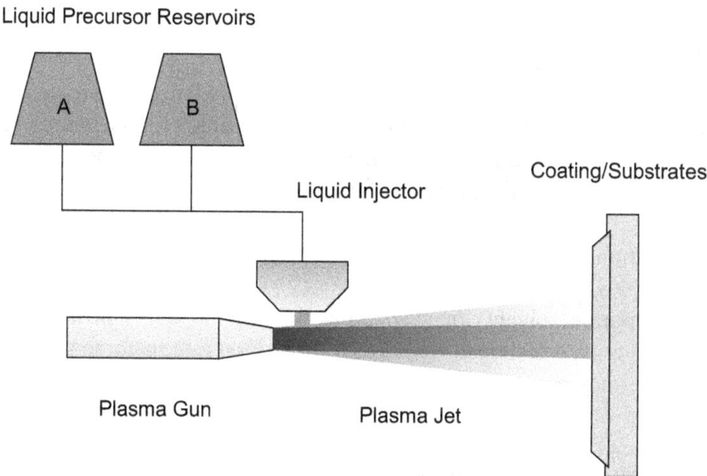


**FIGURE 3.11** Schematic of suspension/solution plasma spray technique.

*Source*: Adapted and redrawn from [23].

process are atomization, particle pyrolysis, solvent evaporation, melting of solid particles and deposition of the coating. The solution precursor contains the metal inorganic salts, organic compounds and solvent, which is organic liquid or water. The process uses the liquid injection method, most commonly being the through injection of a liquid stream. As an alternative, the suspension/solution is atomized before its interaction with the plasma jet. This process leads to the formation of much finer microstructures of coatings than the APS [23].

Factors affecting the quality of the deposited coating include the indirect relation between the particle size and the droplet size, so control of the particle size distribution remains a challenge. Higher deposition efficiency is achieved by projecting a smaller area of the jet into the plasma plume. Characterization of the precursors is also considered an essential factor for the successful deposition of the coating.

#### 3.6.3.1 Advantages of SPPS

The advantage of the SPPS process over other spraying processes is that it is possible to do distinct compositional ratios in a single spray session which is quite a tedious for the conventional process. The other advantages associated with the process are as follows:

- SPPS produces splats ~10 times smaller (1–5 μm) than splats formed by the other plasma spraying techniques, leading to superior properties.
- It produces layered porosity, which helps to reduce the thermal conductivity of yttria-stabilized $ZrO_2$ (YSZ) TBCs.
- It can produce coatings with relatively high segmentation crack density (>7 cracks/mm) with appropriate conditions; fine porosity levels in the coatings can also be achieved using this technique.

#### 3.6.3.2 Disadvantages of SPPS

Despite its advantages, the SPPS process has many challenges which include finding a starting material of low viscosity and then dealing with the complexity during the decomposition stage which affects the quality of the deposited coatings. The other limitations of the process are as follows:

- Deposition efficiency is 4–8 times lower than conventional plasma spray techniques.
- To evaporate solvents or suspending liquid, the SPPS process needs torch energy, which leads to a lesser deposition rate in comparison with conventional processes.
- SPPS involves disintegration and also vaporization of liquid, which controls the process and in turn makes the process far more complex.
- The SPPS process involves a large heat transfer to the substrate due to the short stand-off distance (~30 mm), leading to heat fluxes up to 40 $MW/m^2$ that must be controlled.

TBCs fabricated via SPPS exhibit greater life and lower thermal conductivity in comparison with conventional plasma spraying. SPPS is used to produce various components of solid oxide fuel cells. Bioactive coatings consisting of hydroxyapatite (HA) produced by SPPS technique exhibit a good amount of OH group, which leads to greater structural integrity [24].

## 3.7 COLD SPRAYING

The cold-spraying technique was first invented in the mid-1980s at the Russian Academy of Sciences (RAS) by A. Papyrin and his research team [5]. In cold spraying, the metallic/non-metallic materials are heated at comparatively low temperature (much lower than the melting point) and accelerated at a speed of 500–1200 m/s—depending on the type of gas used (air, He, N), nozzle design (convergent, convergent-divergent) and spraying parameters—in a highly pressurized gas jet, which allows the powder particles to accelerate and plastically deform as impact occurs on the target surface to form splats adhered with the substrate, resulted in coating [25]. Cold spraying is classified as either high pressure (HP) or low pressure (LP). The difference can be explained by the kind of gas used, the working pressure and the method of powder feeding [25].

### 3.7.1 High-Pressure Cold Spraying

#### OPERATION PRINCIPLE

Generally, nitrogen ($N_2$) or Helium (He) gas is used in high-pressure cold spraying (HPCS) at temperatures up to 1200 °C and pressure up to 5 MPa; this helps to obtain high velocity (up to 1200 m/s) and deposits different materials like Al, Ti, Cu, Ni, Ta, Cu-Sn, Ni-Al and so on. Figure 3.12 shows the working principle of HPCS equipment. The heated pressurized gas is delivered to the convergent

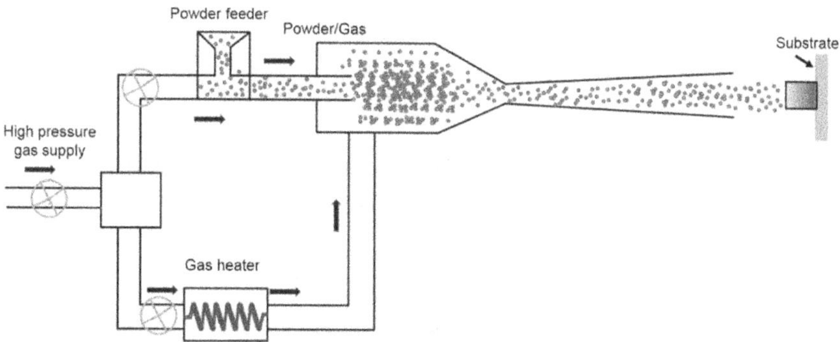


**FIGURE 3.12** Schematic presentation of the high-pressure cold-spraying system.

*Source*: Adapted and redrawn from [27].

divergent nozzle, where it helps to accelerate the pressurized gas to reach its supersonic velocity and simultaneously reduces its temperature. A powder feeder feeds the powder axially with a distinct gas line to gas stream. The powders accelerate with the help of gas, and at the exit of the nozzle it impacts fully deposited on substrate, with the system managed via a control unit [25, 26].

### 3.7.2 Low-Pressure Cold Spraying

Figure 3.13 shows the low-pressure cold-spraying (LPCS) process. Many features are common with HPCS, however the difference occurs in the temperature and gas pressure. In LPSC, $N_2$ gas and air are used with a gas pressure of 0.1 to 1 MPa, and temperature varies from room temperature (~25 °C) to 600 °C. Because the powder particle velocity is relatively less in LPSC, low-strength materials can be sprayed (e.g., Al, Cu, Sn, Zn). To increase the deposition efficiency of LPCS, the ceramic particles (alumina, SiC) are added. Apart from pressure and temperature, the powder feeding system is different in low-pressure and high-pressure cold spraying. In LPSC, the powder is fed radially, whereas in HPCS the powder is fed axially. The different process parameters used in both cold-spraying processes are shown in Table 3.2 [26].

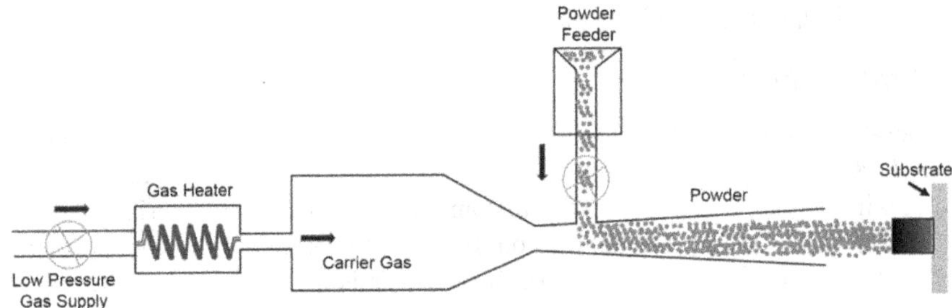


**FIGURE 3.13** A diagram representing the low-pressure cold-spraying system.

*Source*: Adapted and redrawn from [27].

**TABLE 3.2**
**Different Process Parameters Used in the Cold-Spraying Process**

| Process Parameter | HPCS | LPCS |
|---|---|---|
| Gas | Nitrogen, helium, mixture | Nitrogen, air |
| Rate of gas flow ($m^3/h$) | 0.8–2.5 ($N_2$), 4.2 (He) | 0.3–0.4 |
| Gas pressure | 1.5 | 0.1 to 1 |
| Gas temperature (°C) | 20–1100 | 20–600 |
| Rate of Powder feeding (kg/h) | 4.5–13.5 | 0.3–3 |
| Spray distance (mm) | 10–50 | 5–15 |

*Source:* Adapted from [27].

The important process parameters used in cold spraying are shown below (Table 3.2).

### 3.7.3 Advantages of Cold Spraying

The main merit of the cold-spraying technique is that it is a solid state process, which results in unique coating characteristics. Several advantages of the cold-spraying technique are as follows:

- Compared to other spraying techniques, cold spraying has high deposition efficiency up to 95%. It is shown that by minimizing the $O_2$ content in starting material, relieving stress in powder particles, optimizing the spraying parameters and optimizing the distribution of particle size, a very high deposition rate is possible for most of the materials like refractory materials (e.g., tantalum, niobium).
- Thick coatings can be manufactured with a depth rate per pass of 1–2 mm and low tensile residual stress. There is no need or only a slight need for surface preparation of the substrate, as the process is three in one (i.e., grit blasted, cold spray, and shot peened).
- Good microstructure stability is achieved (i.e., no phase changes and grain growth). Minimal heat input during deposition and high bond strength with compressive residual stress makes it suitable for thicker coating, which is a limitation of other spray methods.

### 3.7.4 Disadvantages of Cold Spraying

The major disadvantage of the cold-spray process involves solid-state plastic deformation, which results in lower ductility of the outside layer that necessitates post heat treatment before further processing. Some more disadvantages are as follows:

- Although composite materials can be sprayed, pure ceramics materials and some alloys are not.
- Substrates should have at least limited ductility to produce good bonded coatings via cold spraying. Thus, these coatings over ceramic substrates have limited bond strength.
- Cold spraying is a line-of-sight process, so complex shapes and internal surfaces are difficult to coat.

All types of materials (polymers, metals, ceramics and metal matrix composites [MMCs]) have been deposited via the cold-spraying process. It has been used in many industries, such as automotive, medical, energy, marine and aerospace. In thermal power plants, it is used to improve the life of superheater tubes. In the current scenario, the cold-spraying technique was developed as an additive manufacturing to synthesize freestanding metallic parts and also refurbishing damaged metal parts [27, 28].

### 3.7.5 Particle Velocity and Temperature in Cold Spraying

The details of the cold-spraying process were shown in section 3.7. In brief, in the cold-spraying process electrical energy is used for heating the pressurized gas to temperatures ranging from 300 °C to 800 °C; then the pressurized gas goes through a nozzle (converging-diverging) to form a supersonic gas jet. The cold spray does not melt the materials; rather, it increases the sonic velocity of gas in the throat region. Gas expansion in the diverging section of the nozzle quickly cools the gas and leaves the gun nozzle at a lower temperature (even below room temperature for some cases). Generally, higher gas velocities end with higher particle velocities. The sonic velocity of gas is directly proportional to the temperature and inversely proportional to the molecular weight of the gas. Sonic velocity ($v$) can be calculated as follows [26, 27];

$$v = \left( \frac{\gamma RT}{M_W} \right)^{\frac{1}{2}} \tag{3.2}$$

where $\gamma = \frac{C_p}{C_v}$ is the specific heat ratio (for air 1.4, for He 1.66), $R$ is the gas constant (8.314 J/mol K), $T$ is the temperature of the gas, and $M_W$ is the molecular weight of the gas. So, a convenient way to increase the gas velocity is to increase the temperature of the gas or use gas with a light molecular weight (e.g., He). On the other hand, with an increase in pressure the flow chokes, so after sonic velocity, the increasing pressure does not increase in the gas velocity.

#### 3.7.5.1 Process Gas Flow inside the Cold-Spray Nozzle

Assumptions made in gas flow in the nozzle (model) are as follows:

1. Gas is approximated as a semi-perfect gas and its flow as quasi 1D with $\Delta s = 0$ (isentropic flow);
2. Initial velocity, temperature and pressure are $u_{gi}$, $T_{gi}$ and $P_i$, respectively.

For optimal designing of the nozzle shape, both throat pressure ($P_t$) and exit pressure ($P_e$) must be compared. By this comparison, it is observed that the optimal nozzle will be a convergent type nozzle ($P_t < P_e$) or a convergent-divergent type nozzle ($P_t > P_e$). In cold spray, a convergent-divergent nozzle (de Laval nozzle) is chosen where $P_t > P_e$ so that supersonic gas flow is generated with inlet gas pressure ($P_i$) reaches up to 2 MPa.

Throat pressure ($P_t$) to nozzle intake pressure ($P_i$) is calculated from the following equations [27]:

$$\frac{P_t}{P_i} = \left( \frac{2}{\gamma + 1} \right)^{\frac{\gamma}{\gamma - 1}} \tag{3.3}$$

At the throat for maintaining the sonic conditions, the pressure at throat ($P_t$) must be greater than the atmospheric pressure. Optimal nozzle throat area ($A_t$) is calculated as [26]:

$$A_t = \frac{m}{\sqrt{\gamma\left(\frac{2}{\gamma+1}\right)^{\frac{(\gamma+1)}{(\gamma-1)}}\frac{P_i}{v_i}}} \tag{3.4}$$

where $m$ is the rate of mass flow of gas (kg/h) and $v_i$ is the specific volume of process gas; $v_i$ is calculated by

$$v_i = \frac{RT_{gi}}{p_i} \tag{3.5}$$

$T_{gi}$ is the gas temperature at the nozzle intake, and $R$ is the gas constant.

Further, the nozzle exit area ($A_e$) in the convergent-divergent nozzle expressed as [28]:

$$\frac{A_t}{A_e} = \left[\frac{\gamma+1}{2}\right]^{\frac{1}{\gamma-1}}\left(\frac{P_e}{P_i}\right)^{\frac{1}{\gamma}}\sqrt{\frac{\gamma+1}{\gamma-1}\left[1-\left(\frac{P_e}{P_i}\right)^{\frac{\gamma-1}{\gamma}}\right]} \tag{3.6}$$

The optimum exit area ($A_e$) is reached when the nozzle pressure at the nozzle exit ($P_e$) is the same as the atmospheric pressure ($P_a$); this is determined by equation (3.6).

Several parameters (i.e., pressure, temperature [$T_g$] and velocity [$u_g$]) at a given cross sectional area ($A$) can be calculated with respect to the nozzle throat area:

1. Pressure

$$\frac{A}{A_t} = \sqrt{\frac{\left(\frac{\gamma-1}{2}\right)\left(\frac{2}{\gamma+1}\right)^{\frac{\gamma+1}{\gamma-1}}}{\left[\left(\frac{P}{P_i}\right)^{\frac{2}{\gamma}}-\left(\frac{P}{P_i}\right)^{\frac{\gamma+1}{\gamma}}\right]}} \tag{3.7}$$

2. Temperature

$$T_g = T_{gi}\left(\frac{P}{P_i}\right)^{\frac{\gamma-1}{\gamma}} \tag{3.8}$$

3. Velocity

$$u_g = \sqrt{u_{gi}^2 + 2\frac{\gamma}{\gamma - 1} P_i v_i \left[ 1 - \left( \frac{P}{P_i} \right)^{\frac{\gamma-1}{\gamma}} \right]} \quad (3.9)$$

### 3.7.6 Particle Behavior inside the Nozzle

The heating of particles and acceleration in the gas flow inside the nozzle is attained with the help of heat transfer and motion equations. The following assumptions are considered before calculating the particle behavior within the nozzle:

- The spray particles are spherical in shape with almost no internal temperature variations
- The specific heat of particles is independent of their temperature.
- Interaction among the particles and gravitational force are ignored.
- The effect of particles on the gas flow is ignored.

Based on the above assumptions, the equation of particle motion for the cold spraying process is expressed as [28]:

$$\frac{du_p}{dt} = \frac{3}{k} \frac{c_d}{D_p} \frac{\rho_g}{\rho_p} \left( u_g - u_p \right) \left| u_g - u_p \right| \quad (3.10)$$

where $u_p$ is the velocity of the particle, $t$ is the time, $C_d$ is the drag coefficient, $\rho_g$ is the process gas density, $D_p$ is the particle diameter and $\rho_p$ is the particle density. $C_d$ is dependent on the particle Reynolds number.

The particle's heating in the gas flow is written as:

$$\frac{dT_p}{dt} = \left( T_g - T_P \right) \frac{6h}{c_p \rho_g D_g} \quad (3.11)$$

where $T_p$ is the particle temperature, $h$ is the coefficient of heat transfer and $c_p$ is the particle's specific heat.

The velocity of the particle is expressed from the gas velocity as:

$$\frac{v_p}{v} = \left[ 1 + 0.85 \sqrt{\frac{D_p}{L} \frac{\rho_p v^2}{P_o}} \right]^{-1} \quad (3.12)$$

where $v$ is the gas velocity, $P_o$ is the inlet pressure of density and $L$ is the length of the barrel.

In the convergent-divergent nozzle, the acceleration of particle predominantly occurs initially in the nozzle throat area and also at the start of the diverging section (first third); now gas velocity ($v_g$) reaches approximately 85% of exit gas velocity. At the nozzle throat, the velocity of gas reaches its sonic velocity; on the other hand, in the diverging section it reaches supersonic velocity. Also, during this time the gas temperature reaches much below the room temperature as gas expands in the nozzle's diverging section.

## EXAMPLE

Calculate the optimum nozzle exit area for a convergent-divergent nozzle that is being supplied with Ar gas at 1 MPa pressure, the length of the provided the length of nozzle is 150 mm, and the throat are is 10 mm². (ii) Calculate the exit gas temperature when the exit gas pressure is 150 kPa (at inlet gas temperature of 600 °C); (iii) calculate the exit velocity of gas for an optimally designed nozzle when the inlet gas velocity is 40 m/s, provided that the specific volume of Ar (0.622 m³/kg); (iv) what is the particle velocity for an Al particle of 30 µm diameter (density of 2.7 g/cm³). (Given: atmospheric pressure = 101.325 kPa.)

## SOLUTION

Given:

$P_i = 3000$ kPa, $P_e = 101.325$ kPa, $\gamma = 1.67$, $A_t = 10$ mm²

$$\frac{A_t}{A_e} = \left[\frac{\gamma+1}{2}\right]^{\frac{1}{\gamma-1}} \left(\frac{P_e}{P_i}\right)^{1/\gamma} \sqrt{\left(\frac{\gamma+1}{\gamma-1}\right) \times \left[1-\left(\frac{P_e}{P_i}\right)^{\frac{\gamma-1}{\gamma}}\right]}$$

$$\frac{10}{A_e} = \left[\frac{1.67+1}{2}\right]^{\frac{1}{1.67-1}} \times \left(\frac{101.325}{1000}\right)^{1/1.67} \times \sqrt{\left(\frac{1.67+1}{1.67-1}\right) \times \left[1-\left(\frac{101.325}{1000}\right)^{\frac{1.67-1}{1.67}}\right]}$$

$$\frac{10}{A_c} = \left[\frac{2.67}{2}\right]^{1.4925} \times (0.101325)^{0.5988} \times \sqrt{3.985 \times (1-0.4002)}$$

$$\frac{10}{A_e} = 1.5391 \times 0.2538 \times 1.5460$$

$\mathbf{A_e = 16.55\ mm^2}$

(ii) exit gas temperature

$$P = 150 \text{ kPa}$$

$$T_g = T_{gi}\left(\frac{P}{P_i}\right)^{\frac{\gamma-1}{\gamma}} \quad T_g = 873\times\left(\frac{150}{1000}\right)^{\frac{1.67-1}{1.67}}$$

$$T_g = 873\times(0.15)^{\frac{0.67}{1.67}}$$

$$T_g = 408.7 \text{ K or } 135.7\ ^\circ\text{C}$$

(iii) at exit pressure of 101.325 kPa

$$u_{gi} = 40 \text{ m/s}$$

$$u_g = \sqrt{2\frac{\gamma}{\gamma-1}P_i v_i\left[1-\left(\frac{P}{P_i}\right)^{\frac{\gamma-1}{\gamma}}\right]+u_{gi}^2}$$

$$u_g = \sqrt{2\times\frac{1.67}{1.67-1}\times1000\times1000\times0.622\times\left[1-\left(\frac{101.325}{1000}\right)^{\frac{1.67-1}{1.67}}\right]+40^2}$$

$$u_g = \sqrt{2\times\frac{1.67}{0.67}\times1000\times1000\times0.622\times\left[1-(0.101325)^{\frac{0.67}{1.67}}\right]+1600}$$

$$u_g = \sqrt{1857889.198+1600}$$

$$\mathbf{u_g = 1363.63\ m/s}$$

(iv) particle velocity

$v$ = gas velocity, $D_p = 30 \times 10^{-6}$, $L = 150 \times 10^{-3}$, $\rho_p = 2700$ g/mm$^3$

$$\frac{v_p}{v} = \left[1+0.85\sqrt{\frac{D_p}{L}\frac{\rho_p v^2}{P_o}}\right]^{-1}$$

$$\frac{v_p}{1363.63} = \left[1 + 0.85 \times \sqrt{\frac{30 \times 10^{-6}}{150 \times 10^{-3}} \times \frac{2700 \times 1363.63^2}{1000 \times 10^3}}\right]^{-1}$$

$$\mathbf{v_p = 735.70\ m/s}$$

## 3.8 SUMMARY

The thermal spraying process is an extensively used technique for deposition. There is probably no limitation available for the combination of substrate and coating materials. So, thermal spraying coatings have a broad area of applications, both for repair and manufacture of new parts. Today, there are many thermal spraying techniques, and many new techniques have been emerging. Every spraying technique has some advantages and disadvantages, and use of these techniques is dependent on the shape and size of the parts being treated, the atmospheric/service conditions, accessibility of thermal spraying equipment, and human knowledge. Thermal spraying is growing scientifically and is connected with many fields (e.g., materials science, tribology, medicine). Recently, a hybrid between vapor deposition and plasma spraying called hypersonic plasma particle deposition is developed, in which vapor precursors are injected with high-velocity plasma to make nanostructured coatings. The new coating materials like nanostructured and quasicrystalline have also been investigated apart from new thermal spraying techniques. Powder mixed from micrometer sized and nanometer sized have been inspected as well. As the thermal spraying is used to preserve components from wear, from corrosion at high temperature, and to restore damaged surfaces, nowadays these coatings are also used in biomedical applications.

### Questions for Self-Analysis

1. What is thermal spraying?
2. Give a classification of thermal spraying processes.
3. How does flame spraying work?
4. What is the difference between the powder flame spraying and wire flame spraying processes?
5. Define the working principle of the D-gun process.
6. Give some advantages, disadvantages and applications of HVOF.
7. What is the principle of twin arc spraying?
8. What is the principle of the air plasma spraying process?
9. List some applications of air plasma spraying process.
10. What are the process parameters involved in vacuum plasma spraying?
11. What is the advantage of the suspension/solution precursor plasma spraying process over conventional plasma spraying?
12. What are the differences between high-pressure and low-pressure cold spraying?

## REFERENCES


[1] Robert C., Thermal spray coatings. In C.M. Cotell, J.A. Sprague, F.A. Smidt (Eds.), *ASM Handbook,* Volume 5: Surface Engineering, Netherlands: ASM International (1994) 497–509

[2] Ang A.S.M., Berndt C.C., A review of testing methods for thermal spray coatings, *International Materials Reviews* (2014) 59:179–223

[3] Amin S., Panchal H., A review on thermal spray coating processes, *International Journal of Current Trends in Engineering & Research* (2016) 2:556–563

[4] Davis J.R., *Handbook of Thermal Spray Technology, First.* Materials Park, OH: ASM International (2004)

[5] Pawlowski L., *The Science and Engineering of Thermal Spray Coatings.* Hoboken, NJ: John Wiley & Sons Ltd (2008)

[6] Vuoristo P., Thermal spray coating processes. In D. Cameron (Ed.), *Comprehensive Materials Processing,* 1st edition, Volume 4. Coatings and Films, Amsterdam, Netherlands: Elsevier (2014) 229–276

[7] Schneider K.E., Belashchenko V., Dratwinski M., Siegmann S., Zagorski A., *Thermal Spraying for Power Generation Components.* Weinheim: Wiley-VCH (2006) 271

[8] Heimann R.B., *Plasma-spray Coating—Principles and Applications.* New York: VCH (1996) 339

[9] Matejka D., Benko B., *Plasma Spraying of Metallic and Ceramic Materials.* New York: John Wiley & Sons (1989) 280

[10] Hermanek F.J., *Thermal Spray Terminology and Company Origins,* First Printing, Materials Park, OH: ASM International (2001)

[11] Lunn G.D., Riley M.A., McCartney D.G., A study of wire breakup and in-flight particle behavior during wire flame spraying of aluminum, *Journal of Thermal Spray Technology* (2017) 26:1947–1958

[12] Lima C.R.C., Libardi R., Camargo F., Fals H.C., Ferraresi V.A., Assessment of abrasive wear of nanostructured WC-Co and Fe-based coatings applied by HP-HVOF, flame, and wire arc spray, *JTTEE* (2014) 23:1097–1104

[13] Lia X., Zhaia H., Lia W., Cuib S., Ninga W., Qiuc X., Dry sliding wear behaviors of Fe-based amorphous metallic coating synthesized by D-gun spray, *Journal of Non-Crystalline Solids* (2020) 537:120018

[14] Martin D.T., Rad M.R., McDonald A., Hussain T., Beyond traditional coatings: A review on thermal-sprayed functional and smart coatings, *Journal of Thermal Spray Technology,* (2019) 28:598–644

[15] Kilic M., Ozkan D., Gok M.S., Karaoglanli A.C., Room- and high-temperature wear resistance of MCrAlY coatings deposited by detonation gun (D-gun) and Supersonic Plasma Spraying (SSPS) techniques, *Coatings* (2020) 10:1107

[16] Reddy N.C., Kumar B.S.A., Reddappa H.N., Ramesh M.R., Koppad P.G., Kordd S., HVOF sprayed $Ni_3Ti$ and $Ni_3Ti$ + ($Cr_3C_2$ + 20 NiCr) coatings: Microstructure, microhardness and oxidation behaviour, *Journal of Alloys and Compounds* (2018) 736:236–245

[17] Doolabi D.S., Rahimipour M.R., Alizadeh M., Pouladi S., Hadavi S.M.M., Vaezi M.R., Effect of high vacuum heat treatment on microstructure and cyclic oxidation resistance of HVOF-CoNiCrAlY coatings, *Vacuum* (2017) 135:22–33

[18] Kiilakoski J., Musalek R., Lukac F., Koivuluoto H., Vuoristo P., Evaluating the toughness of APS and HVOF-sprayed $Al_2O_3$-$ZrO_2$ coatings by in-situ- and macroscopic bending, *Journal of the European Ceramic Society* (2018) 38:1908–1918

[19] Chmielewski T., Siwek P., Chmielewski M., Piatkowska A., Grabias A., Golanski D., Structure and selected properties of arc sprayed coatings containing in-situ fabricated Fe-Al intermetallic phases, *Metals* (2018) 8:1059

[20] Vassen R., Stuke A., Stöver D., Recent developments in the field of thermal barrier coatings, *Journal of Thermal Spray Technology* (2009) 18:181–186

[21] Wanga Y., Wanga D., Yand J., Suna A., Preparation and characterization of molybdenum disilicide coating on molybdenum substrate by air plasma spraying, *Applied Surface Science* (2013) 284:881–888

[22] Vaben R., Kabner H., Mauer G., Stover D., Suspension plasma spraying: Process characteristics and applications, *Journal of Thermal Spray Technology* (2010) 19:219–225

[23] Fan W., Bai Y., Review of suspension and solution precursor plasma sprayed thermal barrier coatings, *Ceramics International* (2016) 42:14299–14312

[24] Kassner H., Siegert R., Hathiramani D., Vassen R., Stoever D., Application of suspension plasma spraying (SPS) for manufacture of ceramic coatings, *Journal of Thermal Spray Technology* (2007) 17:115–123

[25] Assadi H., Kreye H., Gartner F., Klassen T., Cold spraying A materials perspective, *Acta Materialia* (2016) 116:382–407

[26] Assadi H., Schmidt T., Richter H., Kliemann J.O., Binder K., Gartner F., Klassen T., Kreye H., On parameter selection in cold spraying, *Journal of Thermal Spray Technology* (2011) 20:1161–1176

[27] Malachowska A., *Analysis of the Cold Gas Spraying Process and Determination of Selected Properties of Metallic Coatings on Polymers.* Materials. Université de Limoges; Uniwersytet Wroclawski (2016). English. ffNNT: 2016LIMO0012ff. fftel-01741350ff

[28] Champagne V., Ed. *The Cold Spray Materials Deposition Process: Fundamentals and Applications.* Cambridge: Woodhead Publishing Ltd. (2007) 117–130

# 4 Spraying Parameters

*Ariharan S, Rubia Hassan and Kantesh Balani*

## CONTENTS



The deposition of coatings mandates utilizing optimal spraying parameters during thermal spraying. But one aspect is assured, that the process is so complex that even utilizing the same spraying conditions may result somewhat different coating characteristics due to the stochastic nature of the process. The spraying process parameter influences the quality of the coatings and its application (see Figure 4.1). There are plenty of variables, such as power rating, various plasma/primary/secondary/carrier gases, feed rate, stand-off distance, type of substrate and substrate preparation that affect in-flight conditions. These spraying parameters will be discussed in this order. But it may also be noted that it may not be practically feasible to optimize each and every parameter, as changing one parameter may affect many other attributes of spraying. In addition, coupling of the coating quality with electrode usage/life, type of gun/powder injector, powder settling in powder feeder, substrate roughness, coating quality and so on may also alter the final microstructure and phase development even when the same spray parameters are being used. Nonetheless, the particle diagnostic tools, explained in Chapter 9, may also be utilized to achieve spray consistency. In this chapter, the focus will be on various spraying parameters and their effect on deposition characteristics.

## 4.1 IN-FLIGHT CONDITIONS

The powder experiences temperature variations depending on its size, morphology, location and dwell time in the plume/flame, the temperature of the plume/flame and the feed rate of powder, to name a few. Thus, in-flight conditions are very important as the starting powder begins to result as a deposit on the surface of a substrate. The effect of velocity and temperature profile of in-flight

DOI: 10.1201/9781003321965-4

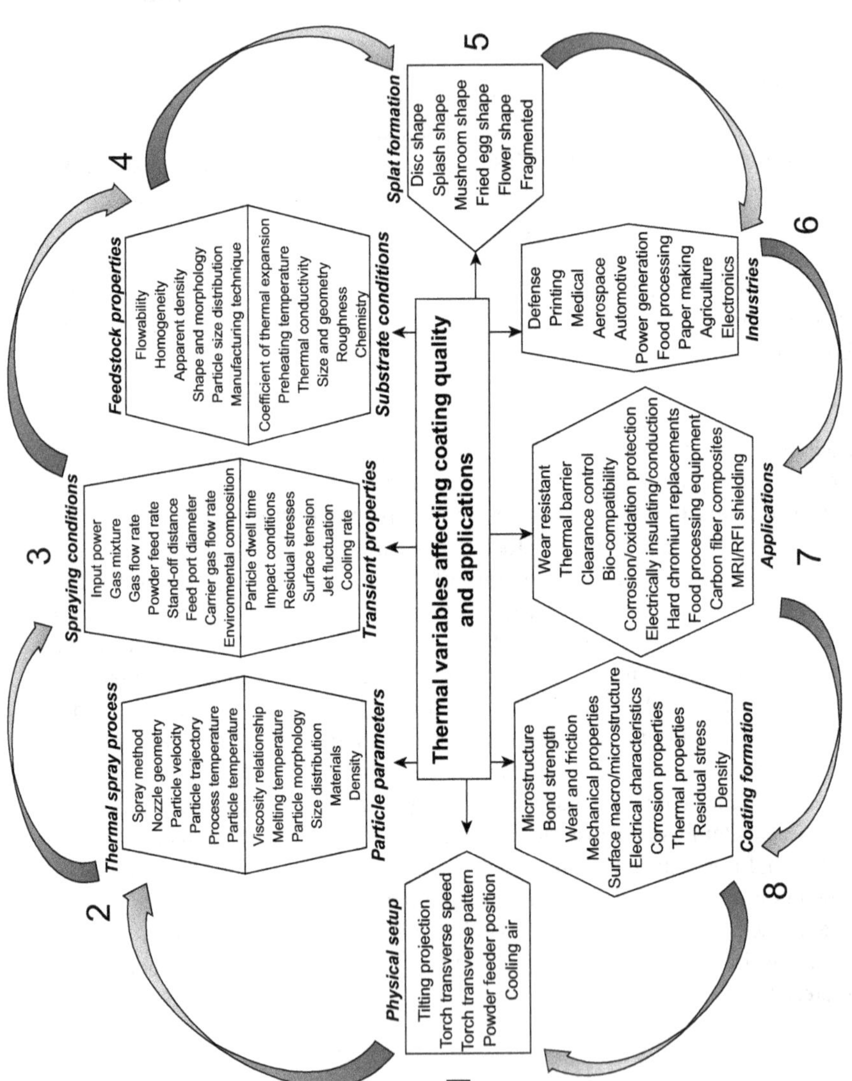


**FIGURE 4.1** Thermal spraying variables that are influencing coating quality and application.

*Source*: Adapted and modified from [1].

particles on the microstructure of YSZ TBCs deposited via plasma spraying show decreased void content as the temperature and velocity increase. As the temperature increased from 2689–2693 °C to 2843–2853 °C, and velocity from 180–185 m/s to 225–231 m/s, the total porosity content decreased from 14% ± 0.5% to 7% ± 1%, respectively. A further rise in temperature to 3242–3524 °C and velocity to 380–450 m/s accounted for 5% ± 1% voids [2].

Under ideal conditions:

1. All the injected powder particles should get deposited (the thermal plume should not throw the particles away);
2. High enough temperature above their melting point (avoids any vaporization if temperatures are very high and also avoids any restricted plastic deformation if temperatures are low);
3. Attaining uniform temperature (even with varied particle size typically between 10 and 45 μm and corresponding mass ratios as high as ~90);
4. Maximum compatible velocity of particles for complete surface coverage overcoming bow shock (with plastic deformation of particles if needed while attaining required temperature during its limited dwell time in the thermal plume).

Though these conditions are preferred, the follow-up phrase in the following list also points toward the issues that restrict achieving these ideal conditions:

1. Overspray is a natural phenomenon that occurs due to Gaussian distribution of sprayed particles. The particles traveling at the center of the plume attain higher temperatures and velocities, whereas the particles at the periphery may wander or bounce off from the substrate surface without deposition due to their low temperatures and velocities. Inherently, in order to attain uniform coating thickness, there is overspray (or the spray gun traverses even ahead the sample edge or till some portion of the masked region) that gets wasted as the in-flight particles do not deposit on the substrate.
2. It may be envisaged that the temperature should just be right (see Figure 4.2) in order to promote complete coverage of particles to coat the substrate. A high enough temperature ($>>T_m$) may lead to evaporation, splashing and fingering, whereas a much lower temperature may require processes dominated by plastic deformation (e.g., cold spraying, high-velocity oxy-fuel or D-gun) to take over and stretch the initial particles to cover the substrate surface as a coating. Otherwise, bouncing off of the hard starting particles may occur when impacting on the substrate surface. The coverage of corners and solidification before reaching the unfilled areas also needs to be minimized in order to ensure complete coverage. Thus, an ideal temperature, just above the melting point of the starting powder, is required to ensure that the coating is formed to uniformly cover the substrate surface.

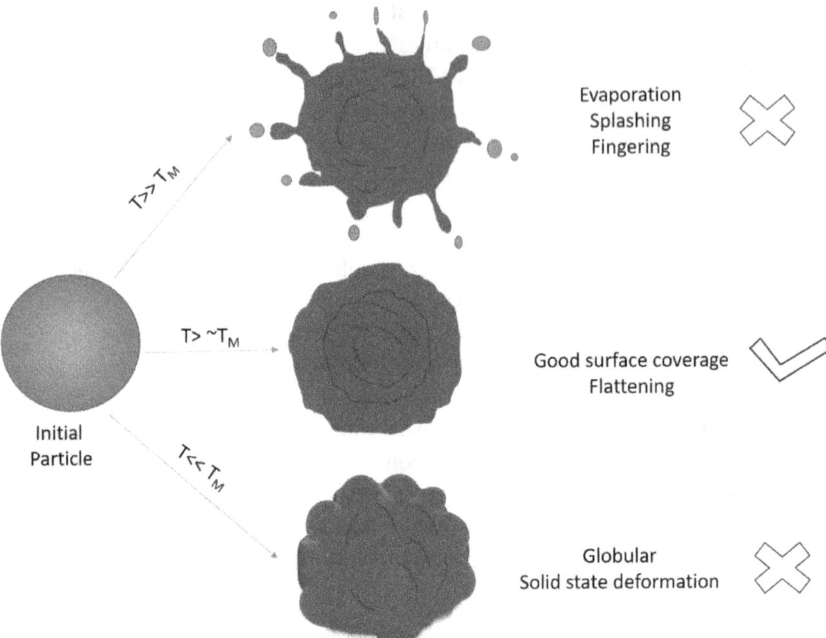


**FIGURE 4.2** The effect of temperature on the final response of the starting particle during coating formation.

*Source*: Adapted and modified from [3].

3. The initial particle size is also critical in dictating the temperature achieved by the particle. As the size increases, the corresponding volume (and the mass contained within) increases. In order to appreciate this effect, if the particle size variations are between 10 and 100 μm, then the mass ratio scales up to the cube of mass ratio (as high as 1000 times in this case; see Figure 4.3a). The corresponding temperature change ($\Delta T$) is given as follows:

$$\Delta T = \frac{Q}{m \times C_P} \tag{4.1}$$

where $Q$ is the heat energy, $m$ is the mass of the particle, and $C_P$ is the specific heat capacity. Please note that *heat capacity* and *specific heat capacity* are two different terms, as the former is mass dependent, whereas latter is a material property (and mass independent). Assuming that the heat flux is constant in a region of plume/flame, the differently sized particles arriving at that region will leave with different temperatures. As the higher mass (even of the same material) requires higher enthalpy to heat up, only a lower temperature may be achieved by a bigger particle compared to that of a smaller particle. Thus, a particle size in the range

(a)

| Initial Particle | Volume (μm³) | Mass ratio |
|---|---|---|
| 10 μm | $\frac{4}{3}\pi \times (10)^3$ | 1 (base) |
| 30 μm | $\frac{4}{3}\pi \times (30)^3$ | 27 |
| 50 μm | $\frac{4}{3}\pi \times (50)^3$ | 125 |
| 100 μm | $\frac{4}{3}\pi \times (100)^3$ | 100 |

(b)

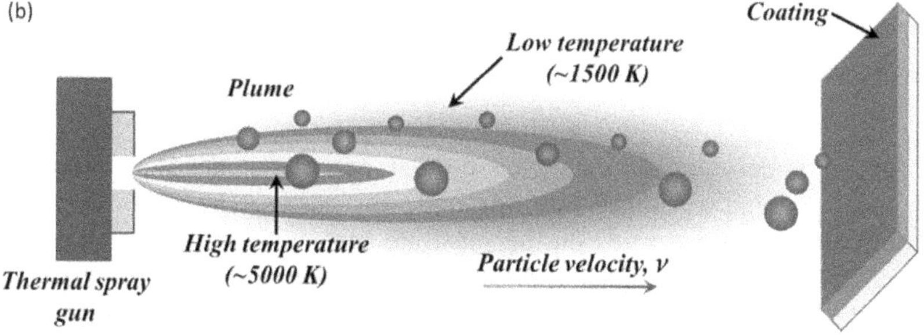


**FIGURE 4.3** Effect of (a) size and (b) location on the uniformity of temperature attained by the starting particle in plume/flame.

15–40 μm is generally preferred for ceramics, whereas a larger particle size of ~150 μm may be used for metals and alloys (due to their high thermal conductivity and plastic deformation capability).

In addition, the location of powder particles (from powder injector to the plume/flame) will also affect its attained temperature due to different particle trajectories and existence of sharp thermal gradients in the plume/flame. The core usually has a much higher temperature than the periphery (in radial direction) of the stretched plume/flame (Figure 4.3b). The cooling effect by the surrounding environment may induce thermal gradients as high as ~1000s K per inch or so. The shrouding effect, due to the shape of plasma plume and the surrounding particle environment (or higher powder feed rate), may also restrict complete melting of the sprayed particles. Therefore, it is preferred that the sprayed particles stay sparsely in the middle/core region of the plume/flame and attain uniform temperature.

4. A high enough velocity is required in order to overcome the bow shock (Figure 4.4); that is, when the carrier gas (carrying the powder particles) strikes the substrate, the consequent return of it creates a shock wave that hinders the arrival of oncoming carrier gas. As the carrier

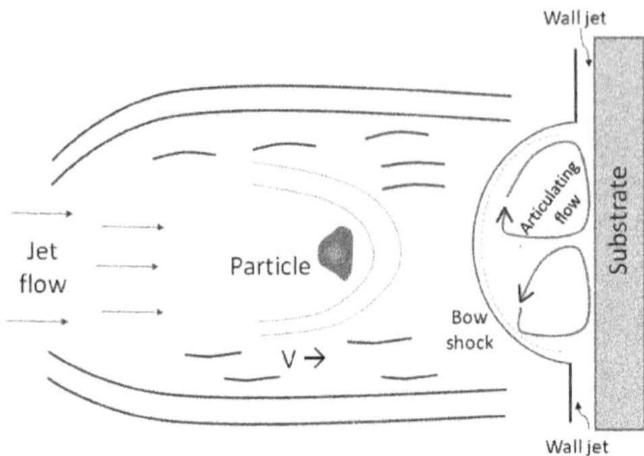


**FIGURE 4.4** Velocity of carrier gas affecting the deposition of particles due to bow shock.

*Source*: Adapted and modified from [4].

gas provides momentum to the powder particle, it needs to pierce this frictional resistance from the returning gas (i.e., bow shock) and attempt to get deposited on the substrate surface. If the particle does not attain enough velocity, then it will fly off along with the bow shock and will not be able to reach the substrate for deposition. Secondly, the impact of particle should occur with high enough velocity to allow its spreading (either in molten or semi-molten state or via plastic deformation) across the crevices and uneven regions to be able to produce dense coatings.

Added to the above conditions, the plasma, primary/secondary and carrier gases also play a dominant role in dictating the eventual microstructural evolution and coating deposition, as discussed in the next section.

## 4.2 PLASMA/PRIMARY/SECONDARY GASES

*Plasma* is defined as a fourth state of matter comprising ions, electrons, and neutral atoms/molecules. Nitrogen is a primary gas which is inert with most materials with a few exceptions, like titanium. It is used either alone or with hydrogen as a secondary gas. It is also the cheapest plasma gas. However, the most favored primary plasma gas is argon [5]. It is the easiest of the gases to form plasma. Typically, Ar is passed through the electric arc created between high potential electrodes in the spray gun to form plasma. Though ions and electrons and excited/neutral states of atoms or molecules exist, the charge of plasma is electrically neutral. Plasma gas is required to transfer the heat to the in-flight particles. The specific gas enthalpy of plasma typically ranges between 15 and 30 MJ/kg to result in a temperature exceeding ~10,000 K. The typical 10%–20% ionization in the core may drop down to as low as 0.1%, resulting in a sharp temperature gradient within a few inches of the plasma plume.

In order to attain plasma, typically an electric arc is struck between a cathode made of tungsten (W) and a water-cooled anode made from copper. This electric arc serves as a source to excite the plasma gas (typically Ar) into higher energy states, which upon recombination transfers the energy to the in-flight particles. It is mainly the current that is kept high (~400–1000 A), and the voltage is controlled typically within 75 V to keep the plasma stabilized in the conventional gas-stabilized plasma spray process. The earlier gun design primarily uses high power (up to 200 kW) to develop coatings with a high feeding rate, deposition efficiency and plasma stability for uniform melting of the spray particles. Later, the plasma power has been brought down to the range of 10–100 kW by modification in the gun and process design. Recent development in the spraying gun design with power ~40 kW showed improved stability of the plasma arc with optimum powder utilization.

It may be observed (see Figure 4.5) that diatomic gases (e.g., $H_2$, $N_2$) have higher energy content than monoatomic gases (e.g., He, Ar). In addition, the heavier gases (e.g., $N_2$ or Ar) have higher energy content than lighter gases (e.g., $H_2$ or He, respectively). In addition, a sharp change in the energy content of diatomic gas at an intermediate temperature highlights the dissociation of gases (thus some energy is consumed in the process), and the dissociated atoms can diffuse faster and attain a higher temperature. Therefore, dissociated diatomic gases recombine and release heat locally at the in-flight particle region to provide a higher temperature and easily melt materials with a higher melting point. On the other hand, monoatomic gas (mainly Ar) helps provide stable plasma and a protective environment for deposition of materials with a lower melting point.

The gas required to initiate the plasma (typically Ar) is called a primary gas. Though diatomic gases possess higher enthalpy, their plasma is unstable due to gas dissociation and requirement of high arc voltages and sharp voltage fluctuations. Thus, a secondary gas (e.g., $H_2$, He) is mixed with primary gas to provide higher thermal conductivity or enthalpy. The enthalpy of commonly seen gas used in plasma spraying ranks in this order: $Ar < (Ar + He) < (Ar + H_2) < N_2 < (N_2 + H_2)$ [6]. On one hand, where primary gas affects the in-flight particle velocity (as it also directly interferes with the carrier gas for in-flight powder), secondary gas has a greater influence on the temperature (affects enthalpy of primary gas) attained by in-flight particles.

It may also be noted (see Figure 4.6) that the addition of hydrogen in Ar enhances the enthalpy due to its higher thermal diffusivity. This increased thermal conductivity also increases the arc voltage and results an enhanced gun efficiency, but at the cost of rapid electrode degradation (due to increased arc voltage). With restricted plume size, the radiation losses at the anode are also minimized. It may be noted that ionization starts early at reduced pressure (e.g., 100 mbar), as might be required in vacuum plasma spraying compared to that at 1 bar of Ar (e.g., in atmospheric plasma spraying).

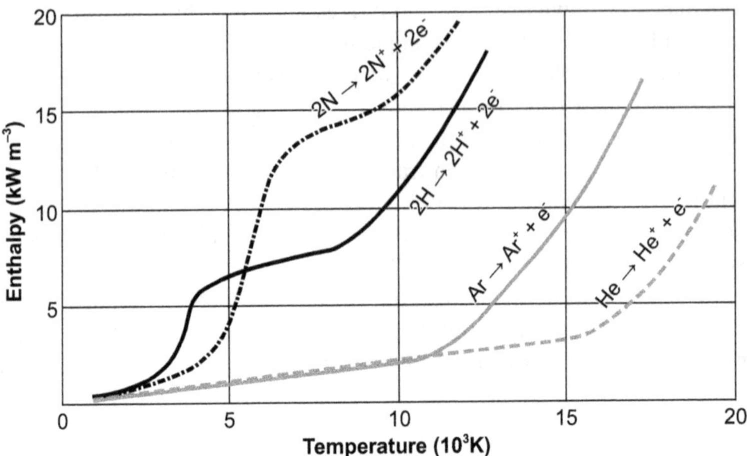


**FIGURE 4.5** The enthalpy per unit volume and the corresponding temperature of various plasma gases.

*Source*: Reprinted with permission from [7].

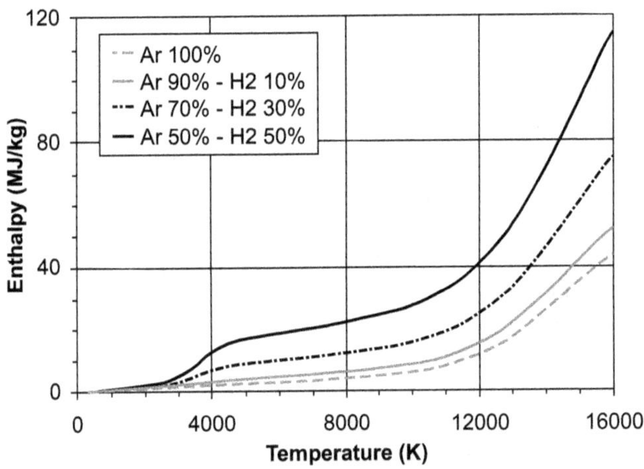


**FIGURE 4.6** The effect of environment (pressure) and addition of secondary gas on the plasma gas enthalpy.

*Source*: Open access from Springer [8].

## 4.3 CARRIER GAS

Carrier gas serves as the medium to carry the powder from the powder feeder to the flame/plume via an internal powder injector (placed in the spray gun) or through an external feeder. The purpose of the carrier gas is to accelerate the powder particles (and gain momentum) through the flame/plume (to gather enough heat) with consequent impact on the substrate to spread the powder particle (via plastic deformation or melting) and deposit as a coating. The role of the carrier gas is to transport the powder particle so that it impacts strongly on the substrate to make it spread (either in partially melted or fully melted condition). It may be noted that the higher the acceleration provided by the carrier gas, the lower the time the in-flight particle spends in the flame/plume, thus gathering a lower temperature (but higher kinetic energy); therefore plastic deformation will dominate to allow spreading of the particle onto the substrate. Further, lower velocities may allow the particles to attain higher temperature and melt (or vaporize). During the cold-spray process, powders are merely dragged and accelerated by the compressed carrier gas, and the in-flight particle velocity is highly governed by the shape of the gas flow field both inside and outside the spray nozzle. It is observed that the temperature and the velocity of driving gas increases with increasing carrier gas temperature at the nozzle throat and in the divergent part of the nozzle. As gas temperatures increase, cold-sprayed powder is accelerated and particle impact velocity is improved [9].

Generally, higher velocities (~1 Mach, where Mach is the speed of sound in air, or ~310 m/s in plasma spraying) are preferred, as they tend to generate denser and refined coatings with fine-grained microstructure (Figure 4.7a). When the particle

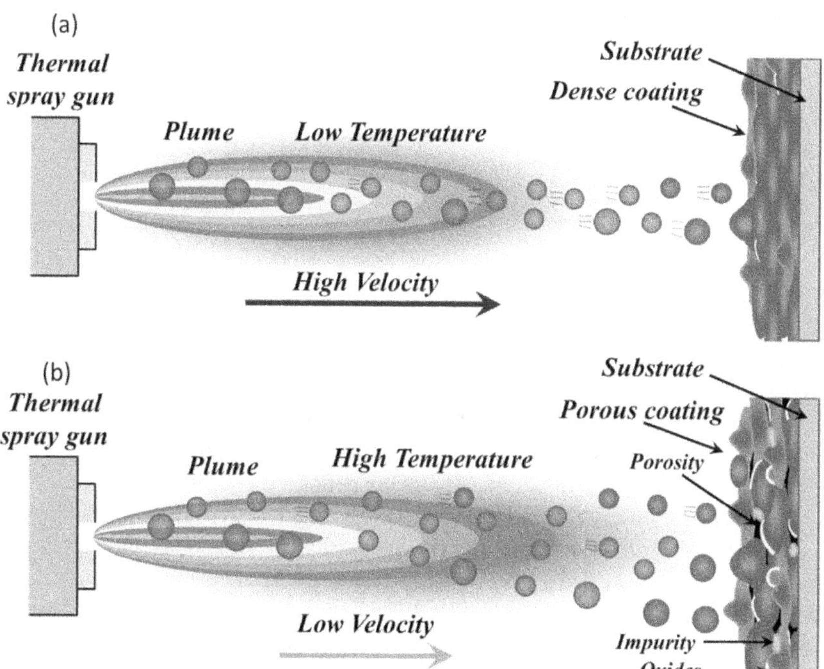


**FIGURE 4.7** The effect of carrier gas velocity: (a) at high velocity, a dense microstructure is obtained, whereas (b) at low velocity, impurity entrapment and porosity is obtained.

velocities are lower, there may be higher particle temperatures, and that may lead to its evaporation or swaying away from the flame/plume. Nonetheless, oxidation becomes inherent (as impurity), or splashing at high temperatures or spheroidization may lead to impurity entrapment or porous microstructure (Figure 4.7b). Thus, velocity requirements dominate over temperature in the spray process towards attaining dense coatings and refined microstructure.

It may also be added that the non-uniform particle sizes used in the powder feed allow smaller particles to attain higher velocities, but due to their low volume/mass, these may also attain high temperatures compared to those of bigger particles. Thus, the optimized selection of powder size may also allow obtaining a bimodal microstructure (i.e., allowing fine particles to melt whereas the larger particles may get entrapped as is in the deposited coatings). Or there can be multiple powder feeders attached to a spray gun to allow deposition of composites or allow their in-situ melting and alloy formation as a coating.

The axial feeding of powders is preferred, as the in-flight particles will stay in the core of the plume. But attaining the same is very difficult as the flame/plume has to be initiated, sustained and also maintained under non-turbulent conditions for retaining a stabilized zone (of core, envelope and overall shape); thus, any disturbance along the axial direction will deter it from those dynamic equilibrium conditions. Hence, a radial/spiral injection or external injection of powders is more popular.

Typically, a powder feeder is used with a pressurized inert gas (called carrier gas) to carry the powder (and provide acceleration/momentum). And usually, argon or nitrogen is used as a carrier gas, as it protects the powder from oxidation under in-flight conditions by forming an envelope around the particles.

## 4.4 FEED RATE

The rate of powder being fed through the injector to the flame/plume is called the *feed rate*. The feed rate decides how much of powder will reach the flame/plume for deposition. But some powder may not penetrate the flame/plume (e.g., in external feeding) or may just move across the flame/plume, or the powder may just evaporate when attaining high temperature or may just bounce back (if not properly heated) after impacting the substrate. A higher feed rate (Figure 4.8a) shrouds the powder from attaining higher temperature, whereas a lower feed rate (Figure 4.8b) may waste the generated heat and provide lower deposition rates. The result is reflected in the properties of the deposited coating. The effect of the powder feed rate on micro-hardness and bond strength of Al-12Si plasma-sprayed coatings, keeping all other parameters constant, showed an increase in microhardness from 115 $HV_{0.3}$ to 130 $HV_{0.3}$ by increasing the feed rate from 30 g/min to 45 g/min, due to the dense level of packing between the splats and low level of porosity. However, a further increase in feed rate to 60 g/min reduced the hardness value to 97 $HV_{0.3}$ because of higher porosity level. The bond strength followed the same trend, increasing from 8 MPa to 12 MPa and then going down to 5 MPa [10].

Higher powder feed rate pours in the powder quantity increase the enthalpy per unit cross-sectional area of flame/plume to heat up the in-flight particles.

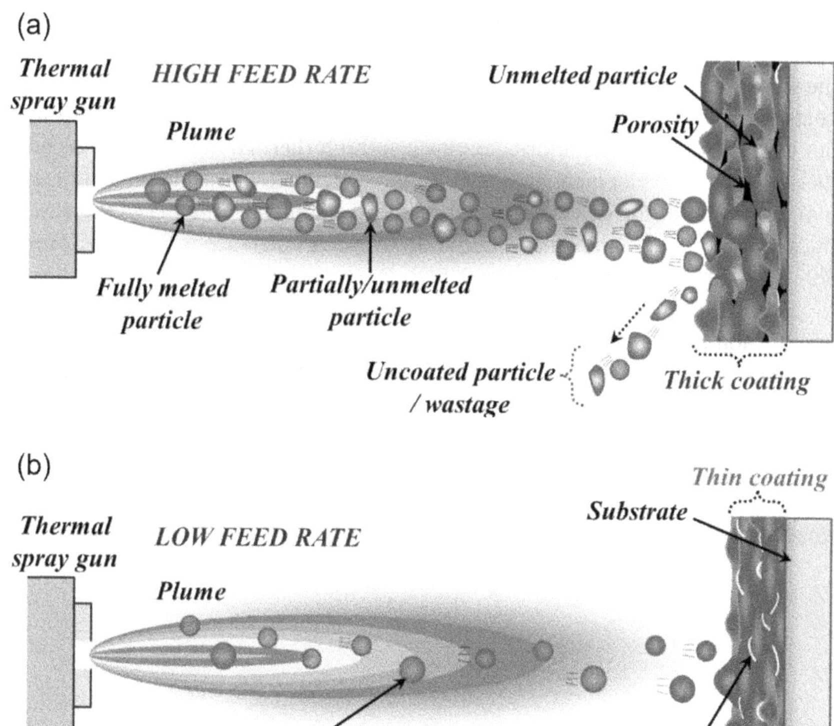


**FIGURE 4.8** (a) A higher feed rate results in a poor quantity of coating with entrapment of un-melted particles; (b) a lower feed rate results in higher particle temperatures and permits oxide entrapment. Thus, an optimal feed rate must be adapted.

Thus, some particles may get shrouded and end up with not acquiring enough temperature to plastically deform and bounce off from the substrate. Though some un-melted particles may remain embedded within the surrounding molten particles (to result in thicker coatings), the coating quality may not be very good. So, apart from poor deposition efficiency, a porous and poor coating quality may be expected. Lower feed rates also may not be well suited, as they heat up the particle to a much larger extent and may result in its vaporization, thus encouraging its oxidation. Therefore, oxide and impurity entrapment may be anticipated in coatings with a low powder feed rate.

## 4.5 STAND-OFF DISTANCE

The *stand-off distance* is the distance between the spray gun to the substrate that dictates the optimal temperature and velocity to deposit on the substrate surface. If the stand-off distance is too low (Figure 4.9a), then the particles hit the substrate before spending enough time in the flame/plume, thus not attaining high enough temperature or velocity. Thereby, the particles mostly bounce off from the substrate without getting deposited. In addition, closer stand-off distance preheats the substrate, which may be good for initial deposition (and thin coatings), but temperature accumulation may result in high thermal damage and eventual thermal stress development arising from the difference in the coefficient of thermal expansion of the deposited material and the substrate (mainly in the case of thicker coatings). On the other hand, if the stand-off distance is too large (Figure 4.9b), the particles may have attained high temperature and velocity but will need to travel the extra distance that, eventually, cools it down and lowers its velocity (or momentum) and, again, it does not deposit effectively on the substrate. Coarse-grained feedstock needs longer spraying distances. A longer spraying distance provides sufficient dwelling time for the powder to melt properly, hence it results in higher efficiency of coating deposition. However, with an increase in spray distance, the number of droplets impacting normally on the substrate decreases. This is because, for a long distance, droplets deviate during flight; this has also been observed to decrease the hardness as well [11]. In addition, the

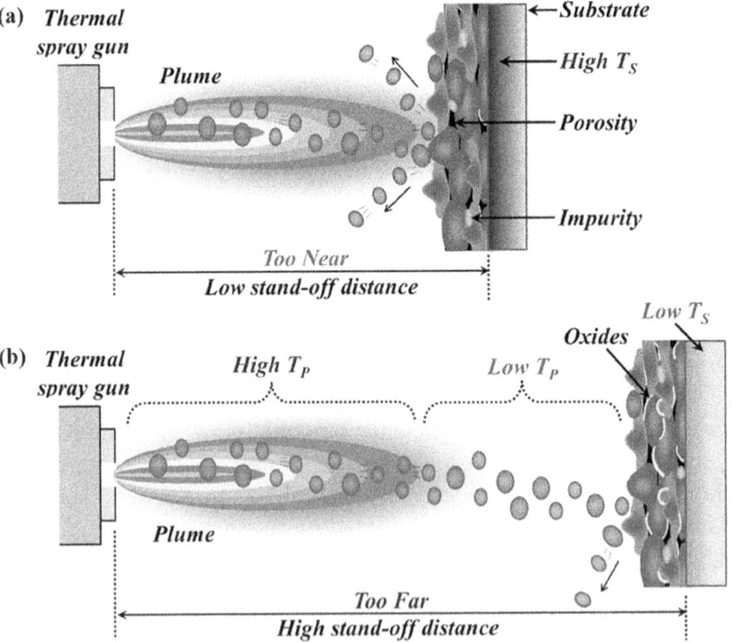


**FIGURE 4.9** (a) Low stand-off distance results in low particle temperature and velocity (and thus poor deposition); (b) high stand-off distance may result in a poor interface between coatings and substrate.

absence of any substrate preheating may result in a poor interface between the coating and substrate.

## 4.6 SUBSTRATE PREPARATION

Substrate preparation is essential because it dictates how the coatings will integrate. Making the substrate surface smoother (finely polished; see Figure 4.9a) will not provide any mechanical interlocking, and the coating may not deposit or may easily spall off. But this smooth substrate preparation can be used to create freestanding structures (e.g., cylinders, conical barrels). As the coatings are mainly used for surface protection (typically against thermal damage or abrasive wear), a rough substrate surface is preferred (Figure 4.10b). Herein, a mechanical interlocking is provided to the depositing splats to stay put and form an integrated surface coating. Surface roughness greatly influences the adhesive strength of the coating. So, most practical cases use grit blasting of the surface to obtain coatings with good adhesion to the substrate [12].

It may also be noted that the substrate surface should be free of any surface oxides, oils/moisture or dust/dirt. Surface oxides are typically brittle and may result in spalling off of the coatings. Similarly, the presence of any oil/moisture will induce its evaporation and entrapment of porosity at the coating-substrate interface, thus compromising its integrity. Also, the presence of any dirt on the surface will be retained in the coating as an impurity, thus surface cleaning is mandated.

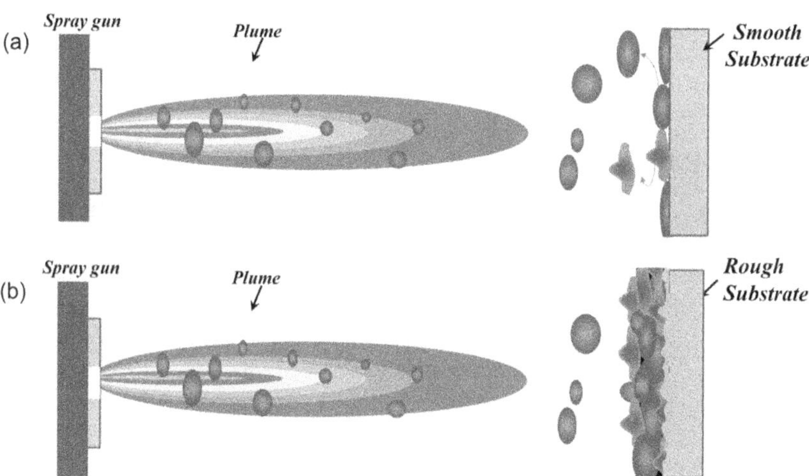


**FIGURE 4.10** (a) Smooth substrate surfaces do not provide interlocking to depositing splats; (b) rough substrate surfaces provide mechanical interlocking to deposited splats and result in an integrated coating.

As a first step, the surface is cleaned with ethanol/acetone to remove any oil and dirt. Cleaning should remove any organic deposits (which can arise from sample handling, or touching with bare hands, or samples may have been stored in oil for protection from corrosion) from the substrate surface.

Then grit blasting is usually carried out to roughen the surface (on the order of tens of μm). Grit blasting uses alumina/silica grit (fine acicular particles) carried in pressurized gas/air to create small dents, scratches and deformation on the surface. This surface roughness serves as a mechanical interlocking site for depositing splats to develop a strong coating-substrate interface. Then, pressurized air is used to remove any surface dust (or remnant grit particles) just before the deposition is to be carried out. That way a cleaner surface is available for deposition of coating. Altogether, a rough and clean surface is mandated in order to allow deposition of mechanically integrated coating on the substrate. Table 4.1 presents some data on flame spraying, plasma spraying and cold spraying, and the resulting coating thickness, their hardness, roughness and porosity with varying processing parameters [13].

As we have observed, multiple plasma parameters influence the quality of coating. This complex setup may, again, be influenced by stochastic nature of the process. Thus, even with the same deposition parameters, a slightly different coating microstructure and coating properties may result. For example, electrode degradation may occur, or an electrical spike during operation may also change the coating quality, or the powder feeding may have an inherent gradient (either in terms of size or composition) to result in a varied coating structure. Interference from powder feeding or carrier gas may also change the plume/flame size/geometry. Substrate heating, external substrate cooling or even the coating thickness may also influence the nature of depositing splats on the substrate. Thus, keeping this multi-parametric condition in mind, spraying parameters guide us as to what may be expected under these complex deposition conditions that finally govern the evolution of the coating microstructure and properties.

## 4.7 SUMMARY

Powder characteristics including powder morphology, powder size and powder size distribution are important, but they are not the only concerns for a good thermal-sprayed coating. Spraying parameters encompass a variety of variables which affect the deposition efficiency and final coating quality during thermal spraying. These variables are associated with the in-flight conditions of the powder (e.g., velocity and temperature), which in turn are dictated by other parameters including the nature and type of plasma gas and carrier gas, distance between the spraying gun and the substrate, feed rate of the powder, flow rate of the gases and so on. Even the substrate preparation has a role to play in determining the quality of the final coating.

**TABLE 4.1**
**Mechanical Properties of Some Flame-Sprayed, Plasma-Sprayed and Cold-Sprayed Coatings at Different Process Parameters**

| Substrate | Coating Material | Particle Size | $p_{O2}$ | $p_{air}$ (bar) | $p_{C2H2}$ | $d$ (mm) | $t$ (μm) | $Ra$ (μm) | $H_m$ | $P$ (%) | |
|---|---|---|---|---|---|---|---|---|---|---|---|
| P.C. steel [14] | FeCr | 100 μm | 3 bar | 0.4 | – | 150 | 865 | ~1.5 | 800 HV | 4.42 | **Flame Spraying** |
| Unalloyed structural steel grade S235JR [15] | 1000 (Al) | 131.64 μm | 4 bar | 4 | 0.7 bar | ~250 | 125 | 29.05 | 32 HV | – | |
| | 1000 (Sn) | 16.79 μm | | | | | 292 | 11.28 | 12 HV | – | |
| | Polymer: Yellow (PA12) | 166.14 μm | | | | > 250 | 1.43 | 1.55 | 9 HV | – | |
| Steel rods [16] | Metco 41C | −140 + 325 mesh | 28 psi | – | 30psi | 150 | – | – | 88–92 Rb | – | |
| | Metco 449P | −120 + 325 mesh | | | | | | | 35–40 Rc | | |
| | Metco 44 | −140 + 325 mesh | | | | | | | 80 Rb | | |
| | Metco 444 | −120 + 325 mesh | | | | | | | 80 Rb | | |

*(Continued)*

**TABLE 4.1 (Continued)**

| Substrate | Coating Material | Particle Size (µm) | $Q_P$ (Ar) Scfh | $Q_S$ (H) Scfh | $I$ (A) | $V$ (V) | $d$ (mm) | $D$ (mm) | $P_p$ | $P_C$ | $t$ (µm) | $Ra$ (µm) | $H_m$ | $P$ (%) | |
|---|---|---|---|---|---|---|---|---|---|---|---|---|---|---|---|
| T91 [17] | NiCoCrAlY | 22–45 | 90 | 90 | 500 | 60–65 | 100 | – | – | 1 bar | – | 7 | – | 2.1 | Plasma spraying |
| Sup. Ni 600 [18] | NiCrAlY | ~45 ± 10 | – | – | 700 | 35 | 90–110 | | 59 psi | 40 psi | bc = 228 | – | 390 HV | 2–4.5 | |
| SF 800H [19] | | | | | | | | | | | bc = 150 | – | – | ~3–4.4 | |
| Substrate | Coating Material | Particle Size (µm) | $T_g$ (°C) | $P_g$ | Carrier Gas | Process Gas | $Q_P$ | $R\ d_f$ | $R_g$ | $n$ | $t$ (µm) | $Ra$ (µm) | $H_m$ | $P$ (%) | Cold Spraying |
| SA 516 [20] | Ni-20Cr | 14 | 450 | 19 bar | $N_2$ | Air | 1.96 m³/min | 113 g/min | 1.96 | – | 400 | 7 | 456–601 HV | – | |
| Copper [21] | Ag/$SnO_2$ | 75 ± 38 | 400 | 1.6 MPa | | – | – | – | 4 $m_3$/h | – | – | | 112 ± 3 HV 0.2 | – | |
| SA516 [22] | Ni-20Cr | -45 ± 15 | | 20.5 bar | | He | 150 m³/h | 40 g/min | | | 250 | – | 221 HV | – | |

*Abbreviations*: bc = bond coat; $d$ = stand-off distance; $D$ = nozzle diameter; $p_{air}$ = pressure of air; $P_C$ = carrier gas pressure; $p_{C2H2}$ = pressure of $C_2H_2$; $P_g$ = gun pressure; $P_P$ = plasma arc gas pressure; $H_m$ = micro-hardness; $I$ = input current; $n$ = number of passes; $P$ = porosity; $p_{O2}$ = pressure of oxygen; $Q_P$ = primary gas flow rate; $Q_S$ = secondary gas flow rate; $Ra$ = roughness; $R_f$ = powder feed rate; $R_g$ = flow rate of gas; Scfh = standard cubic feet per hour; $t$ = coating thickness; $T_g$ = gun temperature; $V$ = voltage generated; Rb = Rockwell hardness B; Rc = Rockwell hardness C; P.C. steel = plain carbon steel; Sup. Ni = Super Nickel; bc = bond coat.

### Questions for Self-Analysis

1. What is plasma? Which is a good plasma gas, and why?
2. What is the role of secondary gases in influencing in-flight powder characteristics?
3. What is the optimal particle size for thermal spraying? What levels of temperature and velocity may be required?
4. What is bow shock, and what is required to overcome it?
5. What is the influence of a shorter stand-off distance of the substrate from the spray gun?
6. What is the issue with high feed rate of powder during coating deposition?
7. What is the effect of substrate preparation on the coating quality?

## REFERENCES

[1] K. Alamara, S. Saber-Samandari, and C. Berndt, "Splat taxonomy of polymeric thermal spray coating," *Surface Coatings Technology*, vol. 205, no. 21–22, pp. 5028–5034, 2011.

[2] Y. Xiao, E. Ren, M. Hu, and K. Liu, "Effect of particle in-flight behavior on the microstructure and fracture toughness of YSZ TBCs prepared by plasma spraying," *Coatings*, vol. 8, no. 9, p. 309, 2018.

[3] V. Kudinov, P. Y. Pekshev, V. Belashchenko, O. Solonenko, and V. Safiullin, *Application of Plasma Coatings*. Moscow, Nauka Publ, 1990.

[4] M. Grujicic, C. Zhao, C. Tong, W. DeRosset, and D. Helfritch, "Analysis of the impact velocity of powder particles in the cold-gas dynamic-spray process," *Materials Science Engineering: A*, vol. 368, no. 1–2, pp. 222–230, 2004.

[5] B. K. Swain *et al.*, "Sensitivity of process parameters in atmospheric plasma spray coating," *Journal of Thermal Spray Engineering*, vol. 1, no. 1, pp. 1–6, 2018.

[6] S. Forghani, M. Ghazali, A. Muchtar, A. Daud, N. Yusoff, and C. Azhari, "Effects of plasma spray parameters on $TiO_2$-coated mild steel using design of experiment (DoE) approach," *Ceramics International*, vol. 39, no. 3, pp. 3121–3127, 2013.

[7] P. Vuoristo, "Thermal spray coating processes," in *Comprehensive Materials Processing, 1st Edition Volume 4: Coatings and Films*. Elsevier, 2014, pp. 229–276.

[8] J. Trelles, C. Chazelas, A. Vardelle, and J. Heberlein, "Arc plasma torch modeling," *Journal of Thermal Spray Technology*, vol. 18, no. 5, pp. 728–752, 2009.

[9] S. Yin, X. Suo, H. Liao, Z. Guo, and X. Wang, "Significant influence of carrier gas temperature during the cold spray process," *Surface Engineering*, vol. 30, no. 6, pp. 443–450, 2014.

[10] M. Mrdak, B. Medjo, D. Veljić, M. Arsić, and M. Rakin, "The influence of powder feed rate on mechanical properties of atmospheric plasma spray (APS) Al-12Si coating," *Reviews on Advanced Materials Science*, vol. 58, no. 1, pp. 75–81, 2019.

[11] S. Khandanjou, M. Ghoranneviss, and S. Saviz, "The investigation of the microstructure behavior of the spray distances and argon gas flow rates effects on the aluminum coating using self-generated atmospheric plasma spray system," *Journal of Theoretical Applied Physics*, vol. 11, no. 3, pp. 225–234, 2017.

[12] C. Kang, H. Ng, and S. Yu, "Imaging diagnostics study on obliquely impacting plasma-sprayed particles near to the substrate," *Journal of Thermal Spray Technology*, vol. 15, no. 1, pp. 118–130, 2006.

[13] S. Kumar and R. Kumar, "Influence of processing conditions on the properties of thermal sprayed coating: A review," *Surface Engineering*, vol. 37, no. 11, pp. 1339–1372, 2021.

[14] S. Harsha, D. DK, and A. Agarwal, "Some studies on wear behaviour of flame sprayed nickel and cobalt base alloy coatings deposited on steel substrate," *Presented at the 1st International & 22nd All Indian Manufacturing Technology Design and Research Conference (AIMTDR)*, IIT Guwahati, India, 2006.

[15] M. Musztyfaga-Staszuk, A. Czupryński, and M. Kciuk, "Investigation of mechanical and anti-corrosion properties of flame sprayed coatings," *Advances in Materials Science*, vol. 18, no. 4, pp. 42–53, 2018.

[16] N. Kahraman and B. Gülenç, "Abrasive wear behaviour of powder flame sprayed coatings on steel substrates," *Materials Design*, vol. 23, no. 8, pp. 721–725, 2002.

[17] C. Sundaresan, B. Rajasekaran, G. Sivakumar, and D. Rao, "Hot corrosion behaviour of plasma and d-gun sprayed coatings on t91 steel used in boiler applications," in *IOP Conference Series: Materials Science and Engineering*. IOP Publishing, England, 2020, vol. 872, no. 1, p. 012092.

[18] H. Singh, S. Prakash, D. Puri, and D. Phase, "Cyclic oxidation behavior of some plasma-sprayed coatings in $Na_2SO_4$–60% V2O5 environment," *Journal of Materials Engineering Performance*, vol. 15, no. 6, pp. 729–741, 2006.

[19] H. Singh, D. Puri, and S. Prakash, "Some studies on hot corrosion performance of plasma sprayed coatings on a Fe-based superalloy," *Surface Coatings Technology*, vol. 192, no. 1, pp. 27–38, 2005.

[20] M. Kumar, H. Singh, and N. Singh, "Effect of increase in nano-particle addition on mechanical and microstructural behaviour of HVOF and cold-spray Ni-20Cr coatings on boiler steels," *Materials Today: Proceedings*, vol. 21, pp. 2035–2042, 2020.

[21] J. Wang, X. Zhou, L. Lu, D. Li, P. Mohanty, and Y. Wang, "Microstructure and properties of Ag/$SnO_2$ coatings prepared by cold spraying," *Surface Coatings Technology*, vol. 236, pp. 224–229, 2013.

[22] N. Bala, H. Singh, J. Karthikeyan, and S. Prakash, "Performance of cold sprayed Ni-20Cr and Ni-50Cr coatings on SA 516 steel in actual industrial environment of a coal fired boiler," *Materials Corrosion*, vol. 64, no. 9, pp. 783–793, 2013.

# 5 Design of Spray Guns

*Priya Kushram, Ariharan S, Alok Bhadauria, Ritik Tandon, Rubia Hassan and Kantesh Balani*

## CONTENTS



Increasing competition in the growing world market has pushed the industry to establish a new and better performance standard for the advancement of advanced materials. The continuous wear or corrosion of a material with time has deteriorated the desired properties of the material. To overcome such issues that cause serious damage to the material, coating and surface treatment is done to retain its thermal, electrical, optical and other properties for a longer period. The coating and surface treatment on the material has served a very important role in various applications to protect the material from oxidation, corrosion and erosion.

DOI: 10.1201/9781003321965-5

This necessitates the formation of coatings achieved through different methods. Different coating techniques can be employed depending on the substrate and material bonding, density variation, type of material to be coated and substrate type. Among the coatings processes, thermal spraying that consists of different coating methods that are evolved is one of the efficient, cost-effective and quick coating techniques. The coating process for thermal spray depends on numerous parameters, such as powder properties, gun design, substrate morphology and preheating. The variation in different spray techniques is because of differences in coating mechanism and predominantly the gun design. Other chapters discuss it in a detailed manner. The current chapter mostly deals with the gun design of different thermal spray methods (plasma spray, electric arc spray, high-velocity oxygen fuel [HVOF], combustion spray and cold spray), as well as how the spraying gun can be designed to achieve more efficiency with minimal loss of material.

## 5.1 THERMAL SPRAY GUNS

The thermal spraying guns are used for powder feed and for acceleration, heating and channelizing the powder flow towards the substrate. The designing of guns plays a dominant role in deciding the governing traits of particle velocity, particle temperature, injection system, feed rate and so on, which are among the most important parameters that decide the quality and quantity of coating deposition. The experience in gun and system design allows engineering-efficient coatings for exposure to extreme environments. For example, the De-Laval converging-diverging nozzle in the cold-spray method increases the powder particle to supersonic speed. The free expansion and exit of compressed gases through the torch nozzle attain supersonic velocity in the HVOF method, and it is responsible for the dense coating. But the high temperature is achieved in the plasma spray technique by the high energy of plasma formed by the arc strike between anode and cathode. So the entire gun design will change accordingly. There are advantages and disadvantages of guns for any particular application. For example, the liquid-fuel HVOF can enhance the bond strength and control the stress of the coatings. It can spray thick coating, minimizes oxide formation, and ensures restricted phase decomposition. On the other side, the high temperature in plasma spray can lead to the formation of oxides of powder particles and also increase the chances of phase decomposition of the in-flight powder when depositing a coating. So an active improvement in designing the spraying gun becomes essential to rectify the drawbacks of guns and improve the characteristics of the coatings. Therefore, an idealized structure to imply that a perfect plasma system should include certain characteristics to produce an efficient coating in a repetitively controlled manner:

1. Stable and unperturbed plasma parameters during spraying;
2. No plasma pulsing during deposition <3kHz;
3. Stability of electrodes (with no erosion);
4. Consistent powder feeding;
5. High deposition efficiency;
6. Simple, easy to maintain, and easy processing.

The plasma generated by an electric arc in DC plasma spraying is maintained between electrodes. Simultaneously, the arc moves along to the anode surface due to electromagnetic forces and plasma gas flow. This gradual movement of arc root leads to substantial plasma fluctuations [1, 2]. Though the correlation between the arc behavior on other parameters is not well understood, the influence of arc root position concerning the anode surface condition is established [1]. It also influences the wear of the electrodes over a period that may considerably influence the coating characteristics and expose the issues in the reproducibility of coatings. These plasma fluctuations exist in the system due to prolonged use with a wide variety of parameters (e.g., arc position, enthalpy, environment. composition, temperature, viscosity, velocity of the particles). Ignoring these parameters will lead to a variation in the important spraying factors. For example, the temperature can go as high as 600 °C, the state of a particle can be superheated to an unmelted state and its velocity can go as high as 200 $ms^{-1}$. The details on the cause and solution of these important characteristics in the design aspect of the advanced system are discussed briefly in this chapter. In this regard, the plasma systems are divided into the following four major groups based on the design of torches and features:

1. Moderately low voltage (40–90 V), high current (400–800 A) conventional plasma torch;
2. Relatively high voltage (>200–250 V), low current (< 500 A) conventional plasma torch;
3. Multi-torch (e.g., triple torch) plasma systems;
4. Cascaded and segmented plasma torches.

### 5.1.1 DC Plasma Spray Gun

Plasma spraying is one among the widely used thermal spray processes where an electric arc (heat source) ionizes the gas that heats the powder material and melts them. The plasma spray technique provides a very high-temperature (12,000–15,000 K) environment resulting in complete or partial melting of the materials used for coatings and accelerating them to a high velocity to hit and deposit on the substrate. The peak temperature can be attained with the subsonic velocity range of 500–2,500 m/s depending on gun design, plasma gas composition and other operational parameters. The power control system permits the modification of the main parameters, powder feed and flow rate (plasma and carrier gas). It controls the safety interlock to avoid arc initiation without coolant flow and primary gas flow. Further, it controls the secondary gas flow when starting the arc. So, the design of the DC plasma spraying gun should possess the following essential systems to prepare a competent coating:

1. Plasma gas supply
2. Cooling facility
3. Power supply
4. Powder supply.

In a plasma gas system, there are several cylinders under high pressure. The control console has a separate controlled gas flow rate with a mass flow controller for each. Gases of specific composition are mixed and pass through the torch. Ar is often the preferred primary gas because of its low energy density, hence the low rate of electrode erosion and, thus, the longer life of the torch. Along with primary gas, a secondary gas (He and $N_2$) is used to improve the power density and powder feed rate by increasing the gas velocity. The plasma gases flow through the water-cooled constricting nozzle-shaped electrode. The plasma is formed between the electrodes by the high-voltage discharge that leads to the local ionization. The local ionization helps the DC arc that is formed between the electrodes to travel through the local ionized region due to its conductive nature. The resistance heating triggered by the DC arc raises the surrounding temperature that causes the ionization of gases. The unstable gas ions formed are then recombined, and the enthalpy released is used to heat the powder particles. Once the plasma is stable, the DC arc extends down the nozzle during the spray instead of shorting out of the electric arc near to the anode edge due to the thermal pinch effect. For good and non-contaminated coatings to form, the powder particles must completely melt (i.e., it should have enough residence time in the plasma plume) and there are no "bumps" formed in the coating due to spitting and plugging.

The coating material in plasma spraying can be injected externally, as well as internally through a powder hopper which is slightly heated and kept in vibration motion to avoid agglomeration. So, the design of the spraying gun or nozzle has the arrangement to provide an effective coating. Accordingly, the external powder feeder (explained in section 5.2) can be of three types:

1. Gravity
2. Rotating wheel
3. Fluidized-bed based powder feeder.

Internally, it can be fed axially or radially. The internal axial injection is considered to be the best among the powder feeders, but it is hard to implement during the application. In the external type, the injector's position is ~10 mm and several centimeters from the nozzle exit for metals and polymers, respectively, since lower temperatures are required to melt the polymers in comparison to that of metallic materials. The powder feed and injection strongly affect the coating quality. The powder particle trajectory is dependent on the momentum imparted on the individual powder particle, which is influenced by mass flow rate. Particle velocity is characterized by the gas and powder flow rate and the type of the carrier gas. For the flow of powder particles along with the plasma plume, the dynamic pressure of carrier gas should be less than that of the plasma jet. Else, the particles will shoot through the jet and will not get dragged by plasma. Then the powders are effectively exposed to the high-temperature environment and conditioned for coatings. The carrier gas provides sufficient velocity to the powder particles. Thus, it penetrates in the plasma and does not get deflected by it.

The extreme heat of the plasma jet heats the external injector tip and the internal part of the powder feeder exit to a maximum of 700–1000 °C. The particles coming out of the injector gets stuck at the hot surface of the injector due to adhesive forces. The stuck particles further hinder the powder flow and block the motion of incoming particles. If the stuck particle is ceramics, it is called *plugging*, which can only be removed by mechanical means. In the case of materials with a low melting point (metallics) stuck at the hot walls, they melt down after attaining a critical size and fall directly into the plasma stream, which is referred to as *spitting*. The spitting results in the formation of "bumps" in the coatings and thus deteriorates its performance. So, the system should be arranged accordingly to get rid of the spitting effect with the designing conditions, such as

1. Low lateral speed of powder particle;
2. Use of high conductivity cathode with the good cooling system;
3. High carrier gas to detach particles from the injector wall surface;
4. Polishing the internal surfaces of the injectors.

As a design of cooling the electrodes, an intensive cooling system is also incorporated. It consists of distilled water in a closed loop at a pressure in the range of 1.2–1.7 MPa. The pressure is kept high to avoid the film boiling due to high heating at the torch anode. Also, the cooling water should be

1. "Bubble-free," to avoid the loss of cooling;
2. Of a low electrical conductivity, to avoid the electrolytic chemical reactions;
3. Highly pure deionized water, to avoid the salt deposition.

The malfunctioning of the cooling system can result in overheating of electrodes, and thus it reduces the lifetime of the system. These are the common design criteria for the different plasma torch guns.

Among the major type of spray torches (DC and RF) used in the plasma spray technique, DC plasma torches provide a gas velocity close to 600–2300 m/s, where the injector is placed radially (Figure 5.1). Conventional DC plasma torches (Figure 5.1a) are designed with a stick kind of cathode. The arc is struck between a stick-type cathode and anode (cylindrical) nozzle. This arc heats up the plasma gas which is introduced through the cathode base and exits at the gun nozzle. The gas exit at the nozzle will possess a very high temperature and velocity jet. Temperature as high as 15,000 K is achieved at the nozzle region. Also, the maximum velocity attained by the gas near the nozzle will be in subsonic velocity (500–2,500 m/s) which depends on the design of the torch, the composition of the plasma gas and other spraying parameters. But there is no electrode in the RF-coupled plasma torch, and the discharge cavity is formed inside a cylindrical insulator. The powder feed can be axially injected into the discharge core and stays for a while; this depends on the time taken by the powder for melting. Then, the plasma torch tip placed at any place inside the discharge cavity will be directed for spraying with the help of carrier gas. The peak particle velocity achieved is lower in the RF plasma torch than the DC plasma torch. The RF plasma torch does not compromise to achieve a high-density coating. Instead, it melts powder particles of relatively big and thermal insulating materials.

The cathode in the plasma spray provides enough electrons to the arc through thermionic emission. The current due to thermionic emission ($J$) is related to the cathode temperature ($T_c$) and work function ($\phi_c$) using the Richardson-Dushman equation (5.1).

$$J = AT_c^2 \exp\left(-\frac{\phi_c}{k_B T_C}\right) \tag{5.1}$$

where $A$ is an empirical constant (~$6 \times 10^5$ A/m$^2$K$^2$ for most of the cathode materials), $e$ is the elementary charge and $k_B$ is the Boltzmann constant. The work function of the metal (including tungsten) is >4.0 eV. However, the arc generation requires a current density of ~$10^8$ A/m$^2$ for which the temperature of cathode and its work function should be ~4500 K and 4.5 eV, respectively. Tungsten has a high melting temperature of 3422 °C, which offers suitability for the cathode materials with high structural stability. To overcome the issue of stability at high temperature, a <2 weight percentage of material (thoria) with lower work function is mixed with tungsten to help provide a high current through thermionic emission. Figure 5.1b and 5.1c present the PT4 and 3MB plasma torches, respectively, along with their nozzle dimensions.

The anode is made up of the water-cooled Cu channel or lined with W/W-Cu sleeves, which serve as a gun nozzle. The nozzle design controls the gas velocity, plasma temperature, arc length and so on. A nozzle diameter of 6–10 mm is required to achieve an arc current of 1,000–3,000 A. It is observed that temperature increases as the diameter of the nozzle decreases. This is attributed to the increased local temperature of the plasma gas due to increased heat loss resulting from the increased electric field strength. The anode collects electrons and allows them to flow into the plasma. Further, the non-equilibrium condition of the cold gas boundary layer, which lies between the anode and the plasma, leads to high electron temperatures compared to that of heavy particle temperatures that are required for the flow across the cold gas boundary layer. Thus, the flow is dependent on the electron density gradients ($J$) as per the expression in equation (5.2).

$$J = \sigma E_x - (\sigma / en_e)\left(\frac{dp_e}{d_x}\right) \tag{5.2}$$

where $e$ is the elementary charge, $E_x$ is the conductivity of electron in the plasma, $\sigma$ is the electric field (which is normal to the anode), $n_e$ is the number density of electrons, and $p_e$ is their partial pressure in the plasma. The high

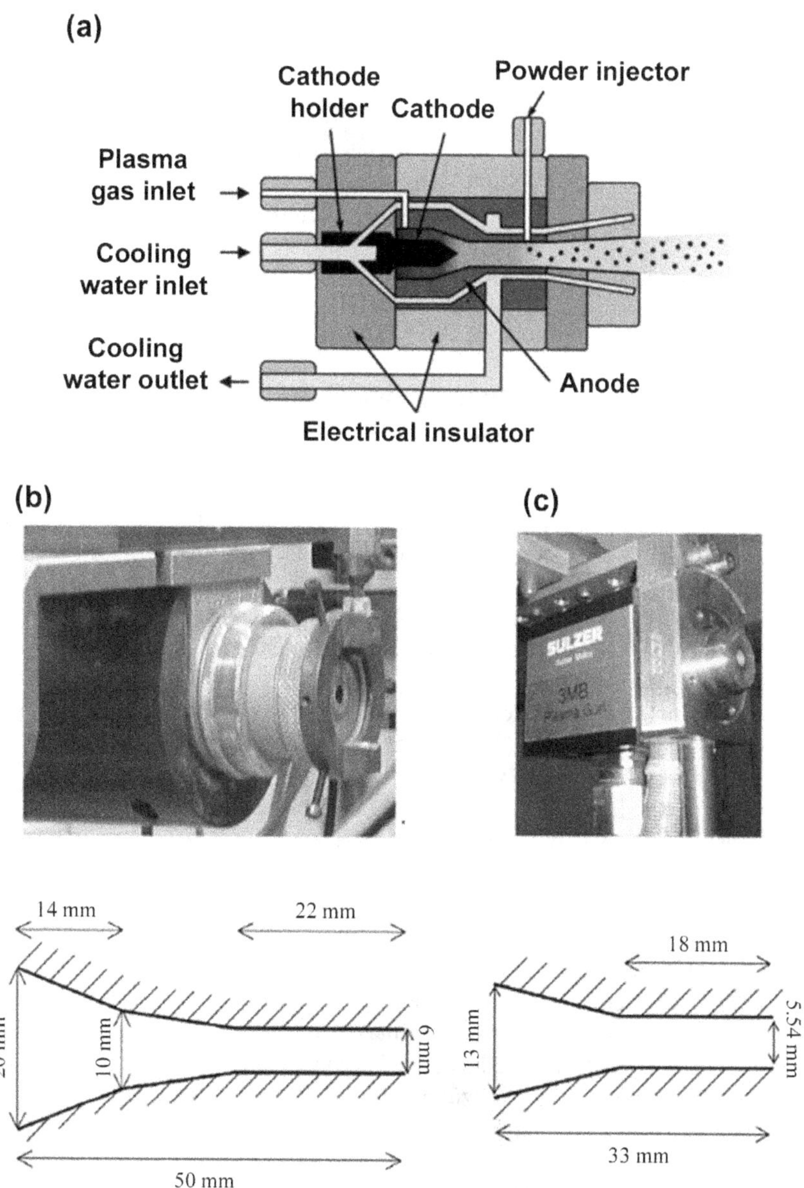


**FIGURE 5.1** (a) Schematic of the commercial DC plasma torch; (b) and (c) plasma torches of PTF4 and 3MB, respectively, and their corresponding internal nozzle schemes.

*Source*: (a) Reprinted with permission from [3]; (b) reprinted with permission from [4].

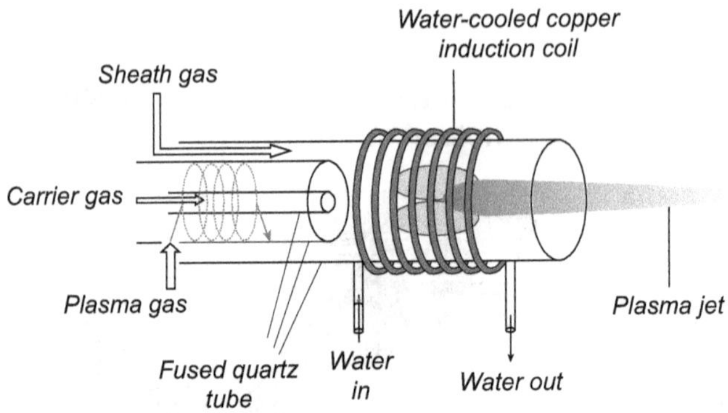


**FIGURE 5.2** Schematic of RF induction plasma gun.

*Source*: Reprinted with permission from [5].

gradient of partial pressure at the cooled surface of the anode resulted in high diffusion flux towards the surface that established the major segment of the electric current flow.

RF torches provides flow up to 100 m/s, and the water-cooled injector is placed axially in the middle of the torch without coupling to the coil (Figure 5.2). Coupling of the coil with plasma occurs near the wall that heats the gas. The gas close to the torch axis gets heated up by convection-conduction. It utilizes a high frequency and power (3.6 MHz and 100 kW, respectively). Generally, the plasma-forming gas velocity will be around 100 m/s while the particle velocity is around 60 m/s. It provides the residence time of tens of milliseconds, and it is sufficient for the powders to melt partially or completely. It uses Ar as primary gas in combination with He/$H_2$ as a secondary gas which increases heat transfer, hence ensuring the complete heating of powder particles. In RF plasma, the powder particles are exposed in the coil region using Ar, and it is partially or completely melted at these regions before spraying. The melted powder particles are then accelerated in a divergent coil, thus increasing the gas velocity up to 2500 m/s and particle velocity up to a maximum of 600 m/s. Correspondingly, the residence time is further reduced, and such parameters are recommended for materials that are prone to oxidize in specific environmental conditions. Due to economic reasons, the RF technique is limited to highly advanced material and composite forming technology.

Usually, the power supply used in the plasma spray is a current-controlled rectifier. The thyristor power system is also one of the frequently used technology for cost-effective production. It is suitable for high current and high voltage. But the low dynamic control, high ripple and low effectiveness limits the use of thyristor power system in technology. The power supply depends on high-frequency switching (>15 kHz for inverter type) to provide a smooth current for the production of uniform coating. Also, it is preferred to work with twice its working voltage in an open-circuit voltage (150–200 V) mode. High voltage and low current are chosen to start the plasma torch with minimum electrode damage/erosion. Primary and secondary power supply systems are mostly used to provide high dynamic control, the possibility of modulation, low ripple and high efficiency/accuracy.

#### 5.1.1.1 Multiple Cathode Gun

The multiple cathode gun spray technique is an advanced development by utilizing multiple cathodes. The cathodes are placed concentrically, symmetrically and radially oriented with a common anode nozzle (Figure 5.3). The arcs in the multiple cathodes coalesce at the common chamber and form a common cylindrical plasma jet. Also, the powder is fed coaxially with the spraying gun. Despite its promise, the multiple cathode gun spray is limited in application, primarily by metallic vapor created when feeding metal powders; a lower operating cost is also a consideration.

The spraying gun should have long-term stability for the perfect plasma system to form with the least damage/erosion of electrodes, stable powder feed, high deposition efficiency and low maintenance cost. Cascade plasma torches provide a steady plasma arc for a wide range of plasma gas (flow rate, composition and pressure). Also, it has high increased productivity throughout and can spray almost all kinds of powders with different specifications. Also, it supports with no or low helium spray parameters. SinplexPro provides similar features, such as a cascaded arc with the efficiency advantage of a single cathode spraying gun. In the case of the Triplex series of Sulzer-Metco, it is based on three cathodes that are separated by an insulator. Also, the electrical energy is distributed by independent sources through three arcs that are sticking at the electrode preceded by insulated rings. The formation of arcs in the Triplex series permits the reduction of the voltage fluctuations percentage (four to five times) compared to that of conventional DC plasma spray torches.

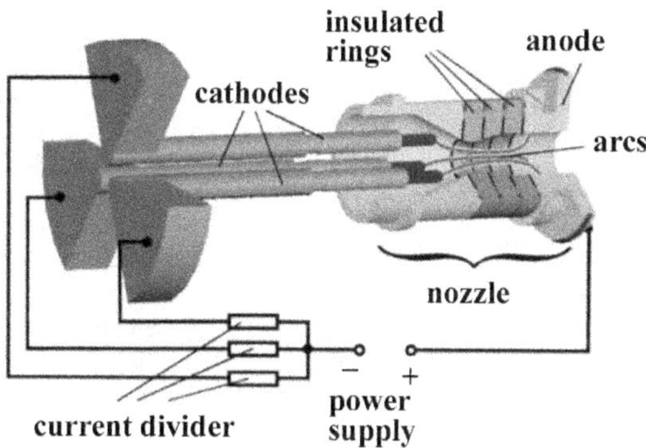


**FIGURE 5.3** Schematic of multi-electrode plasma gun.

*Source*: Reprinted with permission from [6].

### 5.1.1.2 Cascaded Plasma Torches

The cascaded plasma approach is one of the most capable plasma spraying systems to achieve the output as per the theoretical estimation that operates at a high voltage and low current. Correspondingly, it may allow generating plasmas with the following characteristics:

1. Plasma stability with low pulsing (at 3–4 kHz);
2. Extended electrode life;
3. Use of diverse plasma gases while incorporating minor changes in plasma gun;
4. Achieving high-enthalpy plasma at reduced plasma gas flow rate.

Figure 5.4 shows the schematic representation of a cascaded plasma torch. There are insulated sections separating cathode and anode [7]. In the first-generation cascaded plasma gun, penetration of the cascade did not approach to generate sufficient power due to the complex design with adjustable cathode and other elements.

As the second attempt in the design of the cascaded plasma gun, the cathode element in plurality in the cascaded elongated plasma torches is developed with an axial powder injection. Earlier, the powder injection into a plasma plume encountered stability issues and affected the performance of the torches. Subsequently, the alteration in the powder design from external radial powder injection to axial powder injection occurred in the next-generation designs. A different design of the powder injection is discussed in the later section of this chapter. The recent design is with a triple powder injector system which is aligned with the arc root positions (formed using three anodes), which will give a highly effective feed rate with provision to spray on large components. It also can operate using the

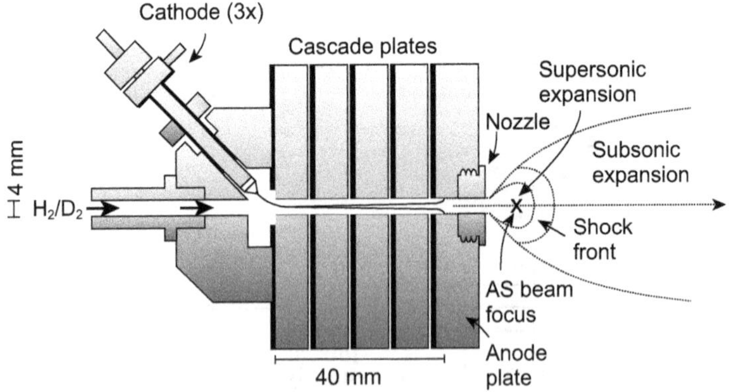


**FIGURE 5.4** Schematic of cascaded plasma torch.

*Source*: Reprinted with permission from [7].

mixture of argon/helium as plasma gas. Significantly low plasma gas flow rate in the Triplex family may provide high plasma enthalpy. Also, a low flow rate of plasma gas through the standard (high-enthalpy) nozzles generates moderate particle velocity, which provides long dwell time in the plasma plume, and it is attractive to produce an efficient coating of relatively low thermal conductivity or high melting point materials.

#### 5.1.1.3 Arc Stability

Arc stability is one of the most important concerns in plasma spraying. The minimum principle states that the arc position will be adjusted itself where there is the least overall voltage drop. It means that the arc will position itself such that energy loss is minimal. The arc stabilization can be done using a cylinder-shaped wall that keeps the arc in a fixed direction, called wall stabilization. Otherwise, the superimposing of the cold gas flow that cools the arc when it goes off-axis is referred to as flow stabilization. In plasma torches, the combination of both torches is used. Usually, the ideal position of the arc is the least distance between cathode and anode. At this distance, the velocity of the cold gas will be maximum, which results in extensive cooling with high arc voltage.

The anode attachment instability is considered as one of the main reasons for arc instability. The movement of the arc-anode attachment can be divided into three categories as shown in Figure 5.5.

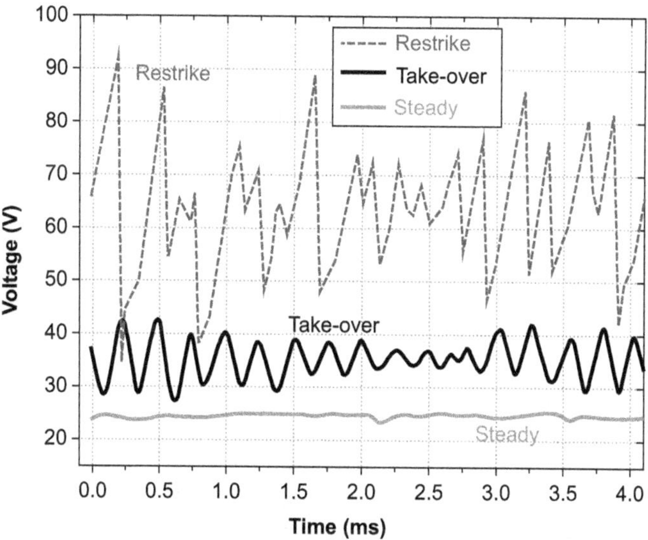


**FIGURE 5.5** Voltage variation during steady, take-over and restrike modes.

*Source*: Adapted and redrawn from [8].

1. *Steady mode*: Operates with very low fluctuations in voltage. There are minimal fluctuations observed in the steady mode (almost constant) which are not linked to the fluid dynamic characteristics of the arc at low frequencies.
2. *Take-over mode*: Operates with arbitrary fluctuations in the voltage traces, and it reaches around 20%–50% of the mean fluctuation in take-over mode.
3. *Restrike mode*: Here, arc connection goes downstream into the anode nozzle. The total amplitude of fluctuation reaches a value of 30%–70% of the average value in restrike mode with a frequency range of 2–6 kHz.

### 5.1.2 Combustion Spray Guns

High-velocity oxygen fuel (HVOF), flame spray and detonation gun (D-gun) are the major types of thermal spraying methods that utilize combustion guns. They also use gas and liquid as the sources for powder feed, velocity, melting and guiding the flow of the spray material to adhere to the substrate. HVOF is one of the widely used internally confined combustion spray techniques that provides an extremely high velocity to particles. The particles with large impact energy deposit on the substrate by means of plastic deformation. A high-pressure gas system or lower-pressure liquid system is used to achieve high velocity of the particle. Subsequently, there are ranges of HVOF guns that differ in the ways of providing high-velocity spraying.

One of the most common methods is the use of a water-cooled long nozzle with a pressurized combustion chamber. When hot gas exits, rapid cooling of the jet occurs due to sudden expansion by the air medium. Also, heat loss is avoided by using a nozzle incorporated with a long De-Laval barrel of ~30 cm. Though some energy is lost, a large quantity of energy is still utilized in providing heat to the particles. The combustion of fuel (i.e., kerosene, hydrocarbons, acetylene) and oxygen produces a hot high-pressure flame when injected under pressure into the chamber. Further, it is forced down to pass through a nozzle which increases its velocity to supersonic velocity (2–6 Mach). Figure 5.6 shows the different designs of guns used in HVOF spraying.

The powder is fed axially in the low pressurized De-Laval type nozzle. Generally, it was found that the spray rate (at least double times) can be achieved using radial injection compared to that of axial injection. The variation in the gas velocity in the HVOF gun design with combustion chamber followed by De-Laval barrel and barrel nozzle is shown in Figure 5.7. Also, the exit gas velocity depends on the ratio between fuel and oxygen due to the reduction in the molecular mass of the combustion gas at high hydrogen flow. Low molecular mass results provide higher gas velocity (at the throat) concerning the velocity at exit. Since the Mach number is constant, the other parameter that can alter the gas velocity is the temperature of the combustion gas. However, the gas velocity is not related to the total gas flow for a constant fuel/oxygen ratio (Figure 5.7). Also, the combustion gas density increases with gas flow and it improves the particle's velocity due to

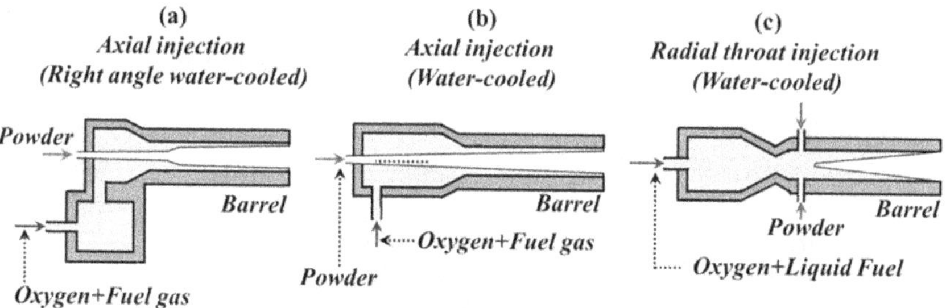


**FIGURE 5.6** Schematic of HVOF spraying gun: (a) early established gun (powder axial injection and gas at normal to the chamber); (b) combustion chamber with axial powder injection; (c) combustion chamber with radial powder injection.

*Source*: Adapted and redrawn from [8].

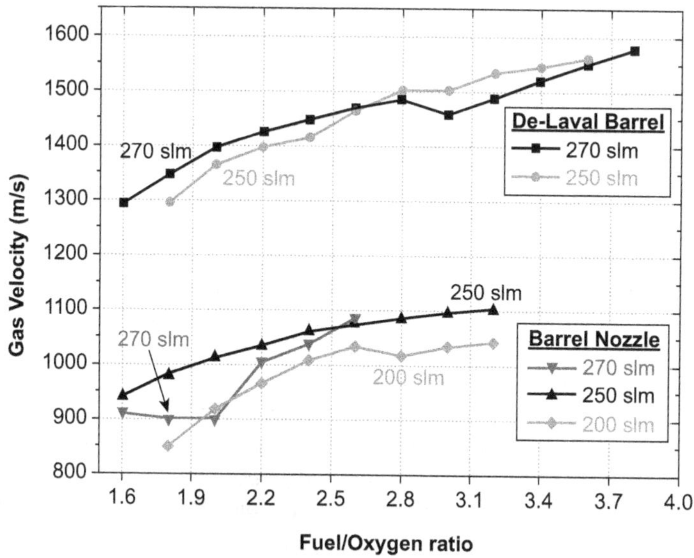


**FIGURE 5.7** Exit gas velocity of the HVOF gun with $H_2/O_2$ ratio for barrel and a convergent-divergent nozzle gun designs.

*Source*: Adapted and redrawn from [8].

increasing the drag force. Thus, the highest particle velocities are obtained, which is recognized by the peak gas velocity and total flow.

Powder to be melted and coated is injected with a carrier gas (He, Ar, and $N_2$). Because of the high back pressure generated by the expansion of combusting gases, there is a requirement for a pressurized powder feeder. So, the spray is injected with rapid supersonic nozzle expansion of gases. The spray in HVOF is controlled by oxy-fuel gas control, powder injector, powder gas flow control, water or air cooling, nozzle cool flow and control, cooling passages, nozzle tube exit and combustion gas injectors. Upon impact with the substrate, the spray forms a thin, dense coating using plastic deformation which produces a homogeneous microstructure. High particle velocity, combined with uniform heating and long dwell time, produces coatings that are very dense and tightly bonded to the substrate. Powder heat transfers and dwell time are greatly increased by nozzle design and hence efficiency. The HVOF technique has the gun design with the provision to use both liquid and gas as the input fuel system, referred to as a hybrid gun (Figure 5.8). Gas fuel systems can use hydrogen, propylene, ethylene and hydrocarbons, whereas liquid fuel systems use kerosene and ethanol.

The advantages of using liquid fuel are controlled stresses, high bond strength, densification of coatings, low oxide content, and minimal or no decomposition of coatings. Although kerosene has a low cost and minimal safety problems when fed with high pressure, large contamination due to its use can result in poor torch performance. Thus, ethyl alcohol is preferred because of its low price and 30% less toxic emissions in most industries. Further, "diamond jet" is one of the most commonly used HVOF guns that operate with a wide range of fuel gases and water or oil cooling systems. Then, it has a low temperature at the convergent portion of the nozzle. The combustion of fuel and air provides supersonic velocity to the powder particle. The coatings using the gun designed with diamond jet provides a high-density, low-oxide content and high deposition efficiency. It is commonly used in the oil and gas, pulp and paper, and steel industries. The WokaStar series uses a spray gun (as shown in Figure 5.8b) that operates with kerosene and oxygen as a fuel with the utilization of profiled nozzle. Subsequently, there are chances of oxidation since the flame temperature is comparatively high. Further, the JP-5000 also utilizes liquid fuel to achieve particles with a velocity range of 700–800 m/s below the melting point and achieves an efficiency of 30%–50%. Generally, the ranking of different HVOF systems (Figure 5.9) will be characterized using the attainment of powder particle temperature and velocity [26].

At present, HVOF systems have fulfilled important needs in developing an efficient coating for different industrial applications. Relatively efficient coating

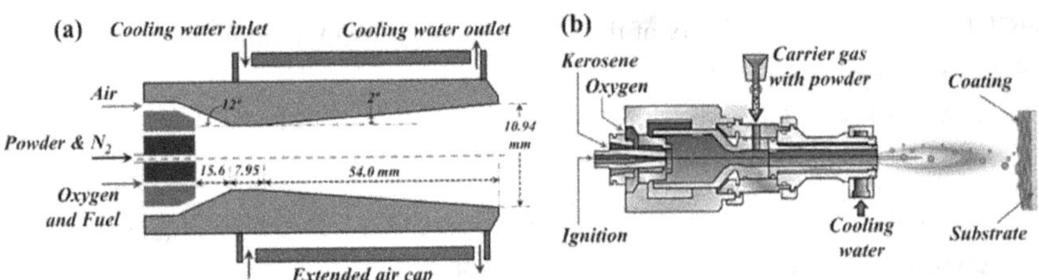


**FIGURE 5.8** Schematic of (a) a jet hybrid thermal spray gun and (b) a liquid-fueled WokaStar series spray gun.

*Sources*: (a) Redrawn from [7]; (b) adapted and redrawn from [8].

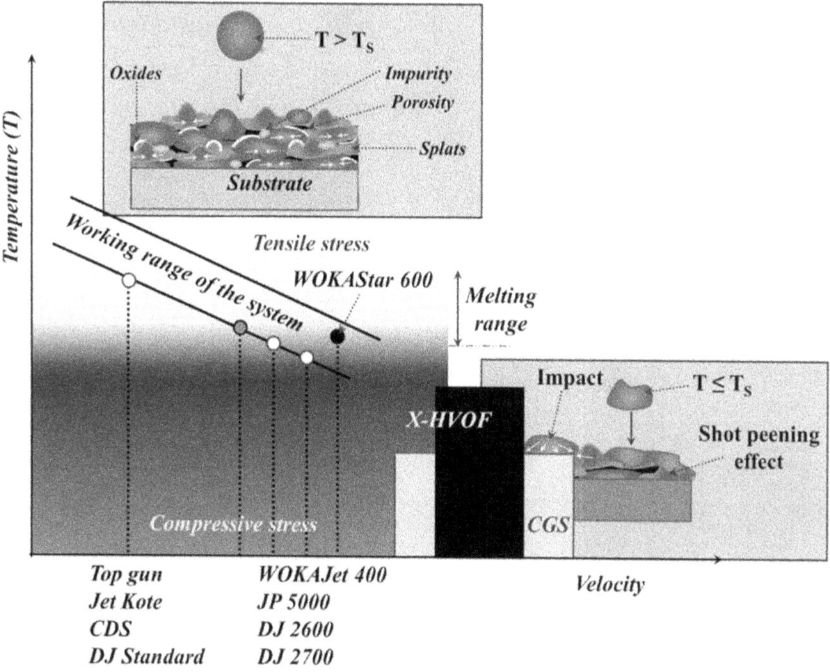


**FIGURE 5.9** Schematic showing particle temperature and velocity in different HVOF systems (CDS: Cold gas-dynamic spraying, DJ: Diamond jet, CGS: cold gas spraying).

*Source*: Adapted and redrawn from [9].

at low operating cost is desirable to fit the actual needs of industries. Improving the performance of the system may be dealing with a prospective optimized parameter of the combustion system, nozzle and barrel geometry, as well as the powder-injection portion of the HVOF system.

### 5.1.3 Detonation Gun

The D-gun utilizes the controlled explosion of fuel gas and oxygen to melt the powder particle. Oxygen and acetylene are introduced into the tube at the closed end, and the fuel and gas are detonated by a trigger. The D-gun is designed in such a way that the nitrogen is injected between the ignition point and the gas mixture to avoid backfiring, because the detonation or ignition can move in every direction and it creates an explosion hazard when backfiring into the gas supply. It confines the combusted gas to the tube and nozzle and thereby produces a high thermal and kinetic energy combustion jet. The different stages of the coating process using D-gun are shown in Figure 5.10. It was the first combustion process to incorporate in the thermal spraying with this concept. Confining the gas can produce a much higher temperature and longer particle dwell time than the conventional spray

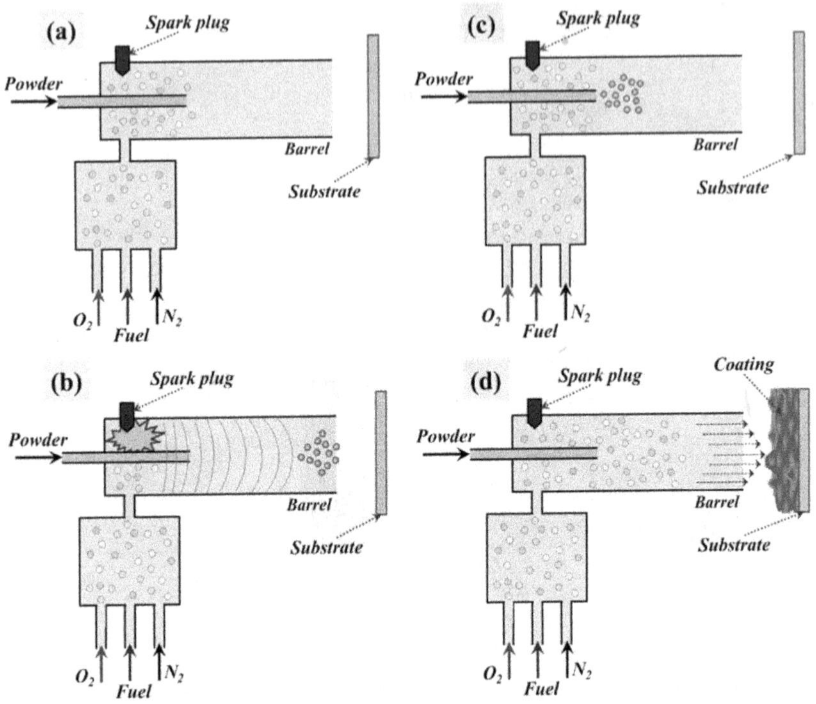


**FIGURE 5.10** Schematic of different stages during coatings process using D-Gun spraying: (a) fuel and oxygen injection in the combustion chamber; (b) nitrogen injection in powder; (c) gas explosion followed by powder acceleration; (d) chamber exhaustion using nitrogen.

*Source*: Adapted and redrawn from [8].

techniques. High particle velocity in the range of 1,000–3,000 m/s and more accelerated combustion flow can be obtained to produce a dense and highly bonded coating. It is predominantly used in aerospace applications.

The inclusion of light gases, such as He and $H_2$, improves the detonation velocity compared to that of heavier gases. Depending on the gas mixture, the velocity of the explosion wave will be in the range of 1000–3000 m/s. Also, it does not depend on the barrel dimension, provided it is beyond the critical barrel diameter. Theoretical velocity, $D$, after the detonation is calculated using equation (5.3). $T_1$ is the temperature attained during the detonation, and $p_1$ of the detonation products, are calculated using equations (5.4) and (5.5), respectively.

$$D = \sqrt{2\left(\gamma^2 - 1\right)Q_0} \tag{5.3}$$

$$T_1 = \frac{2.\gamma}{\gamma + 1} \cdot \frac{Q_0}{c_v} \tag{5.4}$$

$$p_1 = 2(\gamma + 1)\rho_0 \cdot Q_0 \tag{5.5}$$

where $\gamma$ is the ratio of specific heat for the burned gas mixtures and $Q_0$ is the specific energy of ignition.

### 5.1.4 Combustion Powder-Spray Guns

Combustion powder spraying is one of the simplest thermal spraying guns used in flame spraying and it uses powder as the spraying material. The powder is injected axially into the guns with the help of a carrier gas. These powder particles are melted in the oxy-fuel flame and deposit on the substrate. The temperature and enthalpy of the flame are obtained by the oxygen flow rate, fuel gas composition and flow rate. When the molten particles carry forward towards the surface that is to be coated, they rapidly solidify after impinging on the substrate and form a coating. The difference in using powder and wire material is in melting capacity. The restriction in the flame length and lower particle velocity limits the particle temperature after their flight, which is 0.7 times the gas temperature ($T_g$), and thus it prevents the complete melting of ceramics. The flame-spraying technique is limited by the particles whose melting temperature is greater than that provided by flame or if the material decomposes on heating. The 5P-II and 6P-II series spraying guns are some of the most commonly used powder-spray guns.

### 5.1.5 Combustion Wire, Rod and Cord Gun

The gun is designed to melt the feedstock in wire, rod, and cord form by combustion and atomization. Also, it easily incorporates the mixing of material as the tip is melted, and that is followed by compressed air flow for atomization of the molten tip of wire, rod or cord. The typical wire diameter used is 1.2 to 4.76 mm. It is attached to the torch that drives the roll and pulls the wire into the nozzle with the controlled velocity. The atomized particles on passing through the nozzle deposit on the substrate. The gun setup consists of a long nozzle where fuel and oxygen are burned, and the flame extends into the air (Figure 5.11). The rod or wire is fed axially into the nozzle and it extends continuously at the gun exit with an optimum velocity for the tip to melt in the flame. Later, the compressed air blows concentrically with the flame that atomizes the molten material and generates the molten droplets. Using the wire spray process ensures deposition of only molten particles; the other advantage is maximum and uniform heating of material, since the flame is concentric with the wire. Also, the light weight and easy handling of the spray gun supports spraying onto complex shapes. The coating obtained from this technique generally has a porosity of around 10% and can become highly oxidized. The gun has the provision to use nitrogen instead of air, which decreases the oxidation by 2%. It is mainly used in wear-resistant coating for the low-load condition and protection against the extreme environmental condition, and copper coatings for a cooling system. It is easily portable, able to be used in an area without electricity and deposits with high efficiency. One of the disadvantages is

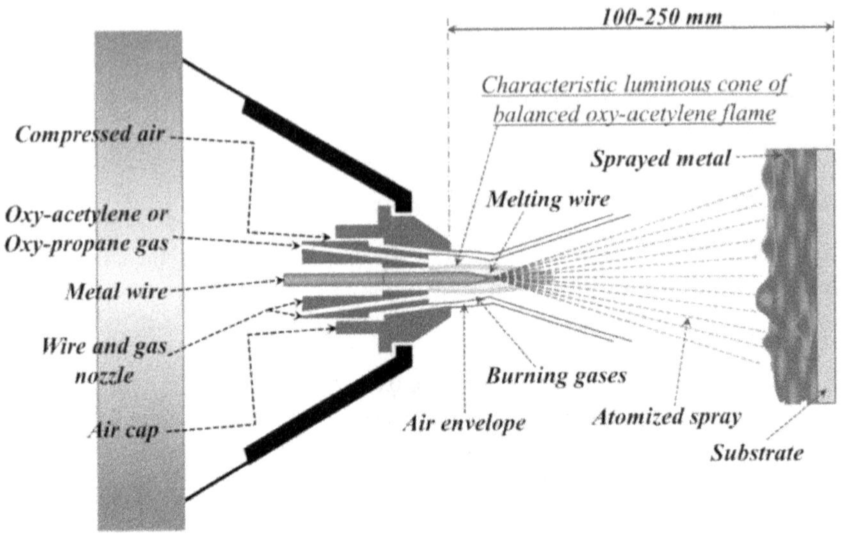


**FIGURE 5.11** Combustion wire spray gun design.

*Source*: Adapted and redrawn from [10].

its low coating strength as compared to HVOF and plasma spraying. A few of the recently used guns are the 16E, 5K, and EGD-K Combustion series.

Apart from the thermal processes, there is a third class of spray techniques, cold spray, that utilizes the kinetic energy of the heated powder particles.

### 5.1.6 Cold Spray

#### 5.1.6.1 High-Pressure Cold Spray

Cold spray exploits the kinetic energy to the spray process, which utilizes compressed gas as the supersonic jets to accelerate the particles to ultra-high velocity (500–1500 m/s). The high dynamic gas acceleration of particles to supersonic velocity is achieved using the convergent-divergent type (Figure 5.12) of nozzle, named the De-Laval nozzle. As thermally exposed powder particles are fed into the nozzle throat, slightly higher pressure is maintained in the powder feeder compared to the upstream pressure at the chamber. Unlike other coating methods like HVOF, plasma spray or combustion, the powder particles are not melted during cold spraying. The particles are preheated (700–800 °C) and avoid condensation while they experiences sonic expansion. The preheating of the particle temperature helps in the deformation of the particle due to impact with a high velocity on the substrate. When helium as a spray gas is used, the spray system is operated in an enclosure to recycle the gas used and reduce the cost. Solid particles on reaching the substrate plastically deform and consolidate to form the coating by Kelvin-Helmholtz instability. This implies that the particles attain a

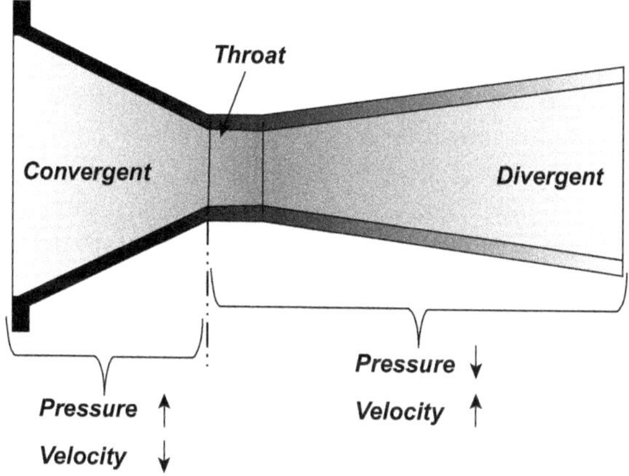


**FIGURE 5.12** Schematic of convergent-divergent De-Laval nozzle used in cold spraying.
*Source*: Adapted and redrawn from [10].

critical velocity that depends on the specification of the particle (size, composition and morphology). The local velocity of the particle at the nozzle's divergent segment of cold spraying can be calculated using equation (5.6):

$$u_g = M\sqrt{T.\gamma.R / M_g} \tag{5.6}$$

where $u_g$ is the velocity of the gas used to carry the powder, $M$ is the Mach number, $\gamma$ is the ratio of specific heat capacity ($c_p/c_v$), $R$ is the ideal gas constant and $M_g$ is the molar mass of the gas. The relation between the temperature at a specific location ($T$) and the initial temperature ($T_0$) of the gas is as per equation (5.7):

$$T_0 / T = 1 + (\gamma - 1)/2 \tag{5.7}$$

For monatomic gases, the ratio of the specific heats ($\gamma$) is 1.66, whereas it is 1.4 for diatomic gases.

Metals, ceramics, polymers, composite materials and nanocrystalline materials can be successfully sprayed using the high-pressure cold-spraying technique. It provides high deposition efficiency, high density and hardness of the coating. The initial properties of the starting materials (physical and chemical properties) can be retained in high-pressure cold spraying. The substrate heating will be minimal and results in cold worked and refined grain microstructure of coating without melting (and solidification). Also, only cold ductile materials can be sprayed without oxidation, as the temperature is kept low throughout the process. Also, the state of the initial powders (solid state) are not altered during the cold spraying, the chances of oxidation, melting and mainly the phase transformation is easily avoided with a deposition efficiency of 70%–90%.

### 5.1.6.2 Low-Pressure Cold Spray

The high-pressure cold-spraying technique is highly cost-effective due to the consumption of a large quantity of gases. Because of its incapability to spray on-site and repair the parts, the low-pressure cold spray (Figure 5.13) has served as an option with relatively lower consumption of spray gases. The spray gun using air has been developed with consumption of gas and the upstream pressure typically about 0.4 $m^3$ and <1 MPa, respectively. Less power (around 3.5 kW) is used to heat the gas. The simplified powder feeder with no pressurization is injected downstream along with the powder into the nozzle throat. The gas velocity of 250–450 m/s obtained in the low-pressure cold spray is insufficient to provide critical velocity to the powder particles to deposit on the substrate. To attain the coating with small particles, it is mixed with big particles. The main role of the big particles is to press the smaller particles further into the substrate. After pressing the fine particles, the big particles rebound on hitting the substrate.

The different thermal spraying techniques provide their pros and cons. Mainly, the design of the gun is changing according to the spraying technique and requirement. Based on the gun design and nozzle used in several thermal spraying techniques, the achievable spraying parameters are in Table 5.1.

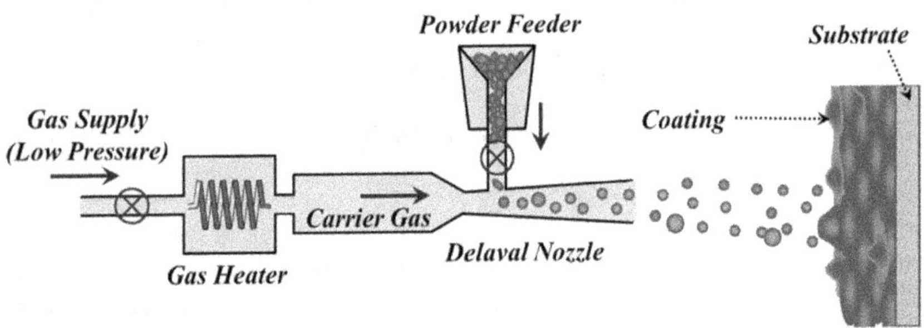


**FIGURE 5.13** Schematic of low-pressure cold spray.

*Source*: Adapted and redrawn from [11].

**TABLE 5.1**
**Particle Velocity and Temperature for Different Thermal Spray Techniques**

| Spray Type →<br>Spray Parameter ↓ | Flame Spray | High-Velocity Oxy-Fuel | Electric Arc | Plasma Spray |
|---|---|---|---|---|
| Temperature (°C) | 3000 | ~3000 | ~5000 | >10,000 |
| Spray rate (kg/h) | 2–6 | 1–9 | 10–25 | 2–10 |
| Particle velocity (m/s) | <50 | 400–1000 | 150 | <450 |
| Adhesion strength (MPa) | 8 | >70 | >12 | 60–80 |
| Coating thickness (μm) | 5–2000 | 5–2500 | 100–5000 | 50–5000 |
| Porosity (%) | 3–10 | <2 | 3–10 | 2–5 |
| Cost | $ | $$ | $$ | $$ |

Among the main causes that affect the major kind of variations in the process, the following should be mentioned to avoid the effect on the process capability:

1. *Specification of coating material*: Coating materials differ in sensitivity during exposure in the high-temperature environment, which affects the process. Production of highly porous materials is more sensitive to the slight variation in the spraying process parameters. In the case of coating thickness, the more precise the specification requirements, then it can be expected with very low process capability.
2. *Size and complexity of the components to coat*: High coating time increases the probability of variations in the coating characteristics during the spraying. The powder feed rate and spray spot size affect the quality of the desired coating and its performance. Further, the weight of the components affects the stability and accuracy of the fixtures in the spraying gun. The essential power required for the torch will depend on the size of the product. The dimension and weight of the spray gun influence the precision of its location in the system, and that will influence the quality of the sprayed part. Components with complex shapes that are difficult to access are usually less stable from the standpoint of coating quality.
3. *Position of processes in the series of the manufacturing process*: Some of the prior manufacturing processes, such as drilling holes (for cooling of gas-turbine blades) make the component more challenging to coat.
4. *Programming approach*: Coating thickness sensitivity with the gun position can be dealt with via sophisticated motion of the gun component using the programming approach.

## 5.2 POWDER FEEDERS

A powder feeder is equipment used for supplying powder in the thermal spray technology. Powder supplied from the feeder needs to have good flowability and regular flow rate irrespective of the powder size distribution, and its morphology and must be able to operate at different pressures ranging from low to high based on the spray technique used. Powder feeder holds the material of up to 10 kg from where it flows through some rate-controlling restriction, mixed with the carrier gas on its way and transporting it to the thermal spraying device. The fluidizing agent, carrier gas carries the powder via a feeder tube which ends in a long metal termination called powder injection port. One of the main problems encountered during thermal spraying is saltation. Saltation is physically manifested by the uneven flow of powder during which powder flow stops after every few seconds. It is thus essential to inspect the powder feeding systems and the requirement for ideal powder feeding. Conventional powder feeders are of three types, as described below [8, 12].

### 5.2.1 Gravity-Fed Hoppers

In gravity-fed hoppers, powder falls freely under gravity in a carrier gas flowing pipe. The powder flow in the feeder can be improved with the assistance of a vibration source to agitate the powder and help it to flow easily into a gas stream and prevent blockages. Figure 5.14 shows a gravity-fed hopper consisting of a compressible rubber grommet. The grommet controls the powder flow into the gas stream. There are no tubes to connect the powder source with the thermal spray source. Near the empty condition of the container, the powder flow starts to become sporadic and irregular, necessitating the full container of powder every time.

### 5.2.2 Rotating Wheel Devices

In these devices, the rotating wheel or disc rotates inside the container with powder. There are holes or slots in the wheel, and the powder from the powder bed fills up the slots. These extra powders are scraped using scalpers and fed into the delivery tube with a specific volume of powder. From the tube, the powder is entrained into the powder injector. The filling of the slots can be improved by introducing vibration into the feeder, especially for powders with poor flowing

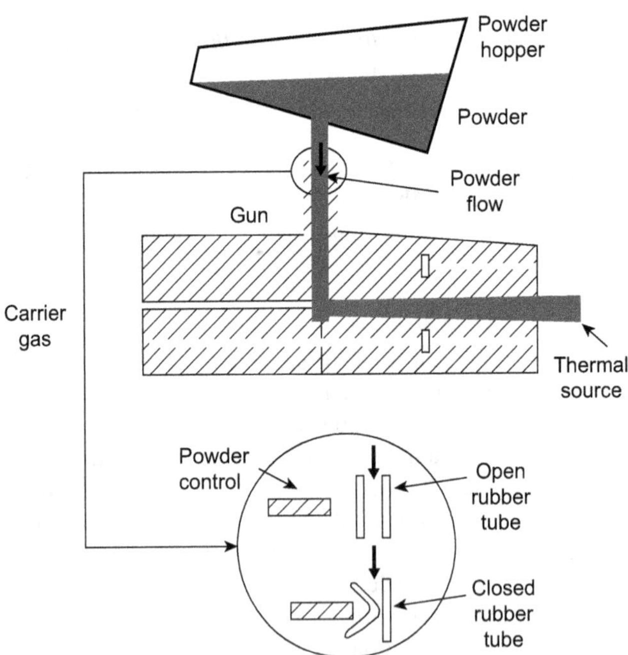


**FIGURE 5.14** Schematic representation of gravity-based hopper.

*Source*: Adapted and redrawn from [12].

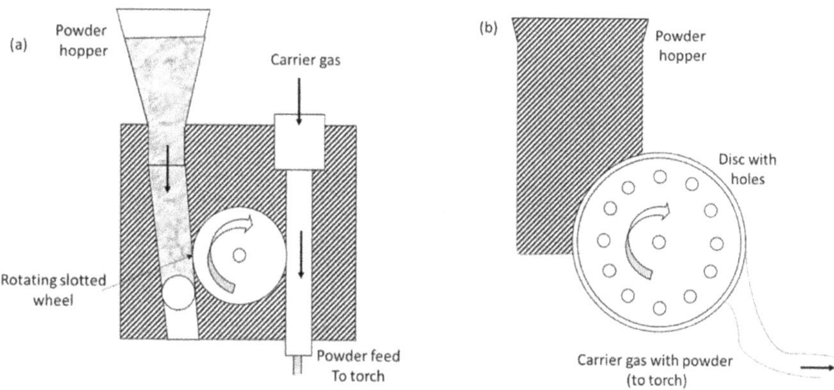


**FIGURE 5.15** (a) Wheel and (b) disc type rotating powder feeders.

*Source*: Adapted and redrawn from [12].

ability. The delivery disc can be mounted either horizontally or vertically. In the vertical wheel type, as shown in Figure 5.15, the powder is dropped into the carrier gas stream. The gas stream tube is connected to both sides of the wheel. So, the gas pressure will be equalized and it prevents the powder from blowing off the slots. In the horizontal type, the slots move through the gas supply system. Then the powder moves away with the slots and is carried by the gas. So, the powder delivery is influenced completely by the physical dimension of the slots, such as the rotating speed of the wheel, the pressure of the injected carrier gas, and the dimension of the delivery tube.

### 5.2.3 Fluidized Bed Hoppers

In these devices, powder feedstock is gas suspended, as shown in Figure 5.16. The carrier gas and a pipe with a cavity cross the powder bed simultaneously. They lead to entrain the powder into the orifice at a uniform rate. The powder bed needs to be fluidized, and gas and powder removed, to be replaced continuously. Powder is fluidized at comparable pressure drop (across the bed) and weight of the bed, making it more sensitive to powder size distribution and specific weight rather than particle morphology. The powder feed rate depends on the gas flow rate and the orifice size and is determined by the Venturi effect (i.e., there will be a reduction in the fluid pressure when a fluid flows through a narrow part of a tube).

These different kinds of feeders are commonly employed in different thermal spraying techniques. Other than these, there are a few other feeders which are limited in use or have been used in the past. There are also some feeders specifically designed for fine powders (<10–15 μm) due to their agglomeration tendency and thus the low flowability.

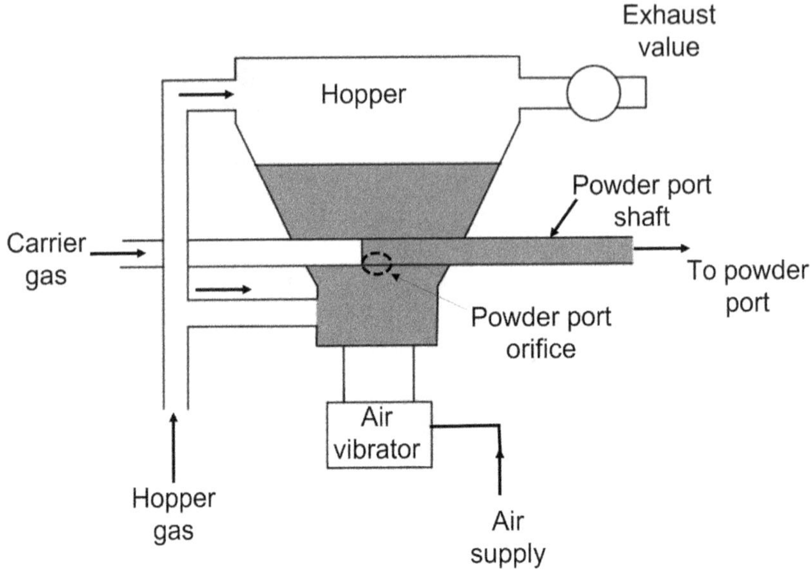


**FIGURE 5.16** Fluidized bed powder feeding system.

*Source*: Adapted and redrawn from [12].

## 5.3 POWDER INJECTORS

Powder particle injection at the heating zone is a major problem for the particle flowing from the powder feeder, such as acceleration, heating, melting and so on. Once the major process parameters (e.g., temperature, gas flow, enthalpy) and composition distributions have been measured, then the heat transfer and momentum of the particle can be calculated. Further, the processing steps that have been made result in the primary assumptions (particle velocity and size distributions) for the powder feed flow modeling. So, the design of the injector for the powder feed is also an important dependent parameter to consider for the utilization in thermal spraying.

There are three general modes for powder injection into the plasma regime: (external radial, internal radial and internal axial. The powder in the radial injector can be fed axially or at some angle to achieve a specific particle trajectory. Practically, axial internal injection provides suitable conditions for the particle treatment, and it is reasonably difficult to show in realization. Further, powder particle injectors are simple and are categorized as straight, curved and double flow injectors (Figure 5.17) for radial injection [13]. In the case of axial injection, water-cooled straight injectors are used. A mandatory cooling system is to avoid the damage of injectors due to the exposure of relatively extreme heat flux. Further, the curved injectors are mostly used in DC plasma torches. Since the injected particles are normal to the nozzle exit torch axis, the injector is of a

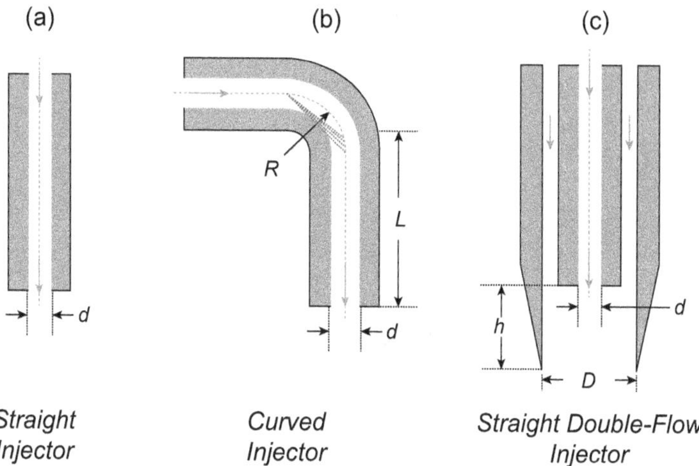


**FIGURE 5.17** Different designs of particle injector (radial) geometries: (a) straight injector; (b) curved injector; (c) double-flux injector.

*Source*: Adapted and redrawn from [8].

curved shape to monitor the torch position. Thus, it allows the pipes to transport the powder and connect it with the torch at the backside. The curved design of the injector interrupts the flow of the powder particles and gas. So, the length of the injector (*L*) is the essential parameter to dilute the effect of the curve on the gas and particle flow. On the other hand, a double-flow particle injector is used to regulate the powder particle stream at the nozzle exit [14].

1. *Straight injector*: The internal diameter affects the particle velocity. The particle velocity at the exit increases with a decrease in the internal diameter (*d* in Figure 5.17a) for a specific gas flow rate. Also, the jet divergence decreases with the injector's internal diameter [14]. But the scenario will change depending upon the carrier gas type. For example, the velocity of the relatively fine particle at the straight injector exit is high with He compared to that of Ar as the carrier gas at a constant flow rate [14].
2. *Curved injector*: The design and geometry of this kind of injector has a great influence on the spray particle behavior. As the spray particle approaches the curved region (Figure 5.17b), it will decelerate and be driven to the outer region of the curve due to the centrifugal forces introduced by the gas flow. At this region, the carrier gas has a low velocity and it results in the decrease of drag force. The drag force is employed on the spray particles due to the carrier gas environment. So, the particle distribution within the curved injector is no longer symmetrical. Further, most of the spray particles (90%) are accumulated or concentrated in the outer part of the curvature [15].

3. *Straight double-flow injector*: Usually, fine and low-density powder particles are difficult to spray. But, this kind of injector is utilized to collimate the relatively fine (<20 μm) and lightweight powder particles.

Placing the injector near the hot gas at the plasma jet is one of the main problems. For example, in the case of the DC plasma jet, the position of the injector at too far a distance from the jet leads to the chances of powders at the edges of the powder injection nozzle or cone, thus it can bypass the plasma jet. Later, it can re-enter the plasma jet farther downstream, creating defects in the processed coatings. As marked in Figure 5.17, as the injector exit is close to the plasma core axis, the probability of high heat that affects the injector tip increases, leading to clogging. Thus, the particle injectors should be water-cooled to avoid the clogging effect. This condition is one of the most requirements for the RF plasma-spraying technique due to the position of the injector at the level of the coil.

### 5.3.1 Powder Injection into the Plasma Jet

In the case of external injection, the powder injectors are placed outside the plasma jet at a certain distance of at least 10 mm from the nozzle exit. Powder injection is determined by the powder mass and powder size, injector diameter and the flow rate of the carrier gas. The penetration of particles into the jet is deflected due to the lower dynamic pressure of the carrier gas than the plasma jet. The role of the carrier gas is to provide sufficient velocity to improve the penetration of the particles into the plasma. But the carrier gas should not exceed a limit to keep the particle being dragged in the plasma. Otherwise, the particles will be bombarded through the plasma jet that will produce the coating with poor characteristics. Before the penetration of the particle with a velocity (radial component) into the jet due to inertia, it needs to cut through the cold and dense boundary layer adjoining the plasma jet. Qualitatively, it can reach some percentage of the jet speed, which is tens of meters per second (several tens to 50 m/s in APS and low-pressure plasma coating system [LPPS], respectively). It suppresses the powder velocity and hinders particle penetration into the plasma core. Hence, the boundary layer acts as "wind" that reduces the radial component of the particle motion and brings the particles towards the jet. In the case of rapidly moving APS guns, particle trajectories could be affected by the surrounding air that makes the process gun motion sensitive. It is due to the injected particles being shaded by the injector during the low injection velocity (<4 m/s). At subsonic APS conditions, the radial velocity produces shear lift in the jet boundary layer, which assists the particles to enter into the plasma core. However, the shear lift induces a force that deviates the particles from the plasma jet of the supersonic LPPS system.

Particles that enter into the plasma jet through the internal radial particle injector will be up-streamed marginally at the APS nozzle exit, or just before the expansion cone in the case of the LPPS gun. Under both circumstances, there is no generation of forces that could hinder the entrance of particles into the plasma jet. Also, the powder injection through the internal mode rises the axial pressure gradient in the nozzle, which causes the back influence of the powder injection factors on the spraying. When the particles are injected radially, it is essential to keep the average force of the particles at the injector exit should be equivalent to that imparted by the heated gas flow. This state leads to an ideal trajectory of the injected particles with controllable acceleration and heating [15]. Also, the divergence nature of the particle jet increases severely with the particle diameter ratio, fine to coarse particles in the particle distribution. Moreover, the diameter ratio should be low (preferably <2) for a uniform melting without overheating.

### 5.3.2 Injector Plugging and Spitting

The tips of external injectors and internal powder feeders lead to high-temperature exposure (700–1000 °C) because of extreme heat fluxes from the plasma. The powder particle that hits at the high-temperature region of the injector or feeder could adhere to the injector wall due to adhesion forces and high contact temperature. Later, it becomes a hindrance to the free powder flow. It depends on the properties of the powder material, alignment and improper cooling of the powder feeding parts during the spraying process.

In the case of fine powders through internal injection, high plasma radiation and thermal conductivity of the anode wall increases the plugging up of powders at the injector. In the case of ceramic particles, the powders are sintered due to excess heat and become stuck at the injector; they can be physically removed regularly to avoid irregularities in the power flow through it. The plugging up changes characteristics of the powder injector including the flow speed and injector direction at the nozzle end, which causes substantial deviation of powder trajectory. Sometimes the plugged material (with a low melting point, such as metallic materials) melts and makes its way into the plasma stream after attaining a critical size of the droplet. Then, the molten metal starts traversing into the plasma plume, which is usual in LPPS and APS, referred to as spitting. It is not beneficial for a quality coating due to sudden appearance of solid bumps on the coating. The injector plugging and spitting is undesirable, and the approaches to prevent these are as follows:

1. No fine powders and better cooling of powder injector;
2. Regular examinations and cleaning of powder injector;
3. Proper implementation of process parameters (powder injector and plasma conditions).

### 5.3.3 Cooling of Spraying System

Without an intensive cooling system, the existing electrode materials cannot survive in the exposure of plasma for a long time. Deionized water, owing to its superior cooling properties and low cost, is used as an appropriate coolant in a closed circulating loop mode. Failure or improper functioning of the cooling system leads to failure of spraying components. The failure of the cooling system leads to failure of spraying gun (in the surface level), such as the formation of oxide scales due to excessive heat, followed by plugging up and spitting in the powder injector and so on, as mentioned earlier. To avoid the malfunctioning of the cooling system, the following major requirements have to be considered:

1. Deionized water free of air bubbles is needed to avoid the loss of cooling inside the torch and minimizes electrode damage.
2. Electrically non-conducting coolant is needed to avoid power losses because of leakage currents. It also avoids any electrolytic chemical reactions that resulted in the formation of gases as the reaction by-product.
3. Highly pure deionized water is needed to avoid the deposition of any kind of salts inside the cooling system.
4. Low inlet temperature is needed to keep the outlet at a suitable temperature. Also, high inlet temperature increases the air in the coolant. Sufficient water flow is maintained to keep the temperature difference at 15–30 K, which depends on the operating power and may reach 40–45 K under extreme conditions. An increase in the high temperature of the spraying gun will lead to the faulty operation or inappropriate design of the system.

The overheating of the anode results in a change of the electric arc root dynamics, affecting the coating properties. So, the proper cooling system is one of the supporting systems to get an efficient coating.

### 5.3.4 Liquid or Suspension Injection

In recent decades, an effort has been made to spray fine particles in solutions or suspensions. Usually, a fine particle that has to be sprayed should have sufficient momentum to impinge on the substrate. It will be achieved with the support of carrier gas that imparts sufficient momentum to drive the fine particles into the thermal zone. Otherwise, injecting the liquid that carries the particles is one of the ways to spray. Atomization and mechanical injection are the two ways to inject the liquid into the thermal zone of the plasma spraying.

- Atomization drastically disturbs the plasma due to the atomizing gas which may cause irregularity in the particle distribution and velocities. Usually, co-axial atomization is used that injects the suspension through a tiny nozzle. Further, the fragmentation of the droplets is facilitated using a high-density gas (Ar) [14].
- In mechanical injection, the suspension is passed through a narrow nozzle (diameter 50–200 μm). It results in droplets of fine size distribution and velocities. Thus, the coating features comprise uniform microstructure and properties.

Whatever the type of the particle injector or injection mode may be, the main issue that needs to be taken care of is what happens when the injector is misaligned with the plasma flow. Though the positioning and design of the injector are proper, the position has to be adjusted accordingly to the electrode. But the electrode tends to change its position due to its prolonged use and the damage that occurred on it. Further, damages at the electrode after the prolonged usage of the spraying system due to wear and environmental conditions affect the gun position and thus it reduce the efficiency of the developed products.

## 5.4 ELECTRODE DAMAGE

The most typical source of gun position drifting is due to "electrode erosion." The erosion leads to a considerable extent of uncertainty in the plasma jet. The electrodes are damaged due to the arc that strikes repeatedly at high temperatures. The wear at the surface of the electrode leads to the initiation of the destruction of the plasma spray process. Also, the formation of the arc is delayed at the worn surface of the electrode, which results in the generation of a non-symmetrical plasma jet profile. Also, when the erosion of the electrode increases, the voltage diminishes continuously and the coating properties and deposition efficiency are significantly modified. Worn electrodes (Figure 5.18) decrease the plasma gun voltage, which is one of the most prominent drifting events that occur during plasma spraying [16–18]. Also, drift in the voltage leads to reduction of the net plasma power. Correspondingly, the particle temperature and velocity decrease, which extends to poor process efficiency. The role of the worn electrode in changing the spraying factors is tabulated in Table 5.2.

It is also important to note that the anode wear leads to a substantial fluctuation in the arc pulsations and its frequency, which in turn broadens the random distribution of particle spraying parameters.

Because of the arc root motion, the arc creates more defects in the anode surface, and it becomes a new point for the attachment of the arc. Thus, the arc becomes smaller and there is a drift in the voltage to the lower side. After a particular extent of anode wear, it accumulates the arc core that fixes itself a definite portion of the anode and then the arc ends swirling. It brings an extreme,

**TABLE 5.2**
**Major Variation in the Coating Due to Damage of the Electrode**

| Corrupt Part of the Gun | Concerns | Affected Spraying Parameter |
|---|---|---|
| Anode | Wear | 1 Voltage drift |
| | | 2 Variation of particle parameters |
| Cathode | Erosion | 1 Contaminated coating |
| | | 2 Variation in pulse characteristics |
| | | 3 Particle characteristics |

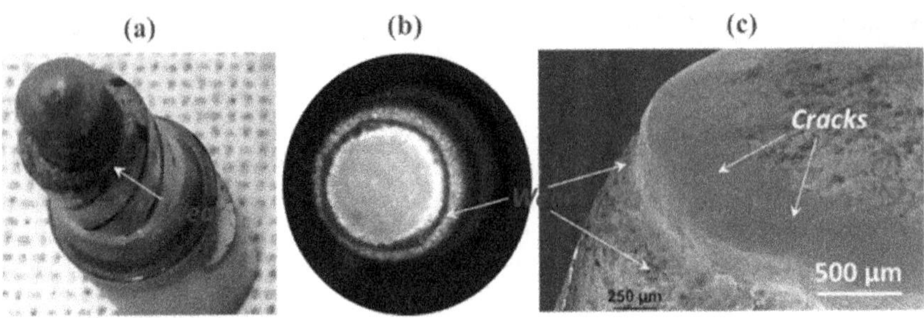


**FIGURE 5.18** (a) Digital, (b) endoscope image, and (c) SEM micrograph of cathode tip used in plasma spraying.

*Source*: Reprinted with permission from [19, 20].

abrupt deviation in the parameters (plasma and particle). Further, it makes the second stage of wear at the anode surface unacceptable for further usage, which should be avoided in terms of quality assurance. The occurrence of the first wear depends on the gun design. All concerns of process capability are acceptable within the limit of the operative window. Then, the wear of electrodes at the torch influences the particle parameters in two ways:

1. Change in the average coating thickness due to voltage drifting: This affects the average power (it decreases) and thus the particle parameters. It should be highlighted that the drift is compensated for with an applicable surge in the current and maintains the appropriate plasma power.
2. Increase in the randomness in the thickness: The wear of the electrode increases the fluctuation in the arc voltage, which is not advisable due to its negative role in efficiency.

For the combination particle velocity and temperature, $P$ ($v$, $T$) that determines the coating adherence with the substrate, the arc-voltage fluctuation against the coating efficiency (otherwise called the probability density function) is shown in Figure 5.19. The figure shows the fluctuations in the time-average density function and its particle-average bonding factor using fresh and worn anodes for spraying. Also, the process deposition is equivalent to the area of the extreme quantity of the distribution function [17, 21]. It shows that the inconsistency in the particle parameters correlates to the voltage fluctuations. Also, the replication of the pulse fluctuation on the efficiency of the coating highly depends on the starting condition of the new nozzle that varies almost linearly.

The addition of $ThO_2$ in W-based electrodes is encouraged to reduce the wear of the electrode in the plasma spraying due to the arc behavior, outstanding electron emissivity, high strength and ease of machinability. Since it is difficult to handle (radioactive nature), $La_2O_3$ is used as an alternative reinforcement material in the electrodes that provides environmental safety [22]. The arc voltage

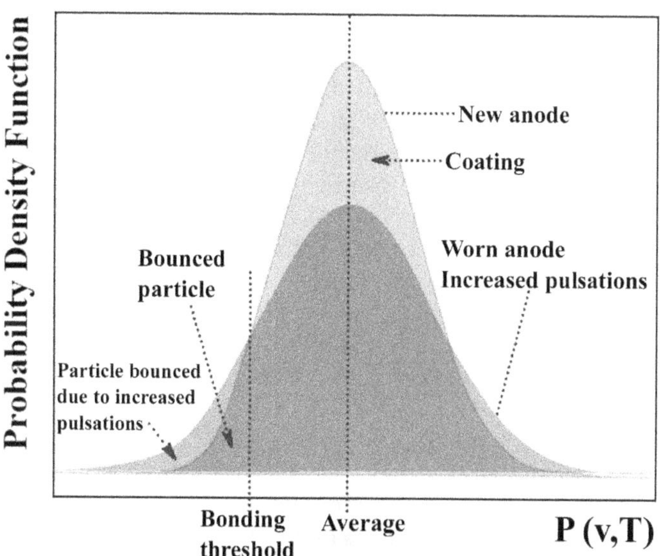


**FIGURE 5.19** Influence of variation in the arc pulsations on the coating efficiency-probabilistic explanation due to worn anode.

*Source*: Adapted and redrawn from [9].

dipping rate and erosion rate of 2 wt% $La_2O_3$-reinforced W cathode is low compared to that of W-$ThO_2$ (2 wt%). Thus, a longer service life of the electrode is expected in atmospheric plasma spraying. Additionally, the lower work function of W-$La_2O_3$ leads to less electrical power consumption to emit the electrons. So, it is not necessary to use traditional electrodes with $ThO_2$ because $La_2O_3$ can be an effective reinforcement in the W electrode to improve the damage (wear) resistance, power utilization and so forth for extended use of electrodes in the plasma spraying.

## 5.5 TYPE OF PLASMA GAS AND PLASMA STABILITY

Among the various parameters to achieve the deposition efficiency, plasma gas composition, features and performance of plasma systems are the major initial parameters to produce the quality coatings with and process repeatability. Gases such as Ar, $H_2$, He and $N_2$ are the major plasma gases used in plasma spraying. Among them, He and $N_2$ are the commonly used plasma gases for plasma spraying. The presence of He in the mixtures of Ar-He and Ar-$H_2$-He [23] improves the conductivity ionized plasma region and that leads to support the high particle heating. Similar to $H_2$, He gas improves the plasma arc voltage and enthalpy that effectively improve the plasma spraying. Plasma arcs with He show excellent ionization potential and plasma velocity, and thus it is experienced with high temperatures. The use of $N_2$ brings an advantage of extraordinary plasma enthalpy with excessive dissociation and ionization energy expenditures. It also reduces the heat losses due to low temperature. Air, $O_2$ and $CO_2$ are not widely used plasma gases due to their interaction with the electrode and corresponding reduction in its lifetime. The water-stabilized plasma spray guns are used with carbon-based sacrificial cathodes, such as graphite, which is restricted to high-throughput ceramics. The required quality of the gas to use is decided by the lifetime of the spraying gun used for the coating processes. Plasma gases and their features, distribution, and including gas connectors with the spraying system can also decide the quality of the coatings. Certain rules that can be applied to optimize the plasma gas distribution system are listed as follows:

1. Gas meter and control devices can be selected based on the necessary accuracy. Otherwise, it becomes one of the major sources in the fluctuations of the gun and particle parameters.
2. There should not be any leakages in the supply of gas. If it is tolerated, it leads to harsh modifications in the coating characteristics, such as chemistry and physical properties.
3. Proper connections are expected.
4. High purity (>99.997%) of plasma gas for better process control is expected.

Plasma arc stability fundamentally depends on the torch power, plasma gas composition and plasma gas flow rate. Plasma arcs with relatively high $H_2$ exhibit high pulsations and they become highly stable with an increase in the gas flow rate and current. The reduction of the plasma gas flow rate lower than the critical value is highly dependent on the design and stabilization of the gun. It makes the spraying process highly unsteady, and the coatings will be inconsistent with irreproducible properties.

## 5.6 SUMMARY

The chapter depicted how different thermal spray processes are employed depending on the application and how modifications are done in the spraying gun to improve the coating quality. There is advanced research going on to improve the gun design to improve the coating efficiency. Surface treatment through thermal spray techniques, being more compatible and easier to carry out has contributed a lot in aerospace, thermal and hydroelectric power plants, petrochemical and other industries. Thermal spray can sometimes be hazardous if not taken with care, and safety precautions should be taken seriously. Nowadays, the design of the spraying system with the automatic operation of machines or by using robots has reduced the chances of hazards. Reduction in noise level, an alternate method to view spraying head, and consistent coating have been achieved.

**Questions for Self-Analysis**

1. What are the essential components to design a DC plasma spraying gun?
2. What is the spitting effect, and how it is avoided in the plasma spraying system?
3. How is anode instability related to arc instability?
4. What are the different issues raised due to the movement of arc-anode attachment?
5. What are the advantages of the multiple cathode gun design for plasma spraying?
6. List different kinds of particle injectors used in thermal spraying.
7. Why should the temperature of the particle injector be maintained low?
8. How is the flowability of the powder improved with a fluidized bed powder feeding system?
9. What is the effect of electrode damage on the injector positioning?
10. How do "plugging up" and "spitting" in plasma spraying affect the coating quality?
11. How does wear of the electrode influence the particle parameters during plasma spraying?
12. What are the main conditions to achieve a better plasma gas distribution system?

## REFERENCES


[1] Duan Z, Heberlein J (2002) Arc instabilities in a plasma spray torch. *J Therm Spray Technol* 11:44–51

[2] Fauchais P, Vardelle A (1997) Thermal plasmas. *IEEE Trans Plasma Sci* 25:1258–1280

[3] Ann Gan J, Berndt CC (2015) Thermal spray forming of titanium and its alloys. In: Qian M, (Sam) (eds). *Froes FHBT-TPM*. Butterworth-Heinemann, pp. 425–446

[4] Nogues E, Vardelle M, Fauchais P, Granger P (2008) Arc voltage fluctuations: Comparison between two plasma torch types. *Surf Coatings Technol* 202:4387–4393

[5] Gabbar HA, Darda SA, Damideh V, et al. (2021) Comparative study of atmospheric pressure DC, RF, and microwave thermal plasma torches for waste to energy applications. *Sustain Energy Technol Assessments* 47:101447

[6] Marqués JL, Forster G, Schein J (2009) Multi-electrode plasma torches: Motivation for development and current state-of-the-art. *Open Plasma Phys J* 2:89–98

[7] Gabriel O, van den Dungen JJA, Schram DC, Engeln R (2010) Nonequilibrium rovibrational energy distributions of hydrogen isotopologues in an expanding plasma jet. *J Chem Phys* 132:104305

[8] Fauchais PL, Heberlein JVR, Boulos MI (2014) *Thermal Spray Fundamentals: From Powder to Part*. Springer Science & Business Media

[9] Schneider KE, Belashchenko V, Dratwinski M, et al. (2006) *Thermal Spraying for Power Generation Components*. Wiley

[10] Maev RG, Leshchynsky V (2009) *Introduction to Low Pressure Gas Dynamic Spray: Physics and Technology*. John Wiley & Sons

[11] Singh H, Sidhu TS, Kalsi SBS (2012) Cold spray technology: Future of coating deposition processes. *Frat ed Integrità Strutt* 6:69–84

[12] Davis JR (2004) *Handbook of Thermal Spray Technology.* ASM International

[13] Vardelle M, Vardelle A, Li KI, Fauchais P, Themelis NT (1996) Coating generation: Vaporization of particles in plasma spraying and splat formation. *Pure Appl Chem* 68:1093–1099

[14] Bouneder M, El Ganaoui M, Pateyron B, Fauchais P (2008) Relevance of a thermal contact resistance depending on the solid/liquid phase change transition for sprayed composite metal/ceramic powder by direct current plasma jets. *Comptes Rendus Mécanique* 336:592–599

[15] Vardelle M, Fauchais P, Vardelle A, et al (2001) Controlling particle injection in plasma spraying. *J Therm Spray Technol* 10:267–284

[16] Bisson JF, Gauthier B, Moreau C (2003) Effect of plasma fluctuations on in-flight particle parameters. *J Therm Spray Technol* 12:38–43

[17] Leblanc L, Moreau C (2002) The long-term stability of plasma spraying. *J Therm Spray Technol* 11:380–386

[18] Fauchais P, Vardelle M, Vardelle A, et al. (1995) Parameters controlling the generation and properties of plasma sprayed zirconia coatings. *Plasma Chem Plasma Process* 16:S99–S125

[19] Zimmermann S, Mauer G, Rauwald K-H, Schein J (2021) Characterization of an axial-injection plasma spray torch. *J Therm Spray Technol* 30:1724–1736

[20] Heißl M, Mitterer C, Granzer T, et al. (2014) Substitution of $ThO_2$ by $La_2O_3$ additions in tungsten electrodes for atmospheric plasma spraying. *Int J Refract Met Hard Mater* 43:181–185

[21] Moreau C, Bisson J-F, Lima RS, Marple BR (2005) Diagnostics for advanced materials processing by plasma spraying. *Pure Appl Chem* 77:443–462

[22] Heißl M, Mitterer C, Granzer T, et al. (2012) *Substitution of Thoria Additions by Lanthanum-oxide Doping in Electrodes for Atmospheric Plasma Spraying.* NA

[23] Janisson S, Meillot E, Vardelle A, et al. (1999) Plasma spraying using Ar-He-$H_2$ gas mixtures. *J Therm Spray Technol* 8:545–552

# 6 Powder Characterization and Synthesis

*Rubia Hassan and Kantesh Balani*

**CONTENTS**



DOI: 10.1201/9781003321965-6

During thermal spraying, the powder feed is subjected to different conditions imposing restrictions on the type and form in which they can be used. In general, the feed material can be supplied in powder form, as a wire or in the form of suspension. For powders, these requirements include desirable chemical composition, proper particle size distribution, particle morphology and flowability, and so on. To test the powder for the desirable characteristics, before feeding in the thermal spraying unit, the powder undergoes various characterizations. Accordingly, it can be modified following a suitable technique if not found proper for feeding. Also, based on these requirements, a wide range of technologies for manufacturing the feed material have turned up. Powder purity, size and distribution are key characteristics in spray processes, as the starting condition strongly influences the quality of the deposited coating. This chapter highlights the size and distribution of powder, powder treatment, and various other attributes that influence the microstructure and quality of deposited coating, which are elaborated here as well.

## 6.1 POWDER PURITY, SIZE AND DISTRIBUTION

A purity level of starting powder of 99%+ may usually be preferred, and in more sensitive applications (e.g., biomedical, nuclear, space and automotive), purity levels of higher than 99.9% may be warranted. Most commercial powders exhibit a purity level of 95%+. It may be noted that the purity of the starting powder particle may be reduced when it is deposited as a coating. Thus, if the starting powder is impure (with oxides and other elements), the final purity of the deposited coating may only be lower than that. The powder may have a carrier media (e.g., organic additive to assist flowability of powder), which is expected to vaporize and not form part of the final coating. But the starting impurity (and probably some reaction with environment as oxide or nitride) may also be retained in the deposited coating. Hence, in order to attain the deposit of high quality, the starting purity of powder must be at least that high while ensuring a processing condition and environment to sustain that.

A powder size of typically 10–150 μm is used in thermal spraying. Due to the high melting point of ceramic particles, powders of larger size may not attain a high enough temperature; thus a smaller size (~10–40 μm) is usually preferred. At the same time, feeding the fine powder particles also becomes a challenge, as their high surface area induces high interparticle friction and restricts their flow (and may clog the nozzle). On the other hand, high thermal conductivity of metallic/alloy powders allow using a higher particle size (~25–150 μm), which allows the powder to conduct heat and completely melt. Herein, high enough pressure also has to be provided to carrier gas in order to ensure enough momentum (i.e., a high enough velocity) to the in-flight particles in order to induce its plastic deformation upon impact with the substrate to deposit as an integrated coating. So in summary, a lower powder size is preferred (e.g., 10–40 μm) during the deposition of ceramic particles, whereas a coarser size (~25–150 μm) is preferred for depositing metallic/alloy powders.

When the term *powder* is used, before dwelling on its size distribution it is important to also appreciate the aspect of powder morphology. The morphology of powder may induce very different in-flight velocity and thermal characteristics, hence it is an important parameter to be characterized and reported. There are different particle shapes, such as disc or coin, fibrous or thread-like, dendritic or branched, acicular or needle, modular or rounded but not regular, irregular or asymmetric, flaky or plate-like, angular or polyhedral, cylindrical or tube-like, crystalline or a certain geometric shape, and spherical or globular (these are presented schematically in Figure 6.1).

Different morphological shapes of powders may result from variations in the manufacturing route. Powders having a similar size distribution and the same chemical composition, but with different particle morphologies may yield structural differences in the final coating. These morphological effects arise during powder feeding and powder deposition through powder mixing, its transport and injection, as well as via aerodynamic forces acting on these surfaces during in-flight and its deposition. Because of difference in the shape of starting powder, the microstructural features like size, shape and amount of porosity in the final deposited coating also are affected. As the particle morphology deviates from the spherical shape, the powder flowability decreases, which results in clogging of the feeder and injection system, thus reducing the coating deposition efficiency. Moreover, it creates irregularities in the feed rate that induces heterogeneity in the deposited coating. It may further cause overheating of the substrate, thermal degradation of the already deposited layers, or even their detachment from the substrate. A spherical shape of powder particles is desirable over other shapes.

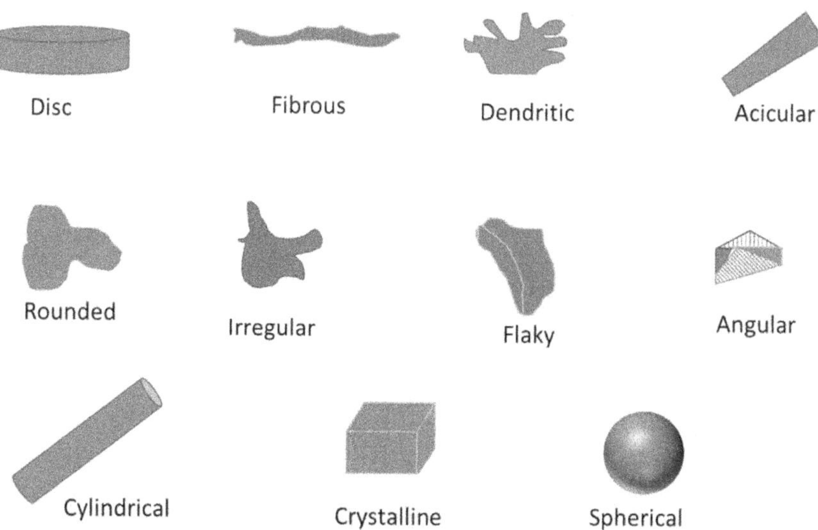


**FIGURE 6.1** Schematic representation of powders of different morphology.

However, not all powders are obtained in spherical morphology, and the additional cost associated with synthesis of spherical powder impels consideration of the use of non-spherical powders as starting feedstock. Their rolling friction is lower than sliding friction, and due to their inherent tendency for rolling than sliding, there is less interparticle friction. As it is less abrasive and has a low surface-to-volume ratio, a spherical shape causes minimum wear of feeding systems and is less prone to react with warm gases, both of which limit contamination in the final coating. Although due to enhanced drag forces in non-spherical particles, the impact velocity is higher in their cold spraying, and the resulting variable drag coefficients due to their irregularity in shape gives rise to inhomogeneous impact velocities resulting in varying microstructure and, thus, structural properties [1]. The cold spraying of Al-$Al_2O_3$ composite coatings with spherical and angular $Al_2O_3$ showed more $Al_2O_3$ retained in the coating with angular $Al_2O_3$ particles because of higher elastic rebound forces in spherical particles due to their constant surface curvature, counteracting the deposition phenomenon [2].

### 6.1.1 Powder Characterization

Properties of final coating are the reflection of various parameters including powder-spray technique used, its optimization and powder characteristics of size, morphology, size distribution, density and flowability, and so on. In order to identify the powder shape and size, the most simple and direct way is visual inspection. However, this may not suffice for the smaller particles, and in case more information is required, we need to consider other characterization methods. Other than direct observation, powder morphology can be ascertained through an optical microscope or an electron microscope. Figure 6.2 shows the spherical morphology of 3YSZ (yttria-stabilized zirconia) and angular morphology of SiC (silicon carbide) powder through scanning electron micrographs (SEMs). Depending upon the powder production technique, the particle surface morphology varies. Powders with different morphological features have different characteristics which offer qualitative ascertainment of flow capability from their synthesis technique. Fused and crushed powders are distinguishable with smooth fracture surfaces and are irregular in shape. On the other hand, spray-dried, sol-gel synthesized, and agglomerated and sintered powders have globular surface features. Regarding size, individual powder particles may exhibit finer particles on their surface (<1 μm) and are indicative of moisture contamination or strong electrostatic forces within the powder body. A wide range of particle sizes in a feed will cause irregular deposition due to inconsistent particle trajectories.

The X-ray diffraction (XRD) technique serves the purpose of identifying the phases and determining the crystallite size of the powder. The XRD pattern obtained from the powder can be compared with any database available, like ICDD (International Crystal Diffraction Data), to determine the crystalline phases in the powder. For multi-phase material, the relative phase fraction can be obtained by the area under the peaks (i.e., integrated intensity). The width of the peaks at half of the intensity (i.e., full width half maximum [FWHM]) can

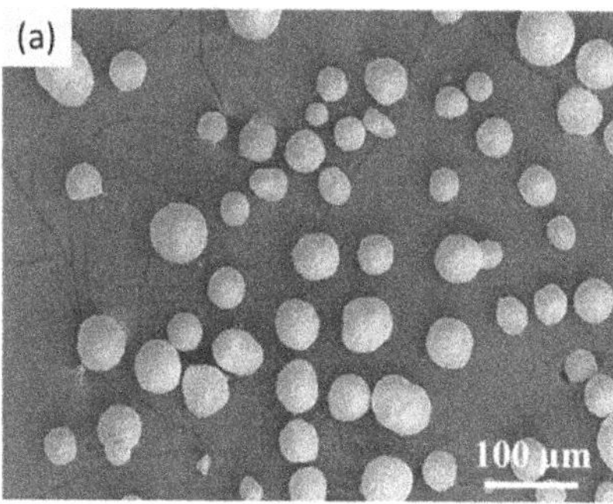

**FIGURE 6.2** SEM micrographs of powders: (a) 3YSZ powder of spherical morphology; (b) SiC powder of angular morphology.

be utilized in determining the crystallite size of the powder using the Scherrer equation.

Elemental composition can be determined using an electron probe microanalyzer (EPMA) or energy dispersive spectroscopy (EDS) equipped in SEM. The technique involves the interaction between incident electrons and matter, analyzing X-rays generating from the sample after bombardment of the electron beam. The *characteristic* X-rays arising from the sample are then used to investigate and obtain the elemental composition.

To calculate the surface area of powder particles, a gas adsorption technique is utilized. Physical gas adsorption can be utilized to determine the surface area of the particles and the volume and size of the porosity in the powder through the BET (Brunauer-Emmett-Teller) method. BET measures surface area through the measurement of the amount of gas adsorbed on the surface as a monomolecular layer. The gas can be nitrogen (usually), argon or krypton. For density (or porosity) measurement of the powder, pycnometry, which is based on the fluid displacement principle, is used but does not provide any information about the pore size. Pycnometry utilizes helium and mercury as working fluids to measure material density and surface area. In order to determine the powder flowability, the Hall flowmeter—a calibrated funnel—measures the powder flow out through this funnel.

For the determination of powder particle size other than the visual approach, many classical and modern methods are used that are based on various physical principles, as discussed in the following sections.

### 6.1.2 Methods for Determining Particle Size

#### 6.1.2.1 Mechanical Method or Sieving

Sieving is one of the simplest methods for particle size measurement. Commonly, sieves are made up of cross-woven wires supported by mechanical vibration or ultrasonic agitation with square apertures and can analyze particles ranging from

5 to 800 μm. The number of openings or apertures per square inch of a sieve is called its mesh number. Mesh number (or sieve size) is determined in terms of the number of wires in one inch. If sieving is utilized in separating the large particles by just passing through one sieve, it is called *scalping*. However, normally a stack of sieves with decreasing aperture/mesh size from top to bottom is used, which is called *classification*. In classification, powder particles of different size ranges can be separated, as illustrated schematically in Figure 6.3. The sieve with the largest opening is placed on the top of a sieve stock. The user should remain mindful of the fact that for non-spherical particles, the diagonal of a sieve or the lower dimension of an elongated powder may also pass through the sieve (i.e., a larger sized particle may also pass through under non-ideal conditions).

The size of openings and sequence of sieves in classification is standardized. Several standards are used for mesh sizes (e.g., ASTM standard, IS standard) providing mesh number, mesh size and wire diameter. Some of these standards along with their mesh number and mesh sizes ranging from 45–4000 μm are tabulated in Table 6.1. These cover size ranges from 37 μm to 5660 μm. However, this range can be widened by means of electroformed mesh and punched-plate sieves which increase the limit by pushing the lower one down to 5 μm and raising the upper limit, respectively. There is no optimum time for sieving. A longer sieving time can help in more recovery of particles for a distribution but at the same time may cause degradation of particles due to wear and attrition. Sieving can be performed

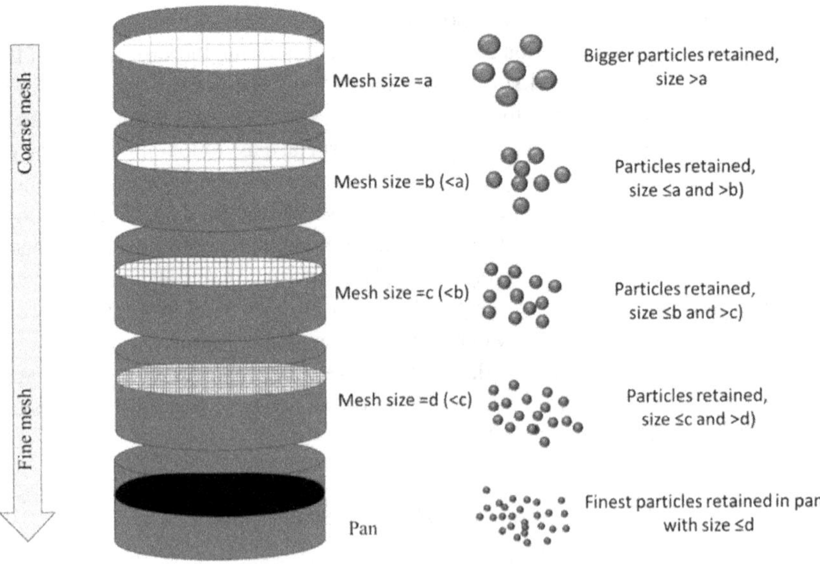


**FIGURE 6.3** Schematic illustration showing the placement of gradually finer sieves from top to bottom, and a collection of powders sequentially downwards during sieving (i.e., from larger to smaller size).

**TABLE 6.1**
**Standard Mesh Size Chart**

| Mesh-Size Standard | | | | Mesh Size |
|---|---|---|---|---|
| ASTM-E11 | TYLER | BSS-410 | DIN-4188 | (μm) |
| – | 2500 | 2500 | 0.005 | 5 |
| – | 1250 | 1250 | 0.010 | 10 |
| – | 800 | 800 | 0.015 | 15 |
| – | 625 | 625 | 0.020 | 20 |
| – | – | – | 0.022 | 22 |
| – | 500 | 500 | 0.025 | 25 |
| – | – | – | 0.028 | 28 |
| – | – | – | 0.032 | 32 |
| – | – | – | 0.036 | 36 |
| 400 | 400 | 400 | – | 38 |
| – | – | – | 0.040 | 40 |
| 325 | 325 | 350 | 0.045 | 45 |
| – | – | – | 0.050 | 50 |
| 270 | 270 | 300 | – | 53 |
| – | – | – | 0.056 | 56 |
| 230 | 250 | 240 | 0.063 | 63 |
| – | – | – | 0.071 | 71 |
| 200 | 200 | 200 | – | 75 |
| – | – | – | 0.080 | 80 |
| 170 | 170 | 170 | 0.090 | 90 |
| – | – | – | 0.100 | 100 |
| 140 | 150 | 150 | – | 106 |
| – | – | – | 0.112 | 112 |
| 115 | 115 | 115 | 0.125 | 125 |
| – | – | – | 0.140 | 140 |
| 100 | 100 | 100 | – | 150 |
| – | – | – | 0.160 | 160 |
| 80 | 80 | 85 | 0.180 | 180 |

*Source*: Adapted from [4].

in dry conditions and in wet form. Dry sieving is usually preferred, and that too for corrosion-susceptible or agglomeration-susceptible powders in the suspension medium. However, it is preferable to use wet sieving for powders that are already in suspension form. Moreover, under high-pressure conditions, wet sieving may result in high throughputs with low toxicity concerns [1, 3].

#### 6.1.2.2 Dynamic Method or Sedimentation

Sedimentation techniques determine the size of powder in a liquid medium by utilizing the concept of particle size dependence on terminal velocity in a gravitational field/centrifugal field. It works on Stokes' principle, stating that the drag force on a particle of spherical morphology in a fluid is proportional to its diameter, velocity and the viscosity of the fluid. At terminal velocity, this drag force is balanced by the gravitational force as:

$$C_D \eta r v_t = \frac{4}{3}\pi r^3 \left(\rho_p - \rho_m\right) g \tag{6.1}$$

where $C_D$ is the drag coefficient, $\eta$ is fluid viscosity, $r$ is the radius of the particle, $v_t$ is terminal velocity, $\rho_p$ is the density of powder particles, $\rho_m$ is the density of the fluid and $g$ is acceleration due to gravity.

The principle is applicable with following assumptions:

- Particles are spherical in shape and possess the same density.
- Particles flow slowly (laminar flow) without affecting one other.
- Terminal velocity is attained instantly.

These assumptions limit the accuracy of measuring the size of non-spherical particles. Thus, the dynamic response under the flow can be studied to determine dynamic diameter, which is the diameter of the sphere with same $\rho$ and $v_t$ as that of the particle flowing in the medium and can be expressed in the Newtonian fluid as:

$$V(\rho_p - \rho_m)g = C_D \frac{\pi}{8}\frac{\eta^2}{\rho_m} Re_t^2 \tag{6.2}$$

$$d_t = \frac{Re_t \eta}{\rho_m v_t} \tag{6.3}$$

Here, $V$ represents volume (particle), $d_t$ is the equivalent dynamic diameter and $Re_t$ is the particle Reynolds number at $v = v_t$.

For a spherical particle, the relationship between $C_D$ and $Re_t$ is given by:

$$C_D = \frac{24}{Re_t}, \quad Re_t < 2 \tag{6.4}$$

$$C_D = \frac{18.5}{Re_t^{0.6}}, \quad 2 < Re_t < 500 \tag{6.5}$$

$$C_D = 0.44, \quad 500 < Re_t < 2\times 10^5 \tag{6.6}$$

Equations (6.4), (6.5) and (6.6) are known as Stokes', Allen's and Newton's equations, respectively. Inserting these relations in equation (6.2), the terminal velocity of a particle can then be related to its diameter as:

$$v_t = \frac{d_t^2(\rho_p - \rho_m)g}{18\eta}, \quad Re_t < 2 \tag{6.7}$$

$$v_t^{1.4} = 0.072\frac{d_t^{1.6}(\rho_p - \rho_m)g}{\rho_m^{0.4}\eta^{0.6}}, \quad 2 < Re_t < 500 \tag{6.8}$$

$$v_t^2 = 3.03\frac{d_t(\rho_p - \rho_m)g}{\rho_m}, \quad 500 < Re_t < 2\times10^5 \tag{6.9}$$

Similarly, in the centrifugal field, dynamic diameter can be determined by replacing $g$ in equation (6.2) with centrifugal acceleration (i.e., $\omega^2 r$).

$$V(\rho_p - \rho_m)\omega^2 r = C_D\frac{\pi}{8}\frac{\eta^2}{\rho_m}Re_t^2 \tag{6.10}$$

$$d_t = \frac{Re_t\eta}{\rho_m v_t} \tag{6.11}$$

where $\omega$ is the angular frequency and $r$ is the distance from the center of the centrifugal field.

#### 6.1.2.3 Optical Microscopy

One of the basic methods in particle sizing is optical microscopy. In optical microscopy, it is the wavelength of light and the objective's numerical aperture which limits the resolution, given as:

$$\delta = \frac{1.22\lambda}{2(N.A)} \tag{6.12}$$

where $\delta$ is the resolution of the microscope, $\lambda$ is the wavelength of light and $N.A$ is the objective numerical aperture. It is also important to select the maximum magnification of the microscope that is useful for any given powder sample; as a rule of thumb, it is 1000 times its numerical aperture. Total magnification can be obtained by the product of the magnification of its eyepiece and objective. The typical resolution of a microscope is ~0.5 μm (it may be as small as ~200 nm), but the size of powders commercially used are in order of tens of μm. As the optical microscopes have very low depth of field, only flat surfaces may come into focus. Thus, the circumference of powder may be focused on the microscope (by dispersing powders on a glass slide) and/or the powders may be compact mounted and polished to obtain the size distribution.

#### 6.1.2.4 Fraunhofer Diffraction

Fraunhofer diffraction measures the particle size without disturbing the medium containing the particles; it is also called *far-field diffraction*. This method is applicable for particles having a size much greater than the wavelength of incident light. In order to obtain a Fraunhofer diffraction, two conditions should be satisfied: (1) the product of the distance of the light source from the particle and its wavelength has to be larger than the particle area (or aperture area), and (2) the product of the wavelength and the distance of the observation plane from the particle should also be greater than the area of the particle. Fraunhofer diffraction by a single particle/aperture is schematically shown in Figure 6.4a, and Fraunhofer diffraction by a suspension of particles in a gas using laser is shown in Figure 6.4b. The intensity of the diffracted pattern from a particle with spherical morphology (diam. $d$) at an angle $\theta$ is then given by [5]:

$$I(\theta) = I_o \frac{\pi^2 d^4}{16F^2\lambda^2}\left(\frac{2j_l(x)}{x}\right)^2 \tag{6.13}$$

where $I_o$ is the incident beam intensity, $j_l$ is the spherical Bessel function of first order, and $x = \alpha\theta$ where $\alpha = \frac{\pi d}{\lambda}$.

By measuring and analyzing the intensity distribution in the diffraction pattern as shown in Figure 6.5, the particle equivalent diameter can be obtained.

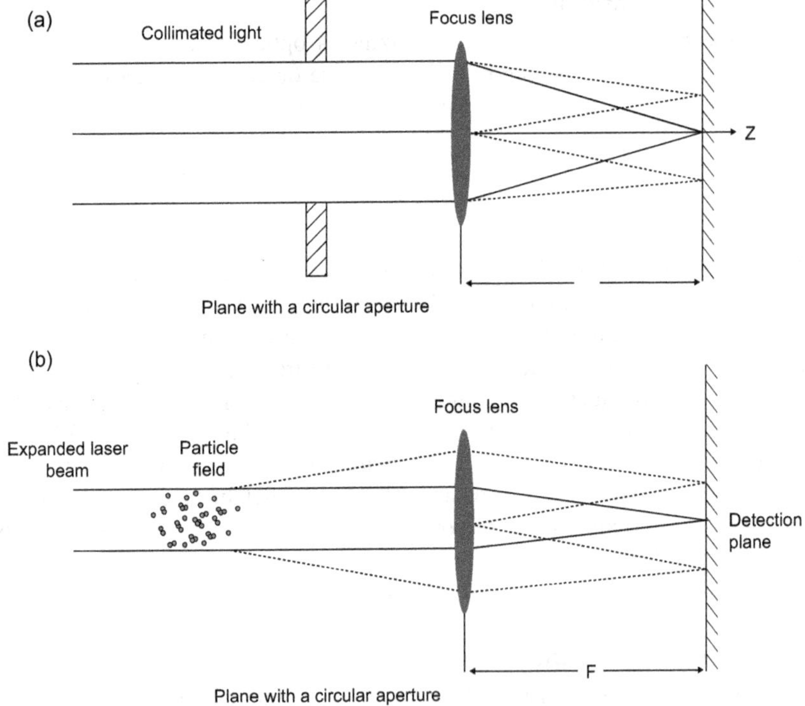


**FIGURE 6.4** Particle size analysis using the Fraunhofer diffraction system (a) from a circular aperture and (b) from a particle cloud.

*Source*: Adapted and redrawn from [3].

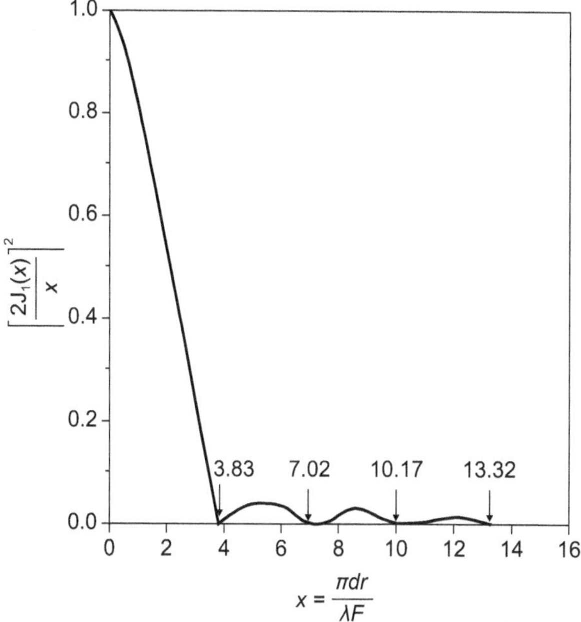


**FIGURE 6.5** Fraunhofer diffraction for a circular aperture of an opaque disc.

*Source*: Adapted and redrawn from [3].

#### 6.1.2.5 Laser Light Scattering

In this method, powder is passed through an expanded and collimated beam of laser, and the light scattered by the powder particles is collected in the forward direction. These particles are made to pass through the beam in front of a lens in whose focal length is placed a photodetector, as shown in Figure 6.6. The angles of diffraction in the forward direction are analyzed, which in general are inversely proportional to the powder size. To yield the particle size distribution, the aspect of its proportionality to scattered-light intensity is analyzed.

#### 6.1.2.6 Electrical Sensing Technique (Coulter Principle)

The electrical sensing technique is based on the Coulter principle, and the instrument is called the Coulter counter, as demonstrated in Figure 6.7. Particles are suspended in an electrolyte in which electrodes are immersed. The electrolyte is made to pass through a small orifice such that one particle traverses through the orifice at a time, which is assured by maintaining the particle concentration low. Once the particle enters the orifice, displacing the electrolyte, a sudden alteration in impedance occurs. This results in a voltage pulse with an amplitude proportional to the particle volume. Particle size and distribution can thus be estimated by the regulation of pulse sizing and pulse.

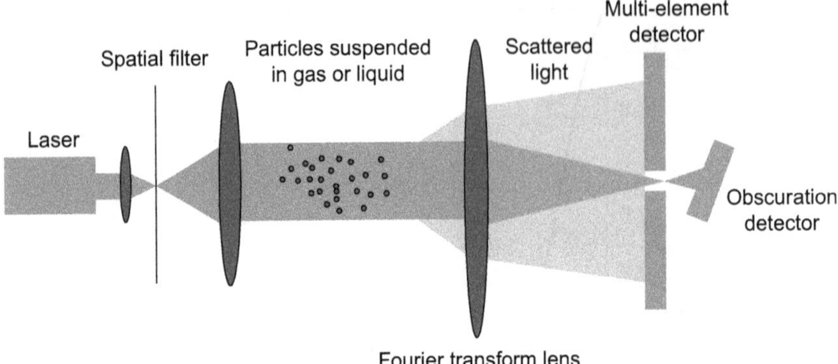


**FIGURE 6.6** Schematic illustration of a laser diffraction system showing scattering of light by powder particles passing through a laser beam.

*Source*: Adapted and redrawn from [6].

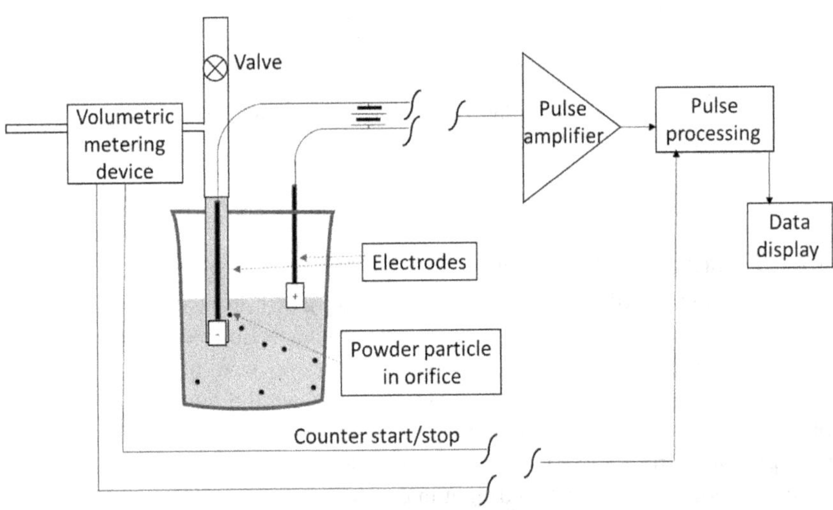


**FIGURE 6.7** Schematic representation of Coulter counter.

*Source*: Adapted and redrawn from [4].

### 6.1.2.7 Electronic Methods

Electronic methods include scanning electron microscopy (SEM) and transmission electron microscopy (TEM), both of which use an electron beam for characterizing the particle size. In SEM, a beam of electrons is scanned over the sample that upon interaction with it produces different signals like secondary and back-scattered electrons (SE and BSE), Auger electrons, X-rays and so on. The detected secondary electrons are converted into an image providing information about the particle shape and size. In TEM, an image is generated by means of the transmission electron beam through the sample on a photographic/detector plate. Although sample preparation in SEM and TEM requires the greatest amount of time and effort, it offers an advantage in terms of viewing particles of different sizes with the same field of focus owing to the high depth of focus of these microscopes [7].

The choice of method for particle sizing depends on factors like the nature of the powder, which includes its flowability, toxicity, solubility, ease of handling, estimated particle size and range, its cost, specific requirements, time restrictions and so on.

#### 6.1.2.8 Particle Size Determination

For a spherical particle, size is easily expressed in terms of its radius or diameter. However, for a particle shape other than spherical, shape is expressed using shape factors and shape coefficients. Equivalent diameters based on several equivalent parameters like volume, terminal velocity and so forth are defined and discussed below:

1. *Sieve diameter*: The minimum aperture size of the sieve that the particle can go through.
2. *Martin diameter* (*M*): The length of a line bisecting the image of a particle or its projected area. The line direction can be along any direction, but it should be maintained constant for each measurement.
3. *Feret diameter* (*F*): The separation between two parallel tangents drawn on any two opposite edges of a particle.
4. *Projected area diameter* ($d_a$): The diameter of a circle with the same area of projection as that of the particle. This can be related to the particle projected area (*A*) as:

$$d_a = \left(\frac{4A}{\pi}\right)^{\frac{1}{2}} \tag{6.14}$$

These diameters, other than the sieve diameter, all depend on the 2D image of a particle; thus they are typically used in optical microscopy and electron microscopy. An illustration of these diameters is shown in Figure 6.8.

### EXAMPLE 6.1

An example illustrating Martin diameter and Feret diameter of a powder particle is cited below in Figure 6.9. If the two diameters are 9 μm and 32 μm, respectively, and the projected area of this particle is the same as that of the area of a circular particle (dotted circle superimposed over the particle), which is 201 μm$^2$, what is the projected area diameter of this particle?

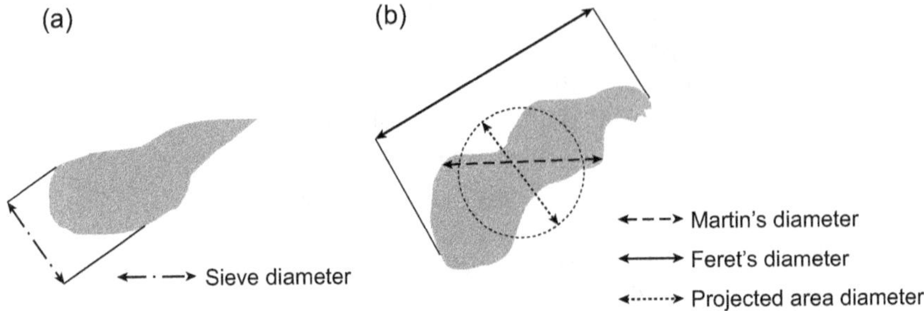


**FIGURE 6.8** Schematic representation of particle diameters: (a) sieve diameter; (b) different particle diameters (Martin, Feret and projected area) based on a 2D projected image.

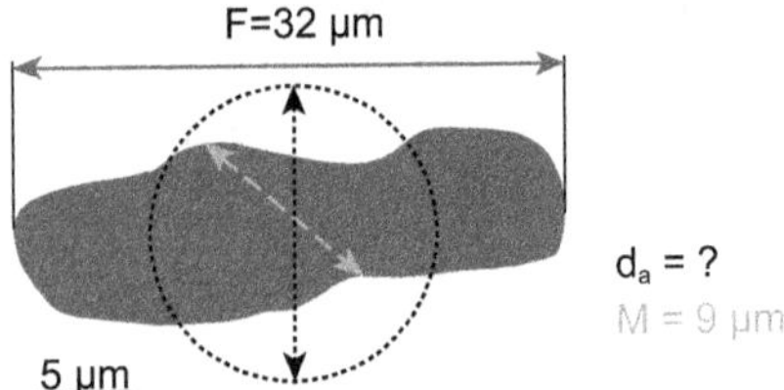


**FIGURE 6.9** Illustration for estimating different diameters of a particle.

## SOLUTION

Projected area of the particle, $A = 201\ \mu m^2$

$$\text{Since, projected area diameter } (d_a) = \left(\frac{4A}{\pi}\right)^{\frac{1}{2}}$$

$$\Rightarrow d_a = \left(\frac{4 \times 201}{\pi}\right)^{\frac{1}{2}}$$

$$\Rightarrow d_a = 16\ \mu m$$

Thus, the three diameters of the particle are $M = 9\ \mu m$, $F = 32\ \mu m$, and $d_a = 16\ \mu m$.

The three-dimensional characteristics of a particle are reflected by diameters discussed below.

5. *Surface diameter* ($d_s$): It is defined as the diameter of a sphere with same surface area ($S$) as that of the particle under consideration, mathematically expressed as:

$$d_s = \sqrt{\frac{S}{\pi}} \tag{6.15}$$

6. *Volume diameter* ($d_v$): It is the diameter of a sphere with same volume ($V$) as that of particle, given as:

$$d_v = \left(\frac{6V}{\pi}\right)^{\frac{1}{3}} \tag{6.16}$$

7. *Sauter diameter* ($d_{32}$): The spherical diameter with ratio of volume to external surface same as that of the particle under consideration. It is also known as surface-volume mean diameter and is expressed as:

$$d_{32} = \frac{6V}{S} = \frac{d_v^3}{d_s^2} \tag{6.17}$$

8. *Dynamic diameter*: Dynamic diameter is characterized by the terminal velocity as discussed in section 6.1.2 and is expressed as:

$$d_t = \frac{Re_t \eta}{\rho_m v_t} \tag{6.18}$$

### 6.1.3 Powder Particle Size Distribution

Powder particle size and its distribution is an important aspect in powder characterization. The weighing factor may be defined either with respect to number of particles or mass of particles in a given size range accordingly called a number density function or mass density function, which are interconvertible. Number density function includes the number of particle fraction in ($b$) to ($b + db$) size range, that is:

$$\frac{dN}{N_0} = f_N(b)db \tag{6.19}$$

where $dN$ is the number of particles in the ($b$) to ($b + db$) range and $N_0$ being the total number of particles. Normalizing equation (6.19), we get:

$$\int_0^{\infty} f_N(b)db = 1 \tag{6.20}$$

For a range $d_1$ to $d_2$, this fraction can be given as:

$$\int_{d1}^{d2} f_N(b)db = \frac{N_{12}}{N_0} \tag{6.21}$$

The mass density function representing the particle mass fraction in size can then be expressed as:

$$\frac{dM}{M_0} = f_M(b)db \tag{6.22}$$

Similarly, over a range $d_1$ to $d_2$, this fraction can be given as:

$$\int_{d1}^{d2} f_M(b)db = \frac{M_{12}}{M_0} \tag{6.23}$$

where $M_o$ is the total mass of the particles. For particles of the same size, mass can be represented in terms of number as:

$$d_M = m \times d_N \tag{6.24}$$

For a spherical particle of mass *m* and diameter *d*, mass can be expressed in terms of density as:

$$m = \frac{\pi}{6}\rho_P d^3 \tag{6.25}$$

Number density can be related to mass density using equation (6.26):

$$f_M(b) = \frac{N_0 m}{M_0} f_N(b) \tag{6.26}$$

Number density is usually obtained through microscopy or Fraunhofer diffraction, while mass density can be acquired via sieving.

Typically, two distributions are used in particle size:

1. Gaussian distribution
2. Log normal distribution.

The density function for Gaussian distribution, also known as normal distribution, is written as:

$$f_N(d) = A_N e^{\left(-\frac{(d-d_0)^2}{2\sigma_d^2}\right)} \tag{6.27}$$

where $A_N$ is the normalizing constant, $d_0$ is the arithmetic mean and $\sigma_d$ is the standard deviation of *d*.

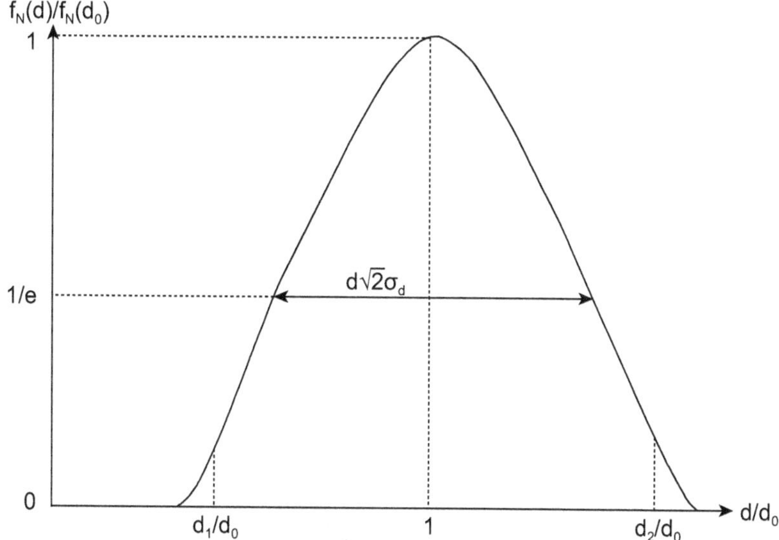


**FIGURE 6.10** Gaussian distribution function of particle sizes bound between $d_1$ and $d_2$.

The range of particle size bounded by $d_1$ and $d_2$ as shown in Figure 6.10 can be expressed as:

$$\int_{d_1}^{d_2} A_N e^{\left(-\frac{(d-d_0)^2}{2\sigma_d^2}\right)} db = 1 \tag{6.28}$$

The corresponding mass density function can be written as:

$$f_M(d) = A_M \frac{\pi}{6} \rho_P b^3 e^{\left(-\frac{(d-d_0)^2}{2\sigma_d^2}\right)} \tag{6.29}$$

$A_N$ and $A_M$ have no simple, exact and explicit solution; however, for a narrow size distribution $\frac{\sigma_d}{\sigma_0} \ll 1$, these are given by:

$$A_N \approx \frac{1}{\sqrt{2\pi}\sigma_d} \tag{6.30}$$

$$\frac{1}{A_M} \approx \frac{(2\pi)^{\frac{3}{2}}}{6} \rho_P \frac{d_0}{\sigma_d}\left(\frac{3}{2} + \frac{d_0^2}{2\sigma_d^2}\right)\sigma_d^4 \tag{6.31}$$

**TABLE 6.2**
**Particle Size Measurements of Powder A and Powder B**

| Powder A | | | | Powder B (μm) | | | |
|---|---|---|---|---|---|---|---|
| S. No. | Powder size (μm) | S. No. | Powder size (μm) | S. No. | Powder size (μm) | S. No. | Powder size (μm) |
| *1* | 0.55 | *26* | 1.15 | *1* | 5.05 | *26* | 3.16 |
| *2* | 0.60 | *27* | 1.12 | *2* | 5.22 | *27* | 4.57 |
| *3* | 0.63 | *27* | 0.53 | *3* | 3.06 | *27* | 2.00 |
| *4* | 2.14 | *28* | 1.22 | *4* | 2.12 | *28* | 3.40 |
| *5* | 1.51 | *30* | 1.45 | *5* | 9.97 | *30* | 4.26 |
| *6* | 2.40 | *31* | 1.15 | *6* | 6.68 | *31* | 4.00 |
| *7* | 1.55 | *32* | 1.32 | *7* | 6.94 | *32* | 3.17 |
| *8* | 1.50 | *33* | 1.33 | *8* | 4.48 | *33* | 3.75 |
| *9* | 1.35 | *34* | 0.91 | *9* | 7.40 | *34* | 2.48 |
| *10* | 0.81 | *35* | 2.05 | *10* | 5.47 | *35* | 2.80 |
| *11* | 0.92 | *36* | 2.10 | *11* | 3.66 | *36* | 7.75 |
| *12* | 0.95 | *37* | 1.82 | *12* | 3.02 | *37* | 6.20 |
| *13* | 1.10 | *38* | 1.22 | *13* | 2.24 | *38* | 3.06 |
| *14* | 1.30 | *39* | 1.23 | *14* | 2.52 | *39* | 3.39 |
| *15* | 1.70 | *40* | 1.12 | *15* | 2.58 | *40* | 3.42 |
| *16* | 2.00 | *41* | 1.65 | *16* | 3.86 | *41* | 7.60 |
| *17* | 1.94 | *42* | 1.85 | *17* | 4.48 | *42* | 4.70 |
| *18* | 1.50 | *43* | 2.35 | *18* | 4.71 | *43* | 4.06 |
| *19* | 2.30 | *44* | 1.60 | *19* | 4.53 | *44* | 4.76 |
| *20* | 1.90 | *45* | 1.40 | *20* | 11.02 | *45* | 5.65 |
| *21* | 1.20 | *46* | 1.75 | *21* | 8.50 | *46* | 4,95 |
| *22* | 0.75 | *47* | 1.56 | *22* | 1.88 | *47* | 5.75 |
| *23* | 1.65 | *48* | 1.75 | *23* | 1.82 | *48* | 3.21 |
| *24* | 1.12 | *49* | 1.55 | *24* | 1.73 | *49* | 6.36 |
| *25* | 1.62 | *50* | 1.82 | *25* | 3.32 | *50* | 10.48 |

In log-normal distribution, the particle size distribution follows the normal distribution with the logarithm of the particle size in semi-log scale, and can be expressed as:

$$f_N(d) = \frac{1}{\sqrt{2\pi}\sigma_{d1} d} e^{\left[-\frac{1}{2}\left(\frac{\ln d - \ln d_{01}}{\sigma_{d1}}\right)^2\right]} \tag{6.32}$$

where $d_{01}$ is the median diameter, $\sigma_{d1}$ is the natural log of the ratio of that diameter to the median diameter for which the cumulative distribution curve has value of 0.841.

Table 6.2 presents the particle size distribution collected from 50 measurements of two different powders; powder A and powder B. The data so obtained is plotted in Figure 6.11. Powder A can be fitted best with the normal distribution (Figure 6.11a), whereas a log-normal distribution fits well for powder

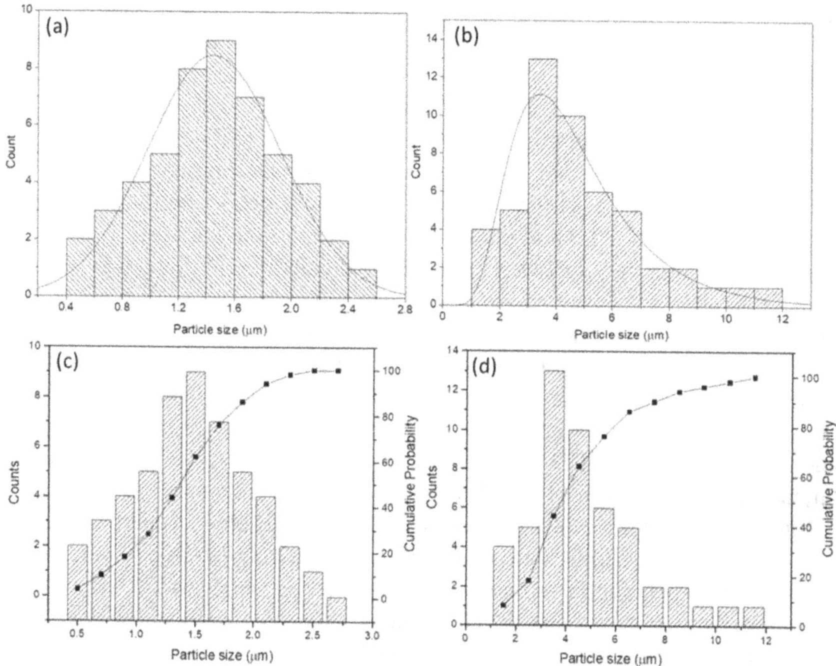


**FIGURE 6.11** Powder particle size distribution: A powder (a) fitted with normal distribution and (b) superimposed with corresponding cumulative percentage of counts; B powder (c) fitted with log-normal distribution, and (d) superimposed with corresponding cumulative percentage of counts.

B (Figure 6.11b). The cumulative counts for both distributions are plotted in Figure 6.11c and Figure 6.11d, respectively.

## 6.2 POWDER MANUFACTURING AND POWDER MODIFICATION ROUTES

The quality of feed material strongly dictates the coating quality and its performance in an application, making it important to manufacture the material in a reliable, efficient, economical and environmentally feasible way, observing the technical specifications of the thermal spraying process. There is a wide range of techniques to manufacture and then modify the powders before using them for the thermal spraying process; these are grouped under the following approaches:

### 6.2.1 Chemical Synthesis

#### 6.2.1.1 Sol-Gel Synthesis

The approach of sol-gel synthesis is wet-chemical. A solution of the chemical compounds is formed first, from which the powder is derived. Sol-gel technique is commonly used for the preparation of high-purity, fine-oxide metal powders. Raw materials are dissolved in a solvent, which may initially result in the formation of a colloidal solution, called *sol*. Reactants are usually the metal alkoxides. The type of alkoxide, pH of the solution, amount of water added and temperature of the reaction determine the condensation and *gel* formation. To promote or avoid the formation of agglomerates, a close control of pH is required. With the addition of water, hydrolysis and condensation occur, promoting the formation of a gel. The formation of a gel-like network is due to the poly-condensation process in which both solid and liquid phases exist. Gelation time may vary from a few seconds to several days followed by drying and calcining of this gel to produce oxide. Powders so synthesized are amorphous and require crystallization step to produce crystalline powders [8]. Processing of powder through sol-gel provides advantages like low synthesis temperature, high purity level and good composition control. The control on size and shape of sol-gel synthesized powder can be improved using variety of approaches including the acid catalyst or base catalyst addition. Preparation of $TiO_2$ nanoparticles with varying morphology like triangular, hexagonal and elongated have been synthesized using titanium alkoxides with tetramethylammonium hydroxide (TMAH). The TMAH base, in addition to catalyzing the reaction, supplies an organic cation stabilizing the polyanionic cores. There occurs a preferential interaction of 101 planes of $TiO_2$ with TMAH, causing differential growth rates in particular directions, controlling the morphology [9].

Gonzalez-McQuire et al. [10] produced hydroxyapatite powders through sol-gel synthesis using $NaCl_2$ and $Na_2HPO_4$ as starting powders. The pH was adjusted through the addition of NaOH. A gel network trapping water within the inorganic structure is formed through the combination of high reactant concentration, fast stirring and electrostatic interaction between the charged species in the solution. Slow drying evaporates water from the structure, leaving behind a porous material network. With no addition of NaOH at pH = 6, crystalline hydroxyapatite plates (5–10 μm) were obtained. However, introduction of NaOH restricted the crystal surface growth resulted in the formation of nanoparticles (at pH = 9, 25–100 nm long and 10–30 nm wide plates are obtained).

#### 6.2.1.2 Self-Propagating High-Temperature Synthesis

Self-propagating high-temperature synthesis (SHS) was initially utilized for producing refractory inorganic powders of carbides, nitrides, borides, silicides and so on. In SHS, reaction is initiated locally, which propagates in a self-sustained manner as a wave throughout the reaction volume. The reaction is highly exothermic [11]. It has been observed from the differential thermal scanning curves that the exothermic peak provides the melting temperature of the lowest melting point component or the reaction product. The product obtained in the form of a cake is crushed and/or ground for obtaining the desirable particle size distribution of the powder.

### 6.2.2 Atomization

Atomization is one among the most common method of manufacturing used to produce a large tonnage of powder, wherein the breakup of a liquid (molten material) is done into a fine spray. Metal or alloy is first melted and homogenized in any type of furnace and then transferred into the reservoir. Molten metal is passed through this reservoir ladle and is made to intercept by an atomizing medium (e.g., air or water) at a high velocity, disintegrating the molten material into droplets which solidify while falling down the chamber. Important factors considerable for the particles collected in the chamber are their size distribution, morphology and the extent of oxidation. Based on the atomizing medium used, atomization is of various types, discussed briefly in the following sub-sections [12].

#### 6.2.2.1 Gas Atomization

Gas atomization, or twin fluid atomization, is known for producing a wide range of powders. In the gas atomization process (GAP), either air (called air atomization) or any inert gas like helium, argon or nitrogen is used (inert gas atomization [IGA]) in compressed form as gas jets to break up the molten metal or alloy. In the case of air atomization, oxygen content in the atomized powders is high (1000 ppm; 1%), while as it is low (100–500 ppm) in IGA. In addition to low oxygen content, IGA results in a high degree of sphericity in the powder in the inert gas atomization process. For the production of more reactive alloys like Fe-based or superalloys required for sensitive applications like aerospace and defense, vacuum inert gas atomization (VIGA) is implemented. In VIGA, before atomization, melting and pouring the metal is carried out in a vacuum chamber. One more variant of gas atomization in which material is not supplied through a reservoir but in the form of an electrode is called electrode induction gas atomization (EIGA). There are two commonly adopted configurations in gas atomization. When the atomizing gas is supplied at the metal delivery nozzle base, it is called a *close-coupled atomizer* configuration. The short distance between the metal exit and the gas injection point causes efficient atomization. The smaller-sized particles are produced at identical energy consumption. However, this configuration faces the problem of premature melt freezing. In the second configuration, called the *free-fall atomizer* configuration, there is a certain vertical distance that a melt travels before being encountered by the gas jets. Due to good separation between the metal exit and atomizing gas, premature melt freezing is considerably low. In this case, the powder PSD is coarser than that in earlier configuration. The size distribution of the molten droplets influences the atomization chamber height, which should be sufficient to ensure solidification of the largest molten droplet before it reaches the bottom. Figure 6.12 shows the schematics of the close-coupled gas atomization system. The process system consists mainly of five parts including gas supply, melt crucible and delivery nozzle, atomizing gas nozzle, atomization compartment, powder collection region and gas extraction system. The parameters that determine the performance of the system include melt temperature, feed rate, gas pressure and flow rate. In gas atomization systems, there are two main parameters, design and operation, which include:

1. Geometry of the atomizing jet, its pressure, velocity and distance from the molten metal/liquid jet;
2. Geometry of the melt nozzle and melt flow velocity;
3. Molten liquid superheat above its melting point;
4. Nature of the gas, its purity and its cooling capacity.

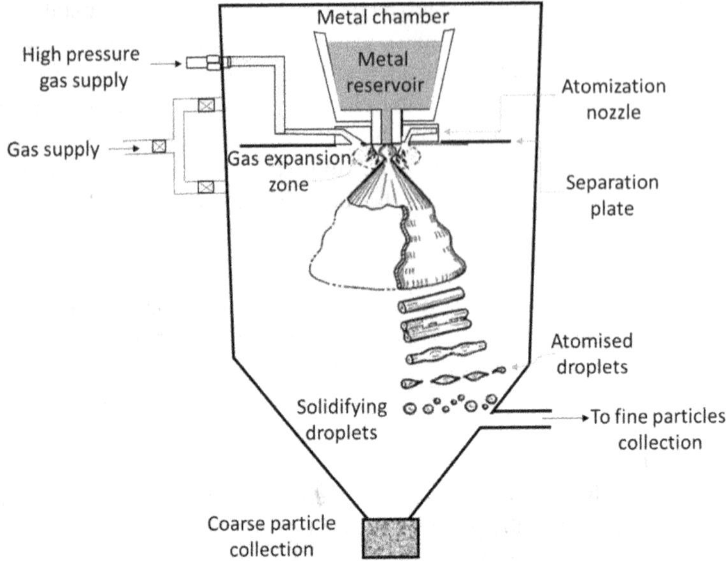


**FIGURE 6.12** Schematic showing different parts of gas atomization setup and its working. The melt exits the tube from the reservoir and is subjected to atomization under high-pressure atomizing gas.

*Source*: Adapted and redrawn from [8, 13].

#### 6.2.2.2 Plasma Atomization

The demand for high-quality powders has been a driving force behind plasma atomization's development. Metal in wire or rod form is melted and atomized to result in spherical, dense powders with good flowability. The process had little production inertia due to the use of rods or wires as starting material, restricting powder production for research and development in small batches. At the same time, cross-contamination-free switchover from one material to another is possible through the plasma atomization technique. This method, though expensive, is usually adopted for Ti and Ti alloy production, providing spherical powders with high purity. Plasma atomization has been applied via three variants:

1. *Plasma rotating electrode process*: To produce reactive metal powders like Ti (titanium), Be (beryllium), Mo (molybdenum), Zr (zirconium) and so on, this technology was developed. Initially, the process used an electric arc that was struck between the stationary tungsten electrode and a rotating metal bar to be atomized. The anode (i.e., the metal bar to be atomized) rotates at a high speed around its axis, hence it is also called rotating electrode process (REP). Molten metal produced at the anode is ejected by the centrifugal forces as spherical droplets. The droplets freeze in-flight and form dense spherical particles. The diameter of these spherical particles depends on the material properties, arc current and

anode rotation speed. In the plasma rotating electrode process (PREP), the tungsten electrode is replaced with a transferred arc plasma torch. The arc stability was enhanced using an inert gas flow of argon or helium, which protects the water-cooled tungsten cathode, improving the cathode life and minimizing tungsten contamination in the produced powder. Figure 6.13 represents a typical PREP installation.

2. *DC-triple plasma atomization process* (*DC-TRAP*): In this technique, material is fed as a wire into the apex of three DC plasma torches, as shown in Figure 6.14. The wire tip melts at the point of convergence of

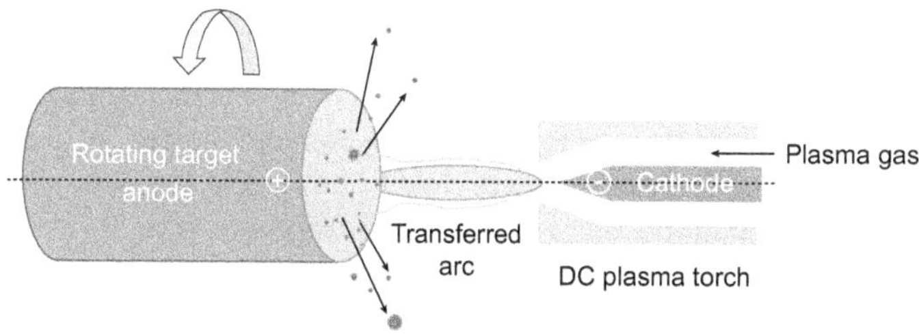


**FIGURE 6.13** Schematic representation of plasma rotating electrode arc setup.

*Source*: Adapted and redrawn from [8].

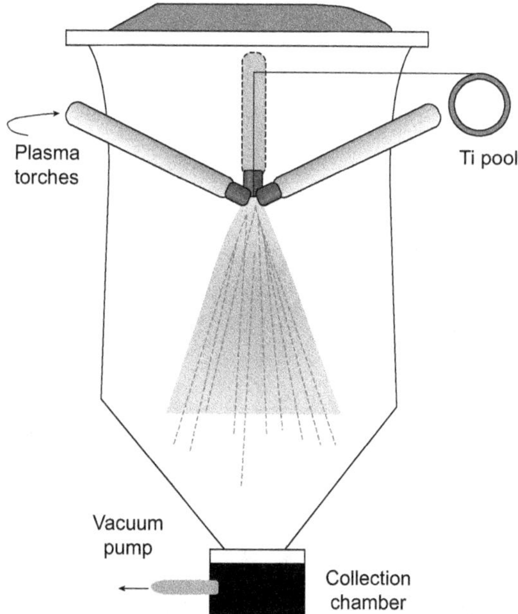


**FIGURE 6.14** Schematic representation of DC-TPAP setup highlighting the placement of three plasma torches; formation of molten droplets and the deposition of atomized powder.

*Source*: Adapted and redrawn from [14].

three plasma jets, which is atomized into fine spray droplets by high temperature and high-speed plasma. The larger particles settle in the bottom by gravity, and finer particles are collected in the filter collector with the help of gases exiting the chamber.

3. *RF-induction plasma atomization process* (*RF-IPAP*): In this atomization process, the preheated wire or rod is further intensely heated using an annular plasma flow of high velocity resulting in melting and a ring of high-speed atomization plasma jet causes atomization of the melt into fine molten droplets. The simple schematic representation of RF-APAP is shown in Figure 6.15.

### 6.2.2.3 Water Atomization

In water atomization, jets of water hit the stream of molten metal and disintegrate it, which then freezes in the atomizer before falling into the pool of water. Water for atomization is applied at a pressure ranging from 3 to 60 MPa. The amount of water supplied should lie between ~4 and 10 L/min/kg of melt to disintegrate the molten metal stream, which results in the particle cooling rates to the tune of

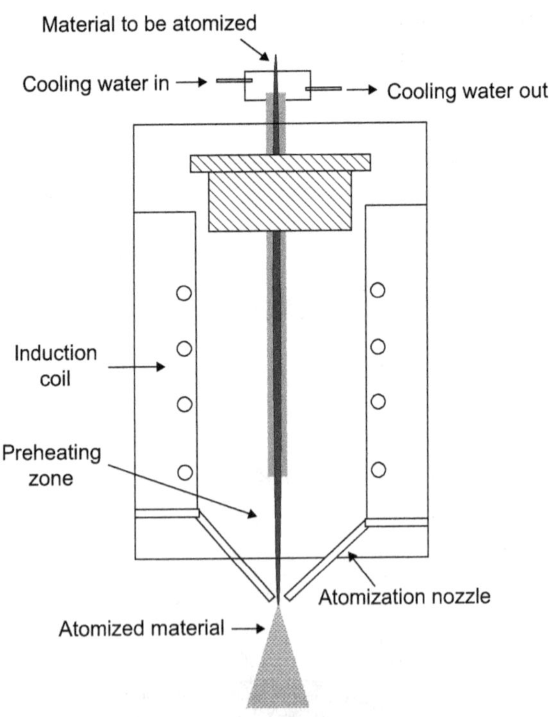


**FIGURE 6.15** Schematic illustration of RF-IPAP setup highlighting the transformation of feed material (wire, cord or rod) into atomized powder through induction heating.

*Source*: Adapted and redrawn from [8].

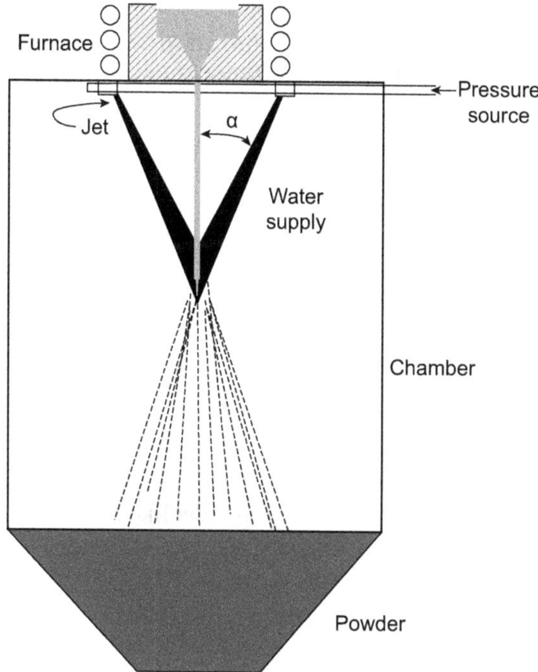


**FIGURE 6.16** Water atomization set up illustrating the use of water stream to atomize the molten stream of material.

*Source*: Adapted and redrawn from [8].

~$10^4$–$10^6$ K/s. For a higher velocity of water jet, finer particles can be obtained. Oxygen content may also be high, which can be reduced by post-treatment of the powder. In ultra-high pressure water atomization (UHPWA), water velocity reaches supersonic level (400–500 m/s) under extremely high pressures (100–200 MPa). Atomization is partly achieved by the shock waves around the jet and results in the production of finer particles. The atomization product needs to be de-watered and dried. The particles obtained are irregular in shape and may need an additional spheroidization step to produce spherical powders. However, partial spheroidization can be obtained by superheating of the molten melt before atomization. Table 6.3 provides some differences in gas atomization and water atomization in terms of different features.

### 6.2.3 Mechanical Size Reduction

Mechanical size reduction is among the popular methods for reducing the size of obtained powder. Some of the techniques used in this method are presented in the next sub-sections.

**TABLE 6.3**
**Comparison of Gas/Water/Plasma Atomization Techniques in Terms of Process Parameters and Obtained Features of Resulting Powders**

| Feature | Gas Atomization | Water Atomization |
|---|---|---|
| Medium | Gas (air, steam, $N_2$, Ar, or He) | Water |
| Velocity | 50–150 m/s | 40–150 m/s |
| Pressure | 30 kPa–65 MPa | 1.4–4.2 MPa |
| Cooling rate | $10^3$–$10^5$ K/s | $10^4$–$10^6$ K/s |
| Powder size | 25–100 μm | 150–400 μm |
| Powder shape | Spherical | Irregular |
| Powder purity | Up to 99.99% (i.e., 100 ppm of oxygen) | Higher oxygen content (1 wt%) |

#### 6.2.3.1 Fusing/Sintering and Crushing

Fusing and crushing is employed for brittle powder like ceramic powder. Ceramics require high temperatures for melting and are usually fused in arc furnaces. The fused solid mass is then crushed using conventional mills like hammer mills, jaw crushers and so on to an appropriate size for spraying.

#### 6.2.3.2 Milling

Milling can be performed in three ways:

1. *Ball milling*: This is a commonly used technique for breaking down the agglomerates and reducing their particle size. The mill contains a barrel made of stainless steel, vulcanized rubber or high-density alumina. At laboratory scale, it is a bowl/container of metal lined with polyurethane or tungsten carbide. The container, partially filled with milling medium, rotates on its axis. The medium contains hard material typically composed of tungsten carbide in a spherical, cylindrical or rod shape. The medium amount and size depend upon the powder to be milled, the final particle size required and the size of the mill. The schematic illustration of a ball mill in Figure 6.17 shows the cascading movement of medium. The main issue with ball milling is the contamination of the powder through the milling medium (during the milling process).

Mill speed is usually expressed in terms of percent critical speed (% CS). At critical speed, the grinding medium will begin to centrifuge; that is, it will rotate with the mill and cease to serve its purpose (of impacting and abrading the powder). So, at critical speed, centrifugal forces equal the gravitational forces at the inside mill wall surface, and no ball will fall from its position onto the shell. For an effective operation, the mill should be operated at 65%–80% of its critical speed. Critical speed ($n_c$) in rpm is given by equation (6.33).

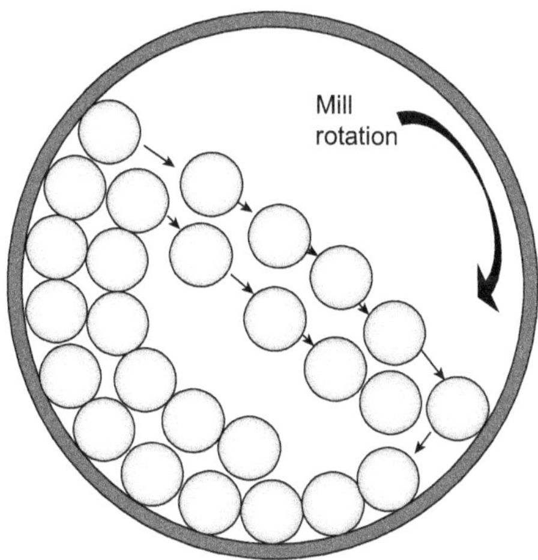


**FIGURE 6.17** Schematic of a ball mill illustrating the motion of balls in the rotating mill.

$$n_c = \frac{60}{2\pi}\sqrt{\frac{g}{R-r}} \tag{6.33}$$

where $g$ is acceleration due to gravity and $R$ and $r$ are radius of the ball mill and ball, respectively.

Optimal performance of the mill is also influenced by the solid loading, since the majority of media motion and milling occurs on the surface of the media bed, where media is cascading. For too few solid particles, media strikes media with little milling effect, which is accompanied with accelerated wear of the media and mill wall. For too much solid, the impact is buffered by a solid bed. Generally, 25% of solid material is required for 50% of the media.

2. *Attrition milling*: In attrition milling, the grinding medium and particles are stirred in the form of a slurry. The chamber in this case is aligned either vertically or horizontally and is usually water-cooled. Media is mixed using a Christmas-tree shaped stirrer, positioned in the center (Figure 6.18). Organic additives can be added in the media to limit agglomeration. The main issue with this technique is the separation of fine powder from the slurry after the attrition milling is completed.
3. *Cryo-milling*: To make the milling process more efficient, cooling of the medium is an effective approach. Through cooling, steady state can be easily achieved and the fracture process is accelerated. It further suppresses the agglomeration of powders and prevents welding with the milling medium. With the nitrogen environment of milling, the oxidation reactions can also be limited.

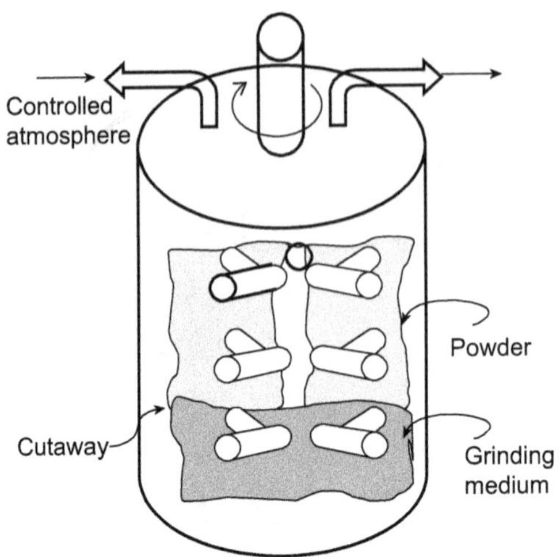


**FIGURE 6.18** Schematic representation of attrition mill.

*Source*: Adapted and redrawn from [8].

### 6.2.4 Powder Consolidation

In powder consolidation, smaller particles are consolidated to produce coarse particles with novel composition and morphology. It can be performed using the following techniques.

#### 6.2.4.1 Mechanical Alloying

Mechanical alloying is a dry and high-energy consolidation technique in which fine powders with controlled microstructure are obtained through repeated welding, fracturing and re-welding in a high-energy ball mill. It was basically developed for oxide-dispersion strengthened (ODS) alloys. It can be used to prepare intermetallic alloys, amorphous alloys, oxide and carbon dispersed metals, and so forth and can be applied to metals, polymers, ceramics and composites. The combination of fracture, welding and strain-enhanced diffusion produces a homogenized powder (i.e., it ultimately forms an alloy). Powders are mixed in the right proportion along with a grinding medium and loaded in a mill. The mixture is then milled for a desired duration till the attainment of steady state, at which the composition of all particles obtained is the same as that of the starting proportion of the individual materials in the mixture. Optimization is required for a number of variables to achieve a desired microstructure and phases in the final powder, including the type of mill, container, milling time, milling speed, nature and amount of grinding medium, milling atmosphere, ball to raw material ratio and so forth. Grinding containers and grinding medium are usually made up of hardened steel, tool steel, tempered steel, WC-Co, WC-lined steel, bearing steel and hardened chromium steel. A process control agent (PCA) is usually added to reduce the effect of cold welding, which is experienced by powder particles, especially ductile powder due to their severe plastic deformation. These PCAs include stearic acid, methanol, ethanol and so on and are also called surfactants or lubricants for this purpose [15].

#### 6.2.4.2 Spray Drying

Spray drying is one of the most versatile practices to produce powders usable for spray processes. Usually, powders obtained through other techniques like crushing, milling and grinding produce fine particles, which cannot be used as such for conventional spraying and require agglomeration before putting into use. Again, for cermets, mixing the fine particles of ceramic and metal/alloy powders is necessary to form agglomerates of their mixture. Nanostructured particles also require agglomeration in the form of micrometer sized particles or sprayed in suspensions. In this regard, spray drying serves the purpose and can be utilized to form agglomerates from suspensions called *slurry*. Although formation of slurry may be easy in certain cases like oxides, which stabilize easily, the high specific density of some materials and widely different properties of their constituents like in the case of cermets, makes it difficult to get a stabilized suspension. In such cases, addition of a stabilizing surface agent that can prevent the sedimentation of powder and ensure the uniformity in its composition is required. Further, to avoid the breakdown of powder before its sintering or calcination, the addition of a binder may be required. It is of utmost importance to choose the composition of these additives carefully in order to avoid any contamination in the final powder product. The suspension prepared is sprayed in the form of drops into a drying chamber with an atomizer, wherein these drops are heated into steam of hot air or by the hot chamber walls. The solvent gets evaporated within few seconds, producing dry powder which is collected in the spray chamber at its bottom. The parameters which affect the production of powder particles include slurry concentration, feed rate of the liquid, air temperature at the inlet, diameter of the nozzle and flow rate of the air.

#### 6.2.4.3 Cladding

Cladding is a process of coating the powder particles with a dense, continuous, porous or heterogenous layer to form composite powder particles. Cladding of ceramic powders with metallic layers improves their wettability and flowability. Cladding may also be required when core material needs protection against the chemical modification like oxidation or decarburization during spraying. A schematic representation of a cladded powder particle is presented in Figure 6.19.

Cladding can be performed in a number of ways:

1. *Mechanical alloying*: In mechanical alloying, the core particles (to be cladded) are attrition milled along with the coating material. The core particles need mechanical strength and stability during milling. The amount of coating material added to the core particles needs

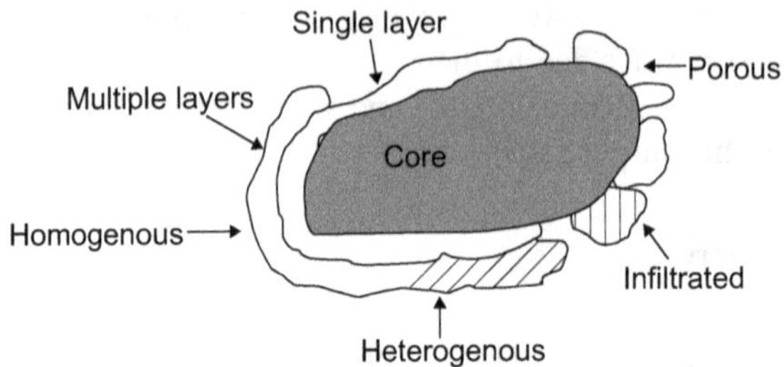


**FIGURE 6.19** Schematic representation of a powder particle showing possible cladding.

*Source*: Adapted and redrawn from [16].

to be optimized. For an insufficient quantity of coating material, core particles may get crushed, while an excessive amount will avoid cladding. Aluminum oxide and chromium oxide powders are coated using mechanical alloying, which are then used for magnetic recording.

2. *Sintering*: By sintering, large particles mixed with the sub-micron-sized particles provide coating of large particles (with material of small-sized powders).
3. *Mechanofusion*: In the mechanofusion technique, different powders are rotated in a rotation chamber and subjected to compression at the same time using two compression pieces. The powders are mixed intensely between the inner wall and the compression pieces, and due to compression, strong mechanical forces result in generation of thermal energy. The thermal energy heats the lowest melting point powder to its plastic state or even melts it. This powder forms the core of the composite powder which is surrounded by a shell of higher melting point material.
4. *Gaseous phase deposition*: The most common technique for gaseous phase deposition is chemical vapor deposition (CVD). In CVD, a fluidized bed reactor of particles and a vapor precursor is distributed throughout the bed. The coating of particles in the fluidized bed occurs as the vapor precursor undergoes thermal decomposition. This technique requires a close control of fluidized bed temperature, and the powder to be coated needs to be stable at that temperature. The distribution of vapor should also be uniform throughout the bed. A maximum coating thickness of 20 μm is achievable through this process.
5. *Electrolytic coating/galvanic plating*: In electrolytic coating, electrically conductive powder is set as anode, whereas in galvanic plating, the coating material to be deposited is made available in its metal salt solution form (as electrolyte). These coatings can be deposited over a wide range of thickness (0.5–2000 μm) but are porous, irregular and spongy.
6. *Chemical precipitation*: The metal powder particles to be coated are mechanically agitated to remain suspended in an aqueous solution of

metal salt, whose coating needs to be deposited. Hydrogen is then introduced in an autoclave containing the suspension at high temperature and pressure. Hydrogen reacts with the solution, precipitating the metal to deposit as a coating when its potential surpasses the metallic ion electrode potential. This is known as the *hyperpoc process* (hydrogen pressure reducing powder coating process) as well [8].

#### 6.2.4.4 Sintering

In sintering, finer particles are either first compacted and then subjected to high temperature in a furnace or subjected to simultaneous application of temperature and pressure in a pressure-assisted sintering machine. The conditions of temperature and pressure should be sufficient to allow their binding by promoting diffusion. The density of the compact depends upon the sintering parameters and the size distribution of powder. With increase in compaction pressure, porosity is reduced. Increase in sintering time and temperature also has same effect on porosity. However, sintering is accompanied by grain growth, which exaggerates with temperature. This exaggerated grain growth has deteriorating mechanical properties, especially in the case of nanostructured coatings. After sintering, the sintered cake undergoes crushing to obtain the desired particle size distribution.

### 6.2.5 Spheroidization

Spheroidization is the process of fine-tuning the powder morphology to improve powder flow characteristics. Dense powder with spherical morphology has many advantages in spraying including better flowability with stable powder flow rate and no clogging of the feeder hose. Due to spherical morphology, only point contact results and restricts wear of pipes and feeding system. Uniform heating of the particles occurs due to their dense nature, which is important in the case of low thermal conductivity of the powder. Although metal spherical powder particles can be obtained by melt atomization, a similar method in the case of ceramics is impractical and can be accomplished through plasma treatment. Flow and packing properties of metallic powders, produced from any method having irregular morphology or low sphericity or in agglomerated form, can be improved by subjecting to plasma treatment. Plasma spheroidization is simple and similar to plasma spraying, where particles are melted by passing through a plasma jet. However, the molten droplets are cooled and resolidified under controlled conditions during flight before falling onto the bottom of the chamber. Figure 6.20 shows the change in tungsten powder morphology after plasma spheroidization.

The materials produced from chemical routes are generally nanocrystalline. It is hard to produce dense powders of spherical morphology using these fine crystallites. In such cases, powder suspension is fed via pump in RF induction plasma, and introduction of atomizing gas produces droplets. Through plasma heating, water is lost producing solid particles. These powders after post-treatment like heating, flash sintering, melting and solidifying form dense spherical powder.

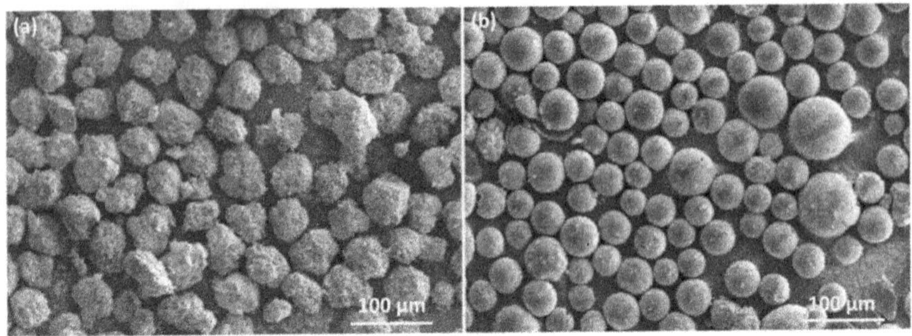


**FIGURE 6.20** Alumina powder (a) before spheroidization and (b) after spheroidization.

*Source*: Reprinted with permission from [17].

## 6.3 SUMMARY

During thermal spraying, the powder feed is subjected to different conditions imposing restrictions on the type and form in which they can be used. In general, the feed material can be supplied in powder form, as a wire or in the form of suspension. For powders, these requirements include desirable chemical composition, proper particle size distribution, particle morphology and flowability, and so on. To test the powder for the desirable characteristics, it undergoes various characterizations before feeding into the thermal spraying unit. Accordingly, it can be modified following a suitable technique if not found proper for feeding. Also, based on these requirements, a wide range of technologies for manufacturing the feed material have turned up.

### Questions for Self-Analysis

1. What is the typical purity in commercial powders, and which applications may warrant the use of high-purity powders?
2. What is the typical size of powders used in thermal spraying, and why?
3. What are the different morphologies of powders? What advantage does spherical morphology play in thermal spraying?
4. What is the limitation of sieving in obtaining a reliable particle size distribution? Highlight its limitations.
5. How the sedimentation method allows determining the size of powder particles?
6. What is meant by "equivalent diameter"? Why is it needed, and how can it be obtained?
7. What additional aspect does the powder size distribution provide?
8. What are the different powder manufacturing routes?
9. Highlight the difference between air and water atomization in terms of powder characteristics.

10. What is critical speed in ball milling, and how is it calculated?
11. What is spray drying, and what is its utility?
12. Compare spray drying with spheroidization, and provide the differences.

## REFERENCES


[1] Nouri, A. and A. Sola, *Powder morphology in thermal spraying.* Journal of Advanced Manufacturing Processing, 2019. **1**(3): p. e10020.

[2] Shockley, J.M., et al., *The influence of $Al_2O_3$ particle morphology on the coating formation and dry sliding wear behavior of cold sprayed Al-$Al_2O_3$ composites.* Surface Coatings Technology, 2015. **270**: p. 324–333.

[3] Fan, L.-S. and C. Zhu, *Principles of gas-solid flows.* 1999. Cambridge University. p. 3–45.

[4] www.bionicsscientific.com/sieve-shakers/test-sieves.html.

[5] Nai-Ning, W., Z. Hong-Jian, and Y. Xian-Huang, *A versatile Fraunhofer diffraction and Mie scattering based laser particle sizer.* Advanced Powder Technology, 1992. **3**(1): p. 7–14.

[6] Shen, S., et al. *Research of laser particle sizer based on scattering theory.* In 2008 International Conference on Information and Automation. 2008. IEEE.

[7] Jillavenkatesa, A., S. Dapkunas, and L. Lum, *NIST recommended practice guide: particle size characterization*, Special Publication (NIST-SP-960-1), Gaithersburg, 2001.

[8] Fauchais, P.L., J.V. Heberlein, and M.I. Boulos, *Thermal spray fundamentals: from powder to part.* 2014. Springer Science & Business Media.

[9] Fornasiero, P. and M. Cargnello, *Morphological, compositional, and shape control of materials for catalysis.* Vol. 177. 2017. Elsevier.

[10] Gonzalez-McQuire, R., et al. *Thermal stability during plasma spray deposition of hydroxyapatite sol-gel derived nanopowders.* in Surface Modification Technologies: Proceedings of the 20th International Conference on Surface Modification Technologies. 2007. ASM International.

[11] Hwang, C.-C., et al., *A self-propagating high-temperature synthesis method for synthesis of zinc oxide powder.* Journal of Alloys Compounds, 2009. **467**(1–2): p. 514–523.

[12] Dunkley, J.J., *Metal powder atomisation methods for modern manufacturing.* Johnson Matthey Technology Review, 2019. **63**(3): p. 226–232.

[13] Davis, J.R., *Handbook of thermal spray technology.* 2004. ASM International.

[14] Kassym, K. and A.J.M.T.P. Perveen, *Atomization processes of metal powders for 3D printing.* Materialstoday Proceedings, 2020. **26**: p. 1727–1733.

[15] Suryanarayana, C., *Mechanical alloying and milling.* Progress in Materials Science, 2001. **46**(1–2): p. 1–184.

[16] Pawlowski, L., *The science and engineering of thermal spray coatings.* 2008. John Wiley & Sons.

[17] Károly, Z. and J. Szépvölgyi, *Plasma spheroidization of ceramic particles.* Chemical Engineering and Processing: Process Intensification, 2005. **44**(2): p. 221–224.

# 7 Coating Formation

*Shipra Bajpai and Kantesh Balani*

## CONTENTS



Plasma arc, HVAF, HVOF and cold spray are the most widely used industrial methods. Irrespective of the process used, thermal spraying basically involves the introduction of a source (particles, wires or rods) into the hot stream of gases. In the case of particles, they are accelerated followed by partial or complete melting, leading to their deposition on the substrate. However, in case of wires or rods, their tips are first melted followed by atomization and then deposition. The kinetic and thermal energy of these molten droplets differ with the thermal spray process used (e.g., plasma spray, cold spray), which in turn results in a coating thickness of 5–5000 μm [1]. Initially a "cold-spraying" process was developed where the powder particles were accelerated at velocity of ~600 m/s at room temperature and the coatings were deposited due to plastic deformation and adiabatic heating induced

DOI: 10.1201/9781003321965-7

by the high-velocity impact of the particles on the substrate [2, 3]. Later, flame-based spray techniques were introduced where the temperature was reduced with increased velocity of the particles. HVOF and HVAF are some examples of flame-based spraying with particle velocity in the range of 650 m/s obtained by converging-diverging type De Laval–type nozzle design resulting in dense coatings with less oxidation [4]. Further development in the coating methods has led to the involvement of electric arc and plasma for the deposition of coatings on the substrate. For plasma spraying the temperature reaches as high as 10,000 K and particle velocity <450 m/s, leading to a dense coating.

During the spraying process, a single particle with accelerated velocity strikes on the substrate surface, flattens and forms a splat. Successive deposition of the splats on the substrate and on the previously formed splat leads to the formation of coatings. Based on the spraying techniques available, coatings/splat formation can be divided into two categories: (a) by completely molten particles and (b) by the partially molten or solid particles. In case of molten droplets, the splat forms due to flattening of fully molten particles of metals, alloys or ceramics upon impact, and the coating formation is due to the layering of such impacting particles. However, for the partially molten or solid particles, plastic deformation at the time of impact leads to the formation of a splat whose diameter is nearly same to the initial particle size followed by coating formation due to pinning effect of the impacting particle. Coating formation on any substrate can be best understood by studying particle flattening or splat formation from the molten, partially molten or solid particle, followed by their deposition on the previously flattened particles on the substrate. Based on this, the current chapter will describe the single splat formation from the fully or partially molten and solid particles, how the heat transfer affects the spreading of splats, the effect of substrate preheating on the splat formation, and the final coating formation due to the deposition of a layer of various splats and its microstructure.

## 7.1 SINGLE SPLAT FORMATION

Whenever a molten/semi-molten droplet impacts on the substrate, a splat forms due to flattening of the droplet, as shown in Figure 7.1. Based on the models suggested by Sampath and Herman [5], the core was suggested to consist of columnar grains with the involvement of maximum particles. However, the core and rim of the splats formed depends on the following parameters:

- Thermal history of the particles;
- Thermal conductivity of the substrate;
- Melting point of particles;
- Kinetics of thermal spraying;
- Thermal/flow properties of the particles.

Based on the available thermal spraying methods, splat formation can occur either by the molten particles or semi-molten particles that are discussed in the following sections.

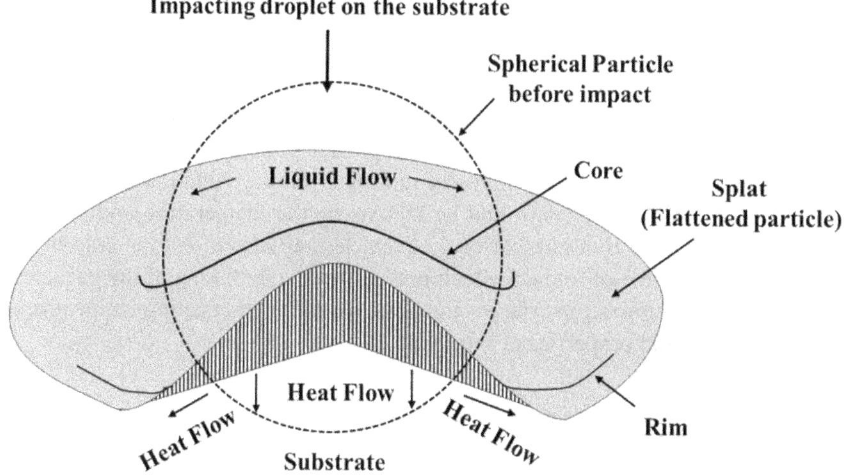


**FIGURE 7.1** Schematic representation of the splat forming upon impact with substrate.
*Source*: Adapted and redrawn from [5].

### 7.1.1 Splat Formation by Solid Particles

When solid particles with temperature less than 0.9 $T_m$ and high velocity are impacted on the surface, they are plastically deformed followed by flattening and splat formation. To achieve the good flattening of the particle, the impact velocity of the particles should be higher than the critical velocity ($v_c$) and slower than the erosion velocity ($v_e$) where particles start to erode. The impact should also be nearly perpendicular to the substrate, otherwise the change in angle may alter the velocity and thus affect the particle deposition. With decrease in the particle impact angle ($\theta$), particle impact velocity also decreases, leading to poor deposition of the particles. When the cold particle impacts the surface with velocity greater than the critical velocity, the deformation of the particle is purely plastic due to higher values of applied stress as compared to the yield stress of the particle. However, in case of hot particles, the plastic deformation of particles become more easier due to lower yield stresses at higher temperatures. There might also be chances of particle and substrate softening at the contact zone due to adiabatic heating induced by dynamic deformation of the particles [6]. This phenomenon has already been reported for the HVOF and HVAF process where the particle temperature is near to its melting temperature; however, no adiabatic heating or material softening was observed for the cold-spraying process [7, 8]. In addition, for the cold-spraying method, no additional in-flight oxidation of the particle occurs, whereas for HVOF or HVAF processes, the oxidation level reaches almost 10 times the initial level.

### 7.1.2 Splat Formation from the Molten Particles

Molten particles with higher velocity when impacting the substrate surface are flattened, forming the splats. Some critical parameters including particle velocity at the time of impact, particle temperature at impact, angle of particle impact at substrate and oxidation tendency of the particles define the degree of splat formation. Impacting particles may get modified due to oxidation during heating stages, which can be controlled by increasing the temperature and velocity of the particles. The Biot number of the particle may also affect the core being solid even at the time of impact, which may affect the flattening of the particles. Substrate preparation (e.g., surface roughness, oxidation, preheating temperature) also define the final properties of the splat.

## 7.2 RAYLEIGH INSTABILITY

Rayleigh instability explains how a molten stream of liquid breaks into small droplets of the same volume but smaller surface area. The practical applications of Rayleigh instability phenomenon include liquid dispensers, inkjet printers and so on. One best example of the everyday application of Rayleigh phenomenon is the formation of rain droplets before hitting the ground. During thermal spraying, a molten metal stream is targeted at the substrate which further disintegrates into small droplets and hits the substrate surface. Formation of the small droplets from the molten stream during thermal spraying is defined by the Rayleigh instability that is caused by the small perturbations present in the stream. No matter how smooth the stream is, there are always some perturbations present, as shown in Figure 7.2a.

If these perturbations are resolved into sinusoidal components, then there will be two possibilities:

1. Growth of some perturbations while others decay with time;
2. Difference in the growth of perturbations.

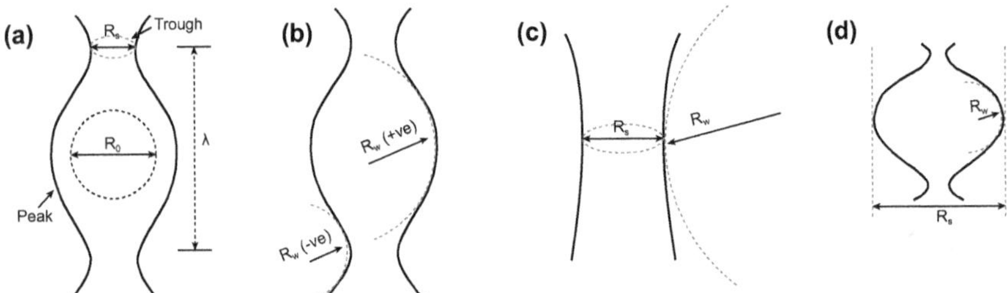


**FIGURE 7.2** Schematic representation of (a) the presence of perturbations in the molten metal stream; (b) arc for radius of curvature in trough and peaks, (c) $R_w > R_s$; and (d) when $R_s > R_w$. where $R_w$ is the radius of curvature of wavefront and $R_s$ is the radius of curvature of molten metal stream.

*Source*: Adapted and redrawn from [9].

Growth or decay of the perturbations is entirely a function of its wave number and radius of the original cylindrical stream.

Wave number is basically defined as the spatial frequency of the wave; in other words, it is defined as the number of waves that exist over a specified distance. Wave number is given by equation (7.1):

$$k = \frac{2\pi}{\lambda} = \frac{2\pi\upsilon}{v_p} = \frac{\omega}{v_p} \tag{7.1}$$

where $v$ is the frequency, $\omega$ is angular frequency and $v_p$ is the phase velocity.

Initially, it is assumed that the amplitude of all the perturbations is roughly equal, and thus the size of droplet is determined by the wave number of the component which grows fastest. With time, the segment with maximum growth rate dominates and pinches the metal stream into droplets.

Perturbations in the molten stream consist of the peak (crest) and the trough (depth), which are shown in Figure 7.2a. At the trough (valley) region, as the radius of stream is smaller, the pressure increases due to an increased surface tension (as per the Young-Laplace equation). Similarly, the higher radius of stream at the crest (hill) possesses lower pressure. With the pressure being higher at the trough, it can be expected that the liquid is squeezed from trough to low-pressure area (i.e., the peak), but this is not the case. There are some other factors which also contribute towards the droplet formation, as the Young-Laplace equation is governed by these two components: (1) radius of stream ($R_s$) and (ii) radius of curvature of wave ($R_w$). The radius of curvature of wave is shown with a fitted arc in Figure 7.2b. At the trough, the $R_w$ is observed to be negative, and thus the pressure decreases, whereas at the crest, the $R_w$ is positive, and thus the pressure increases. The effect of $R_w$ is just opposite to the effect of $R_s$ and, therefore, these are expected to cancel each other; however, they do not. The magnitude of one component will be higher than the other one depending on the wave number and the initial radius of the stream.

When the radius of curvature of wave dominates over the radius of the stream (Figure 7.2c), then it decays and provides a pinching effect (or separates with time). Whereas, when the radius of stream dominates over the radius of curvature of the wave, such components grow exponentially (or become bigger with time). Therefore, on combining all those effects, it is found that the components having product of wave number and initial radius ($k.R_0$) < 1 will be unstable and grow faster. However, the fastest growth of the component was observed when $k.R_0$ is ~0.697 [10].

## 7.3 KELVIN-HELMHOLTZ INSTABILITY

During cold spraying, the particles are subjected to high temperature and contact pressures; in such a case, adiabatic shear instabilities and localized plastic deformation gives rise to the bonding of particles to the substrate via adhesion. Apart from that, some additional nanolevel/microlevel of material mixing/interlocking mechanism may also contribute towards the particle substrate bonding. The Kelvin-Helmholtz instability phenomenon can be used to describe the interfacial instability, resulting in the formation of interfacial roll-ups. This phenomenon can take place when two fluids are moving at different velocities parallel to the interface. In case of the presence of a perturbation with radius of curvature not equal to zero, one of the fluids flows around the other, resulting in the generation of centrifugal forces. This results in a change in the pressure and thus amplification of those interfacial perturbations, as shown schematically in Figure 7.3. Formation of interfacial roll-ups and vortices improves the interfacial bonding and adhesion strength for the following reasons [11]:

1. Vortices formation increases the interfacial area (and consequent adhesion);
2. Fine length-scale mixing at interfacial region;
3. Mechanical interlocking at interface due to roll-ups.

## 7.4 STEPS OF SPLAT FORMATION

Splat formation from the molten/solid droplet can be divided into the following steps:

- Impact of the particles
- Spreading of the splat/particle after impact
- Solidification of the flattened splat
- Fragmentation of the splats

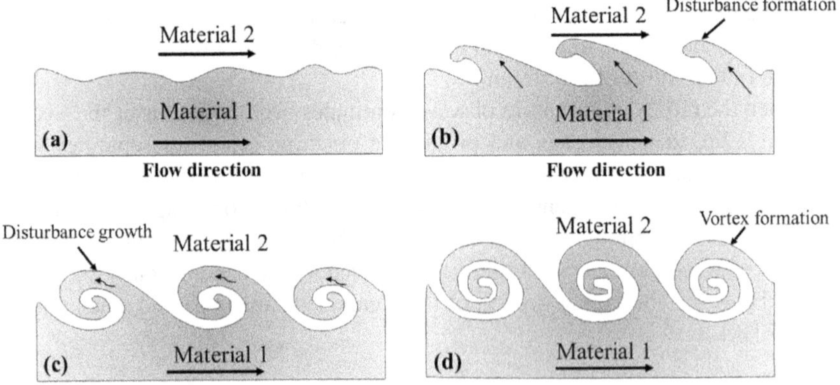


**FIGURE 7.3** Schematic representation of the interfacial instability-based evolution: (a) material 1/material 2 interface, (b) formation of disturbance at interface, (c) disturbance growth and (d) the interfacial roll-ups and vortices formation leading to good interfacial bonding.

*Source*: Adapted and redrawn from [11].

### 7.4.1 Impact of the Particles

During thermal spraying, the particles impact the substrate at higher impact velocity $v_p$ (i.e., 200 m/s for plasma spraying and 500 m/s for HVOF), which results in the generation of the impact pressure as high as 100 MPa and thus the compressibility effects. This generated impact pressure is then propagated as a wave front in the particle with radius $r_m$ (as shown in Figure 7.4c). The radius $r_m$ is generally smaller than the final radius of the splat and is defined via equation (7.2):

$$r_m = \frac{d_p v_p}{2c_i} \tag{7.2}$$

where $c_i$ is the sound velocity of the liquid drop. Figure 7.4 represents the schematic for the successive steps of liquid splashing after impact [12].

### 7.4.2 Spreading of the Splats

After impact of the particle, its kinetic energy is converted into the surface energy and viscous deformation of the particle, leading to the generation of high contact pressure at the interface between substrate and particle. This generated contact pressure is then quickly spread out, followed by its dissipation leading to particle flattening.

The degree of flattening of the impacting particle is given by equation (7.3):

$$\xi = \frac{D}{d_p} \tag{7.3}$$

Where $\xi$ is the degree of flattening, $D$ is the flattened particle diameter and $d_p$ is the diameter of the particle before impacting the substrate. Degree of flattening mainly depends on the particle velocity and state of the particle (partially or fully molten) at the time of impact. Pyrometers, microscopic analysis and so on are some of the techniques to evaluate the temperature variation of the droplets before and after impact. Along with the particle velocity, formation of oxides around the particles and substrate surface characteristics (e.g., preheating temperature, roughness, wettability of the surface by liquid metals) also affect the flattening of the splat.

Impact splashing or spreading of the droplets occurs immediately after the impact, and it is basically governed by the fluid instabilities before the solidification. Figure 7.4 shows the successive events of the particle impact splashing [12]. Many analytical models have already been developed [12] which utilize the dimensionless numbers to define the degree of flattening of the droplets on the substrates. Two-dimensional geometry with good contact between particle and substrate along with the absence of any intermediate oxide layer is assumed in these models to characterize the degree of particle flattening by using the Reynolds number ($R_e$),

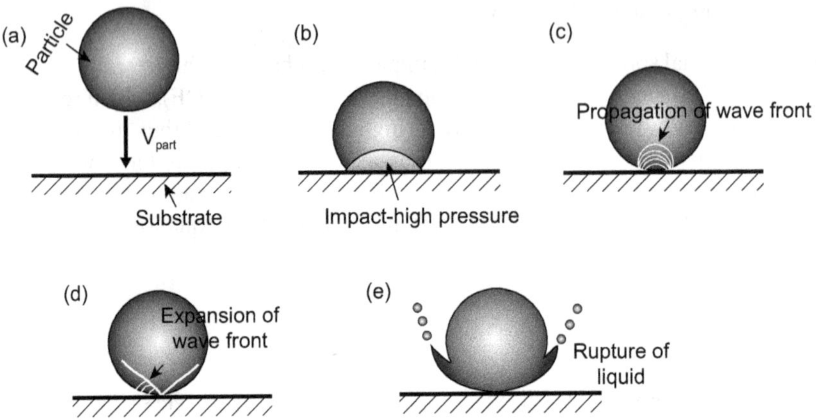


**FIGURE 7.4** Schematic representation of the events of splashing of particles after impact: (a) impacting particle; (b) generation of high pressure at the particle/substrate interface after particle impact; (c) distribution of the pressure in the form of wave fronts; (d) expansion of the wavefront at triple junction between substrate, particle and atmosphere; and (e) final falling-out of the liquid.

*Source*: Adapted and redrawn from [12].

Weber number ($W_e$) and Sommerfeld number ($K$). The Reynolds number is defined as the ratio of inertial forces and frictional forces within a fluid subjected to relative motion due to different fluid velocities, as shown in equation (7.4):

$$R_e = \left[\frac{\rho v d}{\mu}\right] \tag{7.4}$$

The Reynolds number is a dimensionless parameter which helps in predicting the nature of fluid flow (i.e., laminar or turbulent flow), especially when there is change in the behavior of forces, as in the case of a hot stream of particles emitting from the nozzle to the air. When $R_e < 2100$, the flow is laminar; when $2100 < R_e < 4000$, the transition between laminar to turbulent flow occurs; and when $R_e > 4000$, flow becomes completely turbulent.

The Weber number is a dimensional number that is used to analyze the fluid flow in an interface between two different fluids, which is defined as the ratio between the inertia force and surface tension force, as shown in equation (7.5).

$$W_e = \left[\frac{\rho d \mu^2}{\sigma}\right] \tag{7.5}$$

The Weber number indicates whether the kinetic energy or surface tension energy term is dominating.

The Sommerfeld number ($K$) is the dimensionless quantity which is related to the Reynolds number and Weber number, according to equation (7.6):

$$K = \left(R_e\right)^{0.5}\left(W_e\right)^{0.25} \tag{7.6}$$

where $\rho$ is density, $v$ is velocity, $d$ is diameter, $\mu$ is viscosity, and $\sigma$ is surface tension of the droplets at time of impact.

For the partially melted or solid droplets, $K < 3$ corresponds to the rebound of the droplets, a $K$ value in the range of 3–58 results in the particle deposition on the substrate, and $K > 58$ means splashing. However the values of $K$ are different for the fully molten particles [13]. In case of plasma spraying, the value of $K$ is always greater than 58 and it can reach as high as ~1200. For fully molten particles, the solidification starts only after flattening is completed which is related to the Reynolds number and Weber number according to equation (7.7) [14]:

$$\frac{3\xi^2}{W_e} + \frac{1}{R_e}\left(\frac{\xi}{1.2941}\right)^5 = 1 \tag{7.7}$$

For most of the thermal spraying cases, the Weber number of a particle is assumed to be greater than 100 and thus the degree of flattening (i.e., the ratio of the splat diameter and impacting particle diameter) is given by equation (7.8):

$$\xi = \frac{D}{d_p} = A.R_e^{0.2} \tag{7.8}$$

where $A$ is the constant with the value 1.2941 for a perfect disk-shaped particle and 0.83 for the fingered shaped splats [14, 15]. Thickness of the splat can be calculated by considering the complete conversion of particles with diameter ($d_p$) into the disk-shaped splat with thickness $e$ and volume $e.\pi.D_2/4$, where $D$ is the diameter of the splat. In the case of copper particles deposited on the preheated stainless-steel surface, at a low Reynolds number, splats formed were disc shaped, whereas with the increase in Reynolds number to $5 \times 10^4$, splashing of the splat droplets occurred at the periphery of disc [16].

Degree of flattening and spreading of the particles is not only dependent on the impact parameters which include particle velocity and impact angle. Some other parameters, like plume temperature, dwelling time, air resistance, thermal conductivity of the substrate, presence of adsorbents, condensates on substrate surface, and surface roughness also affect the degree of flattening of droplets.

*Particle impact velocity*: Velocity of the particle at the time of impact affects the flattening of the droplet, as well as the splat thickness. Based on the particle velocity, the time required for the flattening of the particle can vary from microseconds to one-tenth of microseconds as reported for the zirconia particles impacting on the substrate with different velocities at 3200 K [17]. Particle with impact velocity of 25 m/s have resulted in the flattening degree of 3.6, whereas with an increase in the particle velocity to 400 m/s, the degree of particle flattening has increased to 6.3.

With an increase in particle velocity, impact pressure increases, which results in increased degree of flattening with reduced splat thickness. Depending on the splat thickness, the particle velocity can be calculated by using equation (7.9):

$$v = \left[ \frac{2\eta d^3}{33\rho S^4} \right] \tag{7.9}$$

where $\eta$ is viscosity, $d$ is diameter, $\rho$ is density, $v$ is velocity and $S$ is splat thickness.

*Plume temperature*: Higher plume temperature leads to the superheating of the particles which reduces the viscosity of the particles and thus better spreading and more flattening.

*In-flight time*: Higher dwell or in-flight time allows the particle to reach a higher temperature and thus better flattening of the droplets on the substrate.

*Air resistance*: Air resistance between the material and plume also affects the spreading or flattening of the particles. Less air resistance provides better thermal transfer and allows the free flow of material, which in turn reduces the viscosity and increases the temperature; both viscosity and temperature contribute towards the spreading of droplets. The mushrooming effect or thickening of the core also occurs in the presence of higher air resistance which does not allow the droplet to spread, and, thus, it solidifies at the substrate.

*Thermal conductivity of particle and substrate*: Lower thermal conductivity of the substrate allows better flattening of the droplets as it avoids the sudden cooling or freezing of droplets upon contact with the substrate. In contrast, higher thermal conductivity of the particle is favorable for the spreading of the droplets. Once the droplets start to flatten, rapid heat loss occurs due to increased surface area. In that condition the core with higher thermal conductivity can provide the heat and thus more spreading. However, if a particle impacts on the still-solidifying splat, further growth of the splat will be impeded, and thus it will not spread anymore.

*Presence of adsorbents and condensates on substrate*: Presence of surface oxides, oxyhydroxides and so forth on the substrate surfaces are considered adsorbents and condensates [18–20]. Presence of these surface oxides, adsorbates and so forth majorly influence the spreading and flattening of the droplets. However, when the particle impacts at higher temperature, the adsorbents may get evaporated, which leads to the generation of pressure and lifting of the droplets as shown in Figure 7.5. In the central zone of highest impact pressure, the vapor pressure can be overcome, and good contact area can be observed. Therefore, based on this, the contact area between splat and substrate increases by 2–6 times with the removal of adsorbates and condensates from the substrate surface. Preheating or heat treatment of substrates prior to spraying may alter the level of these surface oxides and thus flattening of droplets.

*Substrate roughness*: Surface roughness of the substrate affects the spreading of liquid and thus particle flattening. Smooth surfaces result in smaller splat thickness with good spreading, whereas a rougher surface hinders the spreading of liquid.

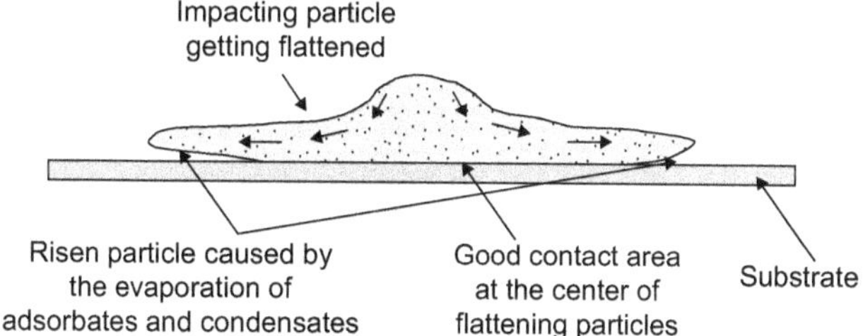


**FIGURE 7.5** Effect of the presence of adsorbents and condensate on the flattening of droplet at room temperature.

*Source*: Adapted and redrawn from [21].

### 7.4.3 Solidification of the Flattened Splats

Another step of splat formation is the solidification of flattened splats which is dependent on its thickness, degree of flattening and thermal conductivity of both the impacting particles and substrate. Impact velocities of the particle affect the flattening of splats and thus the solidification of the splats. Solidification of the droplets starts before the completion of flattening events, and thus it modifies the liquid flow at the surface of flattening droplets. Solidification time is generally 5–10 times higher than the flattening time.

Solidification of the droplet is dependent on the thermal contact resistance ($R_{th}$) between substrate and splat which is defined as equation (7.10) [22]:

$$R_{th} = {}^{\Delta T}\!/_{Q} \tag{7.10}$$

where $\Delta T$ is the temperature difference between the substrate and splat, and $Q$ is the heat flux (W/m$^2$). When the $R_{th}$ value is low, the temperature at the bottom of splat is same as that of the substrate, whereas for higher thermal resistance, poor heating of substrate occurs, which results in almost the same temperature at the top and bottom of the splat. For effective cooling and solidification of the droplet, the thermal resistance between substrate and splat should be as low as possible ($10^{-7}$ to $10^{-8}$ m$^2$.K.W). At the start of solidification, the thermal contact resistance is low; however, at the end of it the resistance may increase.

## HEAT TRANSFER DURING SPREADING AND SOLIDIFICATION

Flattening of the droplet and its solidification are totally dependent on transfer of heat from the substrate or from the previously deposited and solidified splats. During flattening and solidification, the heat transfer occurs through the conduction into the substrate and convection into the molten droplets. However, it has been reported that the conduction contributes 95% of the heat transfer as compared to convection. The heat transfer during flattening and solidification of the splats can be related to the thermal gradient present within the splat (i.e., from the core to the rim), and that can be best understood with the help of the Biot number, defined as equation (7.11):

$$B = \frac{hS}{k} \tag{7.11}$$

where $h$ is coefficient of heat transfer, $k$ is thermal conductivity and $S$ is splat thickness. If the value of the Biot number $B < 0.1$, then the thermal gradient present within the splats can be ignored, and thus the heat transfer can be simplified.

In the case of formation of thin motionless fluid and its solidification, the Newtonian cooling rate is given as equation (7.12):

$$Q = \frac{h(T_s - T_t)}{(\rho C_p e)} \tag{7.12}$$

where $Q$ is the heat cooling rate, $h$ is the heat transfer coefficient (W/m$^2$.K), $T_s$ is the substrate temperature, $T_t$ is the melting point of the splat, $\rho$ is the density of the splat, $C_p$ is the specific heat capacity of the splat (J/kg.K) and $e$ is the splat thickness ($m$). The corresponding solidification rate $I$ is given as equation (7.13):

$$I = \frac{h(T_s - T_t)}{(\rho L_f)} \tag{7.13}$$

where $L_f$ is the enthalpy of fusion.

Splat fragmentation and its solidification of droplets is dependent on the freezing; therefore to predict the likelihood of the splat breakup, a combined single parameter was developed, which is known as the solidification parameter ($\Theta$), defined in equation (7.14) [23]:

$$\Theta = \frac{s}{e} \tag{7.14}$$

where $s$ is the thickness of solid layer formed and $e$ is the splat thickness at the substrate temperatures lower than the droplet melting point. This solidification parameter helps in deciding the final shape and final mechanism of breakup of the splat. Based on the solidification parameter, there can be three possible outcomes during splat spreading:

1. When the formed solid layer is very thin (i.e., $s \ll h$ or $\Theta \ll 1$), the spreading of the splat will remain unaffected. The splat will have a central splat surrounded by the wear debris due to extreme fragmentations.
2. If the growth of solid layer is significant (i.e., $\Theta$ ~0.1–0.3), the spreading of the splat will be restricted, and thus a perfectly disk-shaped splat will form.

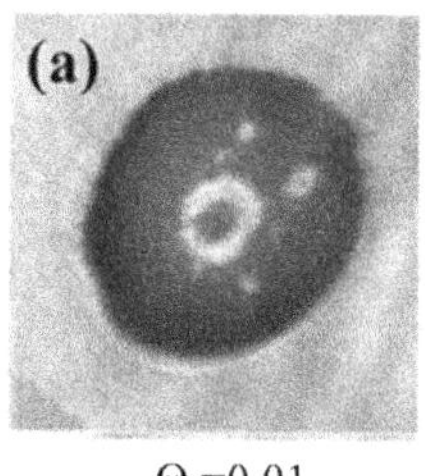


$\Theta = 0.01$
Mo/glass

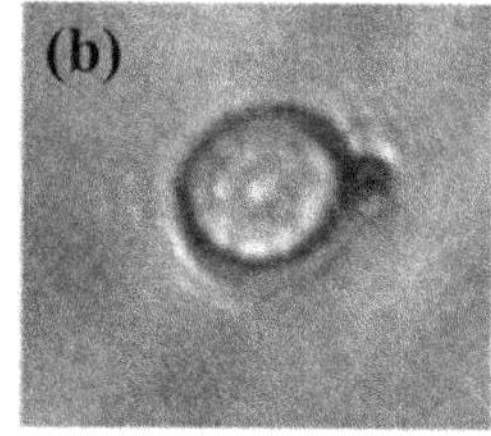


$\Theta = 0.1$
$ZrO_2$/glass, 400 °C

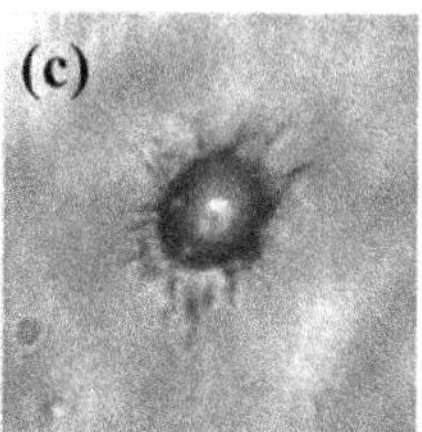


$\Theta = 0.4$
Mo/glass, 400 °C

**FIGURE 7.6** Photographs during and after the impact: (a) fragmentation of the droplets after impact in Mo/glass, (b) formation of disk-shaped splats for $ZrO_2$/glass interface with solidification parameter of 0.1; and (c) freezing induced by the breakup of Mo particles at the glass surface when solidification rate is high (i.e., 0.4).

*Source*: Reprinted with permission from [24, 25].

3. If the solidification is rapid (i.e., $\Theta$ ~0.1), the solid layer formed will restrict the spreading of the splats, and thus the fingers will form at the periphery of the splats.

Some experimental studies have also confirmed the accuracy of the solidification parameter, which is shown in Figure 7.6. Figure 7.6a shows the splats of Mo and Ni particle deposited on glass and steels with high thermal contact resistance, and $\Theta = 0.01$ showing a central splat with surrounded rings of debris. Increase in the substrate temperature resulted in the reduced thermal contact resistance, and thus disk-shaped splats with smooth edges are formed as shown in Figure 7.6b. Further increase in solidification parameter results in the obstruction of spreading, and thus a central splat with fingers is formed (Figure 7.6c).

## EXAMPLE 7.1

With the help of the following table, calculate the Biot number and particle velocity of molten $Al_2O_3$ particle in both the cases.

| Sample ID | Splat Thickness of $Al_2O_3$ (μm) | Splat Diameter of $Al_2O_3$ (μm) |
|---|---|---|
| AO | 2.8 ± 1.7 | 35.5 ± 4.7 |
| AO-20nY | 2.5 ± 1.5 | 36.8 ± 3.6 |

where viscosity of molten $Al_2O_3$/YSZ is taken as 0.15 poise, and theoretical density of $Al_2O_3$ and $Al_2O_3$/YSZ composite are 3.95 g/cm$^3$ and 4.24 g/cm$^3$, respectively. The average heat transfer coefficient is $10^5$ W/m$^2$ for splats and thermal conductivity was taken to be 5 W/m.K.

## SOLUTION

Whenever a molten droplet impacts on the substrate, flattening of the droplet occurs with the mass conservation; that is, the spherical droplet of particle gets converted to the splat of some thickness and diameter which are listed in the given table. Splat thickness and diameter are the parameters which define the Biot number and particle velocity, according to equations (7.9) and (7.11).

The heat transfer coefficient ($h$) defines the heat flux into a particle-to-temperature difference between particle surface and nearby gas. Therefore in the present example, $h = 10^5$ W/m$^2$.$K$

$K$ is defined as the thermal conductivity, which is taken as 5 W/m.K.

Viscosity, $\eta = 0.15$ poise $= 0.15 \times 100 = 15$ g/m.s.

(a) For AO particle:
Splat thickness, $S = 2.8 \times 10^{-6}$ m,
Splat diameter, $d = 35.5 \times 10^{-6}$ m,
Density, $\rho = 3.95$ g/cm$^3$

On substituting the values in the formula:

$$B = \frac{10^5 \times 2.5 \times 10^{-6}}{5}$$

$$B = 0.056$$

Furthermore, velocity for AO particles is calculated by substituting the values in equation (7.9):

$$v = \frac{2 \times 0.15 \times 100 \times \left(35.5 \times 10^{-6}\right)^3}{33 \times 3.95 \times 10^6 \times \left(2.8 \times 10^{-6}\right)^4}$$

$$v = 167.5\ m/s$$

(b) For AO-20nY particles:
Splat thickness, $S = 2.5 \times 10^{-6}$ m,
Splat diameter, $d = 36.8 \times 10^{-6}$ m,
Density, $\rho = 4.24$ g/cm$^3$

The Biot number for AO-20nY particles is calculated as follows:

$$B = \frac{10^5 \times 3.63 \times 10^{-6}}{5}$$

$$B = 0.05$$

Furthermore, particle velocity of AO-20nY particles:

$$v = \frac{2 \times 0.15 \times 100 \times \left(36.8 \times 10^{-6}\right)^3}{33 \times 4.24 \times 10^6 \times \left(2.5 \times 10^{-6}\right)^4}$$

$$v = 186.6 \text{ m/s}$$

In both the conditions, the value of the Biot number is <0.1, which signifies no contact thermal resistance between the particle and substrate, thus effective heat transfer and good spreading of droplets. For AO-20nY particles, higher velocity of 186.6 m/s and lower Biot number (0.05) suggests more spreading of the droplet and thus lower splat thickness of AO-20nY particles.

### 7.4.4 Fragmentation of the Splats

At the later stage of impact, fragmentation of the splats may occur when both the spreading and solidification phenomenon are happening. When the particles are impacted at room temperature, thermal contact resistance is high, thus the fragmentation of splats occurs extensively. However, preheating of the substrate above the transition temperature reduces the thermal resistance, and thus splats formed are disk shaped with reduced fragmentation. Apart from the temperature, there are some other factors which also contribute into the fragmentation of splats. Poor contact between substrate and flattening particle, irregularities in the solidification of core and periphery, and instabilities of the liquid sheet at the splat edge causes the fragmentation of splats during spreading and solidification. Spraying parameters also effect the fragmentation of the splats after impact. Figure 7.7

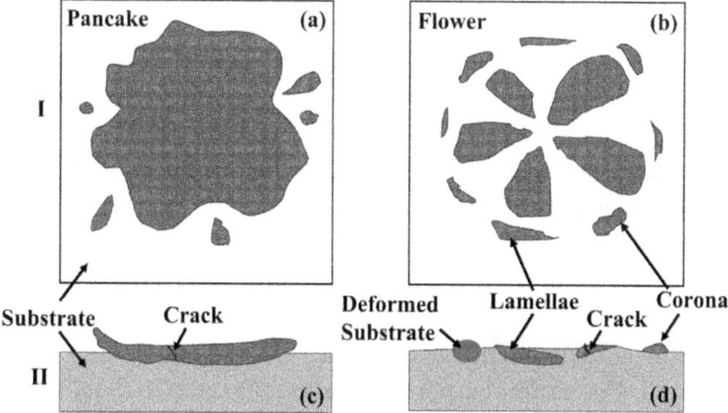


**FIGURE 7.7** Different morphologies of the lamellae after splashing on the substrate: (a) Pancake, (b) Flower, (c) and (d) represent different features formed on the splats including crack, deformed substrate, corona and lamellae. I represents the top view and II represents the cross-sectional view of splats.

*Source*: Adapted and redrawn from [26].

shows the different splats morphologies obtained after HVOF with fully melted particles [26]. During HVOF the particles are fully melted, and at impact the shock waves are generated within the particle, resulting in the formation of either pancake-shaped or flower-shaped lamellae, as shown in Figure 7.7a–d.

## 7.5 EFFECT OF PREHEATING OF THE SUBSTRATE ON SPLAT FORMATION

Preheating of the substrates is done prior to impact to reduce the thermal stresses and modify the level of adsorbents, condensates or any other oxide present on the surface. Preheating of the substrates is usually done above the transition temperature where evaporation of adsorbents and condensates occurs, leading to modified surface roughness and oxide layer thickness.

During preheating, the moisture is absorbed from the substrate, and a dry warm surface is present where the particles are to be impacted. Preheating is usually done in the temperature range of 100–150 °C, followed by immediate spraying by using a gas torch or spray torch (without spraying feed material) moving on the substrate [27]. Preheating of the substrate above the critical temperature $T_c$ (i.e., material/particle property) results in the formation of disk-shaped splats due to reduced thermal resistance, whereas when heated below critical temperature, fingered splats form. However, aluminum, copper, magnesium and its alloys should not be preheated, as the freshly preheated surface may promote oxide formation on the substrate, which eventually will affect the splat flattening. Another way of preheating the substrate is furnace heating under the controlled atmosphere where the oxide layers cannot be produced.

## 7.6 SPLAT LAYERING AND DEPOSITION

Final properties of the coating are dependent on the splat formation and their layering. During thermal spraying, the particles are carried by the gas and injected into the spray jet followed by impact on the substrate, resulting in the final splat formation. Around 100–110 particles of size 30 µm are injected within 1 second. Depending on the particle velocities, the duration of particles into the hot gas plume varies from 0.1 to a few milliseconds. Figure 7.8 schematically represents the time involved during steps of the conventional coating formation, which indicates the wide range of time duration varying from microseconds to hours [28].

Splat layering is dependent on the time-temperature history of the plasma spraying process. For example, particle flattening takes 1–2 µs followed by its solidification within the duration of 2–10 µs. The next particle impacts on the previously solidified splat in the next 12–100 µs, and again its layering takes place within few ms.

During coating formation, three parts of the torch/substrate motions are considered: stationary torch and substrate; 1D motion of torch or substrate leading to beads creation; and pass formation due to covering of beads. In case of both

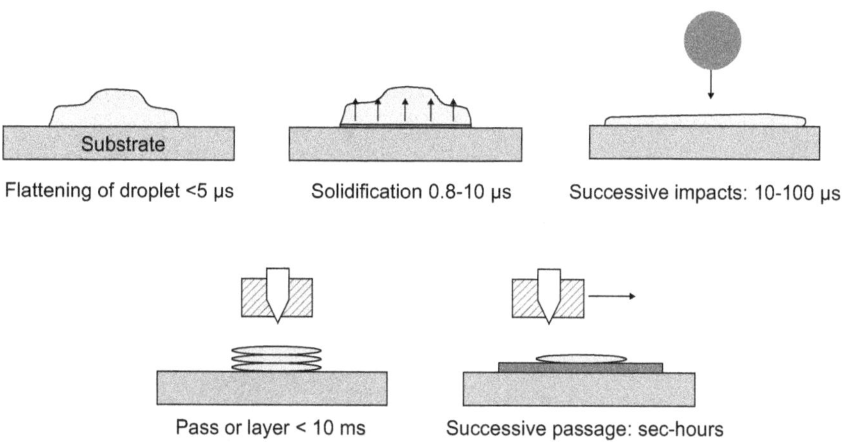


**FIGURE 7.8** Schematic representation of different time ranges involved during various steps of conventional coating process.

*Source*: Adapted and redrawn from [28].

stationary torch and substrate, the layering of the splats appears to be a conical tower shaped due to layering of the Gaussian shape of a successive layer of splats but at an impact angle of 3–4°. Most of the particles impact almost perpendicular to the substrate, whereas some other particles with lower temperature and velocity impact the substrate at an angle range of 70–100°, which is not negligible. Under such conditions, these particles have very few chances to get deposited on the substrate, and thus the center of the already adhered Gaussian layer grows faster than the outer sections, resulting in the tower formation. Formation of tower-shaped layering of YSZ particles sprayed with powder at feed rate of 7 g/min for 86 s and tower height of 21 mm is reported and schematically shown in Figure 7.9a [29]. In another case where the torch is moving whereas the substrate is kept steady, splat layering involves the formation of beads. The height and width of the beads depend on the spray pattern; that is, relative velocity of the torch with respect to substrate $v_r$, mass flow rate of powder $m_p^o$, distribution of particle size in the feedstock, deposition efficiency, injection conditions and spray diameter. Figure 7.9b shows a schematic of the beads formation during splat layering, whereas Figure 7.9c shows the movement of torch to form the pass from the overlapping of beads.

The layering of splat depends on the particle shape and temperature at the time of impact, topology of the deposited layer, accommodation of porosities by the flattening particles and so on. Linear velocity of the spraying torch also affects the splat layering and thickness. Higher torch velocity leads to a lower layer thickness, but at the same time the torch returns to its original position in a short duration, which reduces the contact time with the environment and thus lower oxidation.

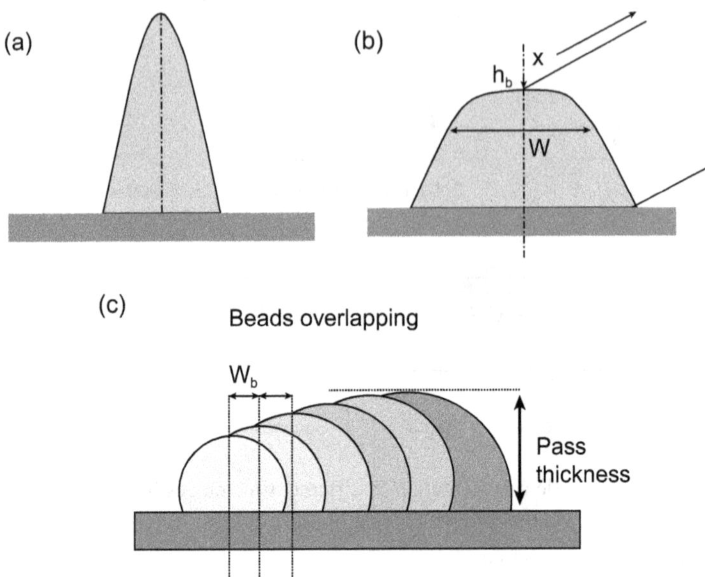


**FIGURE 7.9** Splat layering: (a) tower shaped; (b) bead formation; (c) pass formation due to overlapping of beads.

*Sources*: (a) Adapted and redrawn from [29]; (b) and (c) adapted and redrawn from [21].

## 7.7 ADHESION OF THE SPLATS

Mechanical interlocking, physical interactions, metallurgical interactions involving diffusion and so on are some mechanisms which define the deposition or adherence of the splats to the substate or already deposited splat. Sand blasting or grit blasting on the substrate prior to the spraying introduces roughness on the substrate and thus promotes the mechanical interlocking of the splats. During solidification of the splat, shrinkage of the liquid exerts force which results in the adhesion of splats to the substrate. Epitaxial adhesion of the splats occurs only when the substrate and the splats have the same crystal structure. Physical interaction between the atoms of the splats and substrate occurs because of van der Waals forces which typically requires the surfaces to be clean and activated by the plastic deformation. Metallurgical interactions involve the diffusion of splat atoms or substrate atoms into each other, which may also result in the formation of some new phases that leads to the adhesion of splats on the substrate. Diffusion occurs due to concentration difference and vacancies created by the solidification of high-temperature lamellae. Chemical bonding and formation of new compounds at the interface may also result in good metallurgical bonding and thus proper adhesion of the splats on the substrate, as well as on the previously deposited splats.

## 7.8 COATING MICROSTRUCTURE AND DENSIFICATION

Successive deposition of various layers of splats along with the presence of micro-level and nanolevel asperities result in final coating. A typical microstructure of the coating formed involves the presence of solidified lamellae, unmelted particles, oxides, porosities and various micro-scale defects as shown in Figure 7.10.

During coating formation, the pores are formed due to shadowing, interlamellar spacing between the splats due to unmolten or partially solidified particles. Depending on the size of the pores, they deteriorate the coating properties, therefore this must be avoided. Micro-cracks in the coatings are developed due to quenching stresses, at the interface of the splats due to coefficient of thermal expansion mismatch and the impact angle.

The final microstructure of coating and resulting performance also depends on the following spraying parameters:

1. The final microstructure of the coating is dependent on the contact area between the solidifying splat and the substrate (or deposited splat), as shown in Figure 7.11. The presence of flat interlamellar pores adds up in the porosity of the coating, and thus affects properties like elastic modulus, thermal conductivity and so on. With an increase in the particle impact velocity, the contact area between the substrate and splat increases, leading to the smaller pore size with better thermal conductivity of coatings.
2. Splashing of the impacting particles may also affect the final properties of the coatings. Splashing of the particles results in oxidation and thus poor adhesion of the newly deposited splats.

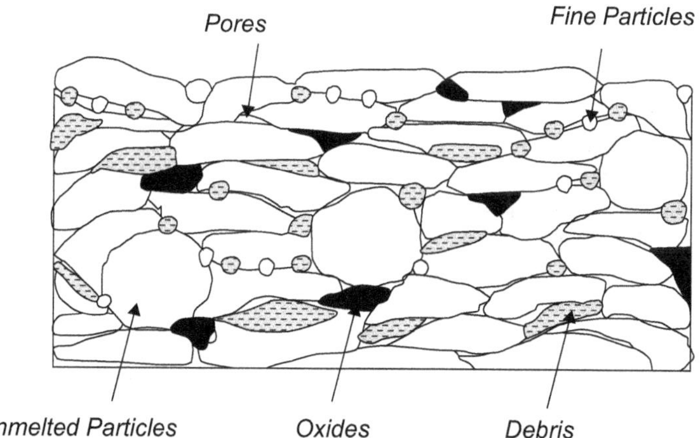


**FIGURE 7.10** Schematic representation of the typical coating microstructure obtained after thermal spraying, which includes unmelted particles, oxides, debris, fine particles and porosities.

*Source*: Adapted and redrawn from [30].

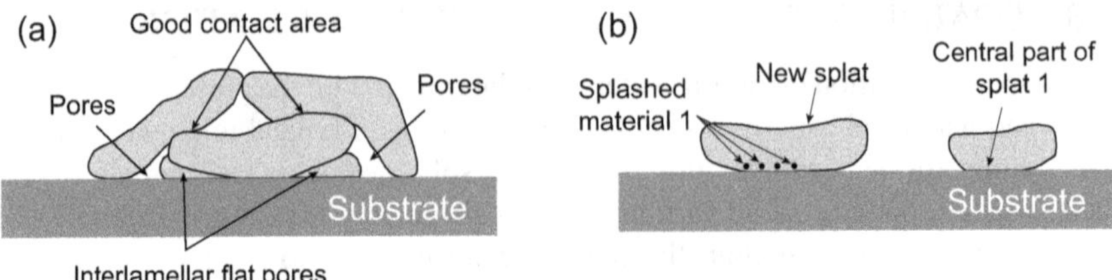


**FIGURE 7.11** Schematic representations of (a) contact between the different layers of splats or between the first splat and the substrate showing presence of interlamellar flat pores and (b) splat deposition on the periphery of already deposited splat, with splashing resulting in poor contact with substrate.

*Source*: Adapted and redrawn from [31].

### 7.8.1 Temperature and Heat during Coating Formation

During thermal spraying, the coating formation receives different heat fluxes from the hot gases and the particles. These heat fluxes are schematically shown in Figure 7.12. Irrespective of the spraying process, the torch and substrate movement are involved during spraying and thus the heat transfer is transient.

The temperature of the substrate and deposited coating is dependent on the type of spraying method and its parameters like spray distance, torch-substrate relative velocity and so forth. Thermal properties of the impacting particles and substrate, which define the heat dissipation, also affect the final temperature of the coatings. Furthermore, different cooling systems used during thermal spraying affects the cooling of the substrate and thus the final temperature of coating and substrate.

The final cooling of the coatings is dependent on the number of splats that form within the location of torch movement. Heat transfer and overall cooling of the splat is dependent on the thermal properties of the substrate and splat, and the thickness of the previously deposited splat. Based on this, a model has been developed [32] which predicts the splat cooling by assuming that the solidification of the splats occurs at the melting point, presence of thermal contact resistance at the splat-substrate interface, the time gap ($t_i$) between the two successive impacts and the presence of various heat fluxes due to hot carrier gas. This model predicted that conductive heat transfer from splat to substrate or previously deposited layer contributes around 95% of the splat cooling, whereas the remaining 5% of cooling is contributed by convective and radiative heat transfer. Therefore, it can be concluded that the splat cooling occurs mainly via conduction of heat from splat to the substrate or previously deposited splat layer. However, during spraying, some additional heat fluxes are introduced by the hot gases, resulting in the particle heating; in such cases some external cooling jets are used to enhance the effective heat transfer and thus promote cooling of the coating. The extent of this heat flux generated by the hot gases varies with the

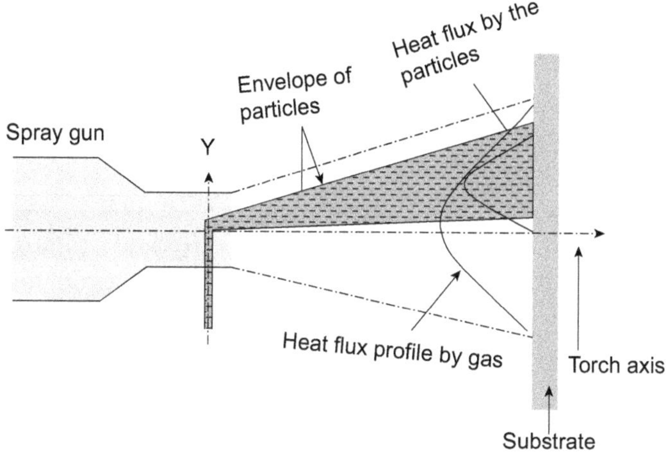


**FIGURE 7.12** Schematic representation of the distribution of heat fluxes obtained from the hot gases and sprayed particles on the substrate and the deposited coating.

*Source*: Adapted and redrawn from [21].

spraying process and spraying parameters like spray distance, velocity, spaying pattern and so on. The heat flux decreases exponentially with the spraying distance.

To take a control on the substrate and coating temperature, various cooling devices like compressed gas jets, $CO_2$/ice/snow, liquid nitrogen and so forth are used. The most-used design of the air jet cooling system is shown in Figure 7.13, which includes two nozzles (R and B) that are attached and move with the spray torch. The R nozzle cools the deposited pass, whereas the B nozzle is used for cooling of the jet of gases and stream of particles to be deposited on the substrate. In case of cooling with air jets, the mode of heat transfer is completely convective, which is strongly dependent on the various parameters including the flow rate of air, diameter of the nozzle, and the distance of nozzle from substrate. However, in the case of an ice/liquid nitrogen/snow cooling system, the latent heat of evaporation contributes towards the cooling of the system.

### 7.8.2 Densification of the Coatings

In general, the coatings obtained are not fully dense therefore some post treatments are given in order to increase the densification, relax the residual stresses generated during the coating formation, improve the coating homogeneity and so on [30]. These post treatments include fusion, heat treatment or annealing, austempering, hot isostatic pressing and spark plasma sintering.

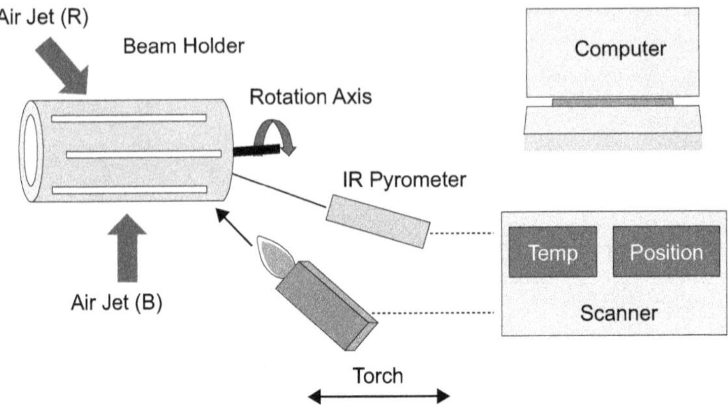


**FIGURE 7.13** Schematic representation of the design of air jet cooling system having two nozzles R and B, used during thermal spraying.

*Source*: Adapted and redrawn from [21].

#### 7.8.2.1 Fusion via Self-Fluxing Alloys

In this process, the coatings are heated at the temperature in between solidus and liquidus, where diffusion of the atoms occurs between the substrate and splat, resulting in increased bond strength and densification of the coatings [30]. This process utilizes the flames or torches to heat the coating up to the fusion temperature. One major limitation of this process is the formation of micro-cracks due to mismatch in the coefficient of thermal expansion between the substrate and coating at the high temperature. To avoid the crack formation, the temperature gradient should be reduced.

#### 7.8.2.2 Heat Treatment or Annealing

During heat treatment or annealing, the coatings are heated to a high temperature but below the melting point of coatings, which results in the changes in the microstructure of coatings. Along with the changes in microstructure, annealing treatment also improves the thermomechanical properties, reduces the residual stresses, improves the bond strength and increases the densification of coatings [30].

#### 7.8.2.3 Hot Isostatic Pressing

Hot isostatic pressing (HIP) involves the simultaneous application of pressure and temperature in the inert atmosphere leading to the shrinkage of pores, increased densification and diffusion between substrate and coatings, which improves the residual stresses and mechanical properties of the coatings. The main industrial application of HIP is for the near-net shaping of the thermally sprayed products with better mechanical properties.

#### 7.8.2.4 Spark Plasma Sintering

Spark plasma sintering involves the simultaneous application of heat and pressure to increase the densification of porous materials. Herein a spark is generated between the particles which causes the plasma or joule heating of the material. These joule heating results in the rapid increase in the temperature and thus faster sintering and increased densification.

There are some other methods like peening or rolling densification, laser glazing and sealing which are used to remove the porosity from the coating microstructure, increase the final densification and thus improve the bond strength, fracture toughness and cohesion of the coatings. Some treatments also helped in the improvement of thermal properties as well.

## 7.9 SOLUTION PRECURSOR PLASMA SPRAYING

Nanostructured coatings are expected to show better mechanical and thermal properties as compared to the microstructured one due to the increased number of interfaces at the nanolevel. However, deposition of nanostructured coatings is quite difficult, as nanoparticles may clog the nozzle. Therefore, a new method has been introduced for producing the coatings of nanoparticles. Solution precursor plasma spraying is used to prepare the highly dense nanostructured coatings with low young's modulus. There are two routes through which nanostructured coatings can be engineers using plasma spraying:

1. Spraying of the nanosized or sub-micron-sized particles on the substrate via suspension. After injection of the suspension, the droplet gets vaporized due to plasma flow and then the heating of the solid particles occurs, followed by their deposition on the substrate. The splats produced by this method have a diameter in the range of 0.1–2 μm and an average thickness in the range of 20–300 nm [33, 34].
2. Another technique involves the spraying of solution of final materials precursor. As the suspension is injected, the liquid gets fragmented and evaporated due to the plasma jet, followed by the precipitation or gelation, pyrolysis or melting of particles resulting in the final impact of molten droplets with diameter in the range of 0.1 to a few micrometers [35, 36].

## 7.10 SUMMARY

In summary, irrespective of the type of thermal spraying process, coating formation involves the basic step of splat formation which includes particle/droplet impact at the substrate followed by its flattening, fragmentation and finally solidification. Droplet formation from the molten metal stream is defined by the Rayleigh instabilities which decide the growth or decay of the droplet based on radius of curvature of wave and stream. Depending on the particle velocity and temperature, splat layering occurs, having a different contact area between the splat and substrate or the previously deposited splat layer. Contact area between the splat and substrate is also dependent on the preheating of the substrate which alters the adsorbates, condensates and other oxides present on the surface of substrate. The final microstructure of the coating involves the presence of melted and solidified splats, unmelted particles, porosities, cracks, oxides and various other asperities and defects. Final properties of the coatings depend on real contact surface resulting in porosities, different defects generated during coating formation like cracks due to quenching stresses and residual stresses, and surface roughness deciding the adhesion of the coating. Other than these, the coating properties also depends on the spray pattern, temperature distribution or heat dissipation, and the average time profile used during spraying. Irrespective of the spraying process, final coating microstructure is never fully dense and consists of various defects like cracks, porosities, oxide particles, residual, and thermal stresses. Therefore some post heat treatments like fusion heating, annealing, hot isostatic pressing and so forth are done on the coating resulting in increased densification, bond strength and thermo-mechanical properties of the coatings owing to diffusion of atoms along with pore shrinkage. Nanostructured coatings with better properties than the microstructured one cannot be produced by conventional spraying technique, and therefore the solution precursor plasma spraying method is used to prepare the nanostructured coatings.

**Questions for Self-Analysis**

1. What factors affect the formation of single splat?
2. How do the perturbations present in molten stream affect the droplet formation?
3. List the substrate parameters which affect the flattening of droplet.
4. How does the substrate preheating affect the spreading of droplets?
5. What are the Reynolds number, Weber number, and Biot number? Mention their significance in thermal spraying.
6. How does the solidification parameter affect spreading of droplet?
7. Which mechanisms define the adhesion of splat on the substrate (or previously deposited splat)?
8. List the factors which affects the final cooling of coatings.
9. Why do the nanostructured coatings not deposit by the conventional method? What alternative method is used for deposition of nanostructured coatings?

## REFERENCES

[1] R. Ghafouri-Azar, J. Mostaghimi, S. Chandra, M. Charmchi, A stochastic model to simulate the formation of a thermal spray coating, *J. Therm. Spray Technol.* 12 (2003) 53–69. https://doi.org/10.1361/105996303770348500.

[2] A.P. Alkimov, *Method of Applying Coatings, Russ. Pat. No. 1618778* (1990). https://ci.nii.ac.jp/naid/10014710108 (accessed July 11, 2021).

[3] A.P. Alkimov, A.N. Papyrin, Ulitsa Vyazemskogo, V.F. Kosarev, N.I. Nesterovich, M.M. Shushpanov, *Gas-dynamic Spraying Method for Applying a Coating, U.S. Patent No. 5,302,414* (12 April, 1994).

[4] F. Gärtner, T. Stoltenhoff, T. Schmidt, H. Kreye, The cold spray process and its potential for industrial applications, *J. Therm. Spray Technol.* 15 (2006) 223–232. https://doi.org/10.1361/105996306X108110.

[5] S. Sampath, H. Herman, Rapid solidification and microstructure development during plasma spray deposition, *J. Therm. Spray Technol.* 5 (1996) 445–456. https://doi.org/10.1007/BF02645275.

[6] T.C. Hanson, C.M. Hackett, G.S. Settles, Independent control of HVOF particle velocity and temperature, *J. Therm. Spray Technol.* 11 (2002) 75–85. https://doi.org/10.1361/105996302770349005.

[7] M. Grujicic, Particle/substrate interaction in the cold-spray bonding process, *Cold Spray Mater. Depos. Process Fundam. Appl.* (2007) 148–177. https://doi.org/10.1533/9781845693787.2.148.

[8] M. Grujicic, J.R. Saylor, D.E. Beasley, W.S. DeRosset, D. Helfritch, Computational analysis of the interfacial bonding between feed-powder particles and the substrate in the cold-gas dynamic-spray process, *Appl. Surf. Sci.* 219 (2003) 211–227. https://doi.org/10.1016/S0169-4332(03)00643-3.

[9] M. Gedvilas, G. Račiukaitis, V. Kučikas, K. Regelskis, Driving forces for self-organization in thin metal films during their partial ablation with a cylindrically focused laser beam, *AIP Conf. Proc.* 1464 (2012) 229. https://doi.org/10.1063/1.4739877.

[10] P.-G. de Gennes, F. Brochard-Wyart, D. Quéré, *Capillarity and Wetting Phenomena*, Springer: New York, 2004. https://doi.org/10.1007/978-0-387-21656-0.

[11] M. Grujicic, Particle/substrate interaction in the cold-spray bonding process, in: *The Cold Spray Materials Deposition Process: Fundamentals and Applications*, Woodhead Publishing: Sawston, 2007: pp. 148–177. https://doi.org/10.1533/9781845693787.2.148.

[12] S.Q. Armster, J.-P. Delplanque, M. Rein, E.J. Lavernia, Thermo-fluid mechanisms controlling droplet based materials processes, *Int. Mater. Rev.* 47 (2013) 265–301. https://doi.org/10.1179/174328013X13789822110946.

[13] C. Escure, M. Vardelle, P. Fauchais, Experimental and theoretical study of the impact of alumina droplets on cold and hot substrates, *Plasma Chem. Plasma Process.* 23 (2003) 185–221. https://doi.org/10.1023/A:1022976914185.

[14] J. Madejski, Solidification of droplets on a cold surface, *Int. J. Heat Mass Transf.* 19 (1976) 1009–1013. https://doi.org/10.1016/0017-9310(76)90183-6.

[15] T. Yoshida, T. Okada, H. Hamatani, H. Kumaoka, Integrated fabrication process for solid oxide fuel cells using novel plasma spraying, *Plasma Sources Sci. Technol.* 1 (1992) 195. https://doi.org/10.1088/0963-0252/1/3/009.

[16] H.L. Liao, P. Gougeon, G. Montavon, C. Coddet, Effect of Reynolds number of molten spray particles on splat formation in plasma spraying—ASM international, in: *Therm. Spray 2003 Adv. Sci. Appl. Tech.*, 2003. www.asminternational.org/home/-/journal_content/56/10192/CP2003ITSC0875/CONFERENCE-PAPER (accessed January 27, 2022).

[17] P. Fauchais, M. Vardelle, A. Vardelle, L. Bianchi, A.C. Léger, Parameters controlling the generation and properties of plasma sprayed zirconia coatings, *Plasma Chem. Plasma Process.* 16 (1995) S99–S125. https://doi.org/10.1007/BF01512630.

[18] H. Li, S. Costil, H.L. Liao, C.J. Li, M. Planche, C. Coddet, Effects of surface conditions on the flattening behavior of plasma sprayed Cu splats, *Surf. Coatings Technol.* 200 (2006) 5435–5446. https://doi.org/10.1016/J.SURFCOAT.2005.07.058.

[19] A.T.T. Tran, M.M. Hyland, T. Qiu, B. Withy, B.J. James, Effects of surface chemistry on splat formation during plasma spraying, *J. Therm Spray Technol.* 17 (2008) 637–645. https://doi.org/10.1007/s11666-008-9237-6.

[20] B. Withy, M. Hyland, B. James, Pretreatment effects on the surface chemistry and morphology of aluminium, *Int. J. Mod. Phys. B.* 20 (2012) 3611–3616. https://doi.org/10.1142/S0217979206040076.

[21] P.L. Fauchais, J.V.R. Heberlein, M.I. Boulos, *Thermal Spray Fundamentals*, Springer Science+Business Media: New York, 2014. https://doi.org/10.1007/978-0-387-68991-3.

[22] P. Fauchais, A. Vardelle, B. Dussoubs, Quo Vadis thermal spraying?, *J. Therm. Spray Technol.* 10 (2001) 44–66. https://doi.org/10.1361/105996301770349510.

[23] R. Dhiman, A.G. McDonald, S. Chandra, Predicting splat morphology in a thermal spray process, *Surf. Coatings Technol.* 201 (2007) 7789–7801. https://doi.org/10.1016/J.SURFCOAT.2007.03.010.

[24] A. McDonald, C. Moreau, S. Chandra, Thermal contact resistance between plasma-sprayed particles and flat surfaces, *Int. J. Heat Mass Transf.* 50 (2007) 1737–1749. https://doi.org/10.1016/J.IJHEATMASSTRANSFER.2006.10.022.

[25] A. McDonald, M. Lamontagne, C. Moreau, S. Chandra, Impact of plasma-sprayed metal particles on hot and cold glass surfaces, *Thin Solid Films.* 514 (2006) 212–222. https://doi.org/10.1016/J.TSF.2006.03.010.

[26] T.S. Sidhu, S. Prakash, R.D. Agrawal, Studies on the properties of high-velocity oxy-fuel thermal spray coatings for higher temperature applications, *Mater. Sci.* 41 (2005) 805–823. https://doi.org/10.1007/S11003-006-0047-Z.

[27] A.A. Syed, A. Denoirjean, B. Hannoyer, P. Fauchais, P. Denoirjean, A.A. Khan, J.C. Labbe, Influence of substrate surface conditions on the plasma sprayed ceramic and metallic particles flattening, *Surf. Coatings Technol.* 200 (2005) 2317–2331. https://doi.org/10.1016/J.SURFCOAT.2005.01.014.

[28] P. Fauchais, Understanding plasma spraying, *J. Phys. D. Appl. Phys.* 37 (2004) R86. https://doi.org/10.1088/0022-3727/37/9/R02.

[29] K. Shinoda, H. Murakami, S. Kuroda, K. Takehara, S. Oki, In situ visualization of impacting phenomena of plasma-sprayed zirconia: From single splat to coating formation, *J. Therm. Spray Technol.* 17 (2008) 623–630. https://doi.org/10.1007/S11666-008-9221-1.

[30] J.R. Davis, *Handbook of Thermal Spray Technology*, ASM International: Netherlands, 2004.

[31] R.W. Trice, K.T. Faber, Role of lamellae morphology on the microstructural development and mechanical properties of small-particle plasma-sprayed alumina, *J. Am. Ceram. Soc.* 83 (2000) 889–896. https://doi.org/10.1111/J.1151-2916.2000.TB01290.X.

[32] H. Koivuluoto, J. Näkki, P. Vuoristo, Corrosion properties of cold-sprayed tantalum coatings, *J. Therm. Spray Technol.* 18 (2008) 75–82. https://doi.org/10.1007/S11666-008-9281-2.

[33] J. Fazilleau, C. Delbos, V. Rat, J.F. Coudert, P. Fauchais, B. Pateyron, Phenomena involved in suspension plasma spraying part 1: Suspension injection and behavior, *Plasma Chem. Plasma Process.* 26 (2006) 371–391. https://doi.org/10.1007/S11090-006-9019-1.

[34] P. Fauchais, V. Rat, C. Delbos, J.F. Coudert, T. Chartier, L. Bianchi, Understanding of suspension DC plasma spraying of finely structured coatings for SOFC, *IEEE Trans. Plasma Sci.* 33 (2005) 920–930. https://doi.org/10.1109/TPS.2005.845094.

[35] E.H. Jordan, L. Xie, X. Ma, M. Gell, N.P. Padture, B. Cetegen, A. Ozturk, J. Roth, T.D. Xiao, P.E.C. Bryant, Superior thermal barrier coatings using solution precursor plasma spray, *J. Therm. Spray Technol.* 13 (2004) 57–65. https://doi.org/10.1007/s11666-004-0050-6.

[36] T. Bhatia, A. Ozturk, L. Xie, E.H. Jordan, B.M. Cetegen, M. Gell, X. Ma, N.P. Padture, Mechanisms of ceramic coating deposition in solution-precursor plasma spray, *J. Mater. Res.* 17 (2002) 2363–2372. https://doi.org/10.1557/JMR.2002.0346.

# 8 Testing of Coatings

*Rubia Hassan, Ashutosh Tiwari and Kantesh Balani*

## CONTENTS



DOI: 10.1201/9781003321965-8

Testing of coatings is performed to ensure that all the requirements are met for any particular application. Through testing, properties of the coatings are evaluated which have a direct impact on their performance. The materials obtained from thermal spray processes differ from the counterparts like forged, wrought, sintered or cast form and the usefulness of these coatings can be assessed after understanding, building and executing appropriate testing and characterization techniques which largely have been borrowed from other disciplines of materials science. Testing of coatings encompasses a wide range of tests including determining mechanical properties, evaluating wear and corrosion resistance, finding coating bond strength, and investigating high-temperature properties. These involve both destructive and non-destructive techniques.

Test methods usually impart some kind of signal or stimulus to the material, having the capability to measure a material's response. The applied stimulus provides a response which should be clear to get measured so that the particular coating property can be characterized. Practically, the output of every test is a number. These generated numbers, directly or indirectly, represent the studied property. It requires some degree of confidence, for these numbers, tests and the materials being characterized, to be useful. A significant number of tests are done to provide a statistical expression of a useful measure of confidence is expressed. A high degree of confidence should exhibit a little standard deviation from the mean.

Initial important characteristics related to the coating microstructure include chemical composition, grain morphology and orientation, defect such as porosity, un-melted particles, cracks and their distribution. Metallographic examination is an important and critical process for these characterizations, discussed in section 8.2 [1, 2]:

## 8.1 STATISTICAL METHODS

When determining coating properties, it is necessary to perform statistical analysis due to complexity of coatings in terms of their structures, anisotropy, and strong dependence on powder characteristics and spray conditions, in addition to a wide scattering of different property values. Statistically, three types of distributions are commonly utilized for fitting the experimental data: normal, log-normal and Weibull distribution. Among these, typically, Weibull and normal distributions are used to characterize thermal spray coatings (TSCs).

### 8.1.1 Normal Distribution

Let us consider the determination of microhardness. For $N$ number of indents with $xi$ being the hardness value obtained from an $i$th indent, $N$-2 values are averaged (discarding the smallest and the largest values). If we define $\eta$ as the arithmetic mean of experimental dataset, and standard deviation $\sigma$ is obtained by taking the square root of the average of $(x - \eta)^2$; see equation (8.1). For $x$ taking the random values of $x_1, x_2, x_3 \ldots, x_N$ from the measured values, each possessing the same probability, $\sigma$ can be obtained from equation (8.1).

$$\sigma = \sqrt{\frac{1}{N}\sum_{i=1}^{N}(x_i - \eta)^2} \tag{8.1}$$

According to the central limit theorem, the probability density function in a distribution of independent random variables, if they are identically distributed is proportional to $exp\left(-\frac{(x-\eta)^2}{2\sigma^2}\right)$, which basically is a Gaussian distribution, and its standard deviation can be defined in terms of a scaling variable reflecting the broadness of the curve.

For approximate normal distribution, 68% of data points lie within ± single standard deviation of the mean (i.e., $\eta \pm \sigma$), ~95% of the data lies within ± two standard deviations (i.e., $\eta \pm 2\sigma$), and 99.7% are within ± three standard deviations of the mean values (i.e., $\eta \pm 3\sigma$).

During statistical distribution calculations, an important issue is the data points to consider for a reliable distribution. In this regard, calculation of variance, which is the mean value of $(x - \eta)^2$ if probability of all the events is same and mathematically represented as $(x - \eta)^2/N$, becomes useful.

Like for an asymmetric data set, Weibull distribution becomes more appropriate, non-homogenous materials like TSCs may not fulfill the criterion for a normal distribution.

### 8.1.2 Weibull Distribution

The Weibull distribution function can mathematically be represented as:

$$F(x,\lambda,m,\delta) = \frac{m}{\lambda}\left(\frac{(x-\delta)}{\lambda}\right)^{m-1} exp\left(-\left(\frac{(x-\delta)}{\lambda}\right)^{m}\right) \tag{8.2}$$

where $x$ is the variable; $m$ is the shape factor (>0), also known as Weibull modulus; $\lambda$ is the scaling parameter (>0); and $\delta$ is the offset.

Plotting $\left[ln\left(ln\left(\frac{1}{1-F(x-\delta)}\right)\right)\right]$ versus $\left[ln(x-\delta)\right]$, these parameters can be determined from the slope and intercept of the best-fit straight line, using a linear regression least square fit. By setting $\delta$ to zero, the Weibull distribution can be simplified. The shape of the distribution is dictated by the value of $m$ and depending on its value, Weibull distribution can be approximated to other distributions; for example, exponential distribution for $m = 1$, Rayleigh distribution for $m = 2$, log-normal distribution for $m = 2.5$, normal distribution for $m = 3.6$ and peaked-normal distribution for $m = 5$.

## 8.2 METALLOGRAPHIC EXAMINATION

To examine the microstructure of thermally sprayed coatings requires metallographic preparation, which should reveal the true, original and undisturbed microstructure. There are various parameters which are involved in the metallographic preparation of coatings. It involves many steps, as discussed in the next sub-sections.

### 8.2.1 Sampling and Sectioning

Specimens selected must be representative of the coating. To display the entire coating thickness, the substrate and all the interfaces, polishing should be done normal to the coating surface. The height of the mounted specimen is kept as what is convenient for handling during polishing, which necessitates the reduction of the size of specimen before encapsulating it in some form of mounting material. This is accomplished by sectioning. TSC specimens need careful sectioning to avoid any effect on the soundness of the coating, and the interface between the coating and the substrate. The standard method of sectioning is the abrasive cutting using rotating wheels or discs. The wheel/disc or specimen is translated into the specimen to form the cut. Proper care should be observed for minimum damage in the coating. Vacuum impregnation of brittle coatings with epoxy mounting compound is recommended, to protect the specimen, before sectioning. Coatings can withstand higher compressive stress than tensile stress. Thus, sectioning should be advanced from the outermost layer through the substrate. This produces compressive stresses rather than tensile stresses within the coating. When sectioning is done in a reverse manner, it may result in delamination or pulling away of the coating from the substrate. This type of damage may be interpreted wrongly while examining the polished specimen. However, if it is unavoidable to place some areas of the TSC in tension, then those areas should be properly marked and excluded during the process of evaluation of the specimen. Quality of the prepared surface is influenced by type of cutting wheel, nature of coolant, cutting conditions, and the type and hardness of the coating and substrate; for example, improper cutting conditions can cause overheating of TSC specimens rendering them unsuitable for proper evaluation.

The abrasive wheels are either silicon carbide or aluminum oxide bonded using resin or rubber. For dry cutting, resin bonded wheels are utilized, whereas rubber bonded wheels are used in wet cutting. In diamond saws, synthetic diamonds are used as an abrasive over a thin metallic disc. Pre-sectioning encapsulation in a protective epoxy reduces further damage to the specimen.

### 8.2.2 Cleaning

Before mounting, it is essential to clean the specimen. While cleaning, make sure that all coolant used during sectioning is removed from the surface and also from any open porosity. Any liquid residual may impede impregnation of porosity and slow down the curing of mounting compounds, which in turn makes grinding and polishing difficult. To accelerate the process of fluid removal, use of an organic solvent and thorough drying is necessary. Drying at low temperature (60 to 80 °C) in an oven can aid this process. It is not advisable to ultrasonicate the TSC specimens for cleaning, especially for fragile or brittle coatings, because during such energetic cleaning process, some coating particles may be lost. In case it is necessary to go for ultrasonic cleaning, minimizing the cleaning time should be considered.

### 8.2.3 Mounting

In mounting, specimen is encapsulated in a cylindrical polymeric mount. Mounting enables the gripping of specimen with the hand or fitting it inside a holder in automatic grinding and polishing machines. The mount must have the capability of being held, should hold the specimen firmly and should not respond significantly differently than the specimen during grinding and polishing. While mounting the specimen, neither the mounting material nor the mounting process should cause any physical damage to the specimen nor should cause any overheating during the process. To retain the original structure of TSC specimens during grinding and polishing, it is necessary to mount them. Mounting processes are basically of two types, discussed below.

*Hot mounting*: In hot mounting, mounting material of thermosetting powder or a preform is placed over the specimen. Under the application of temperature and pressure, the material is softened and is held in such a condition for 5–10 minutes. During this time, the polymeric chains grow and crosslink. The assembly is cooled and finished mount ejected.

*Cold mounting*: Here no heat or pressure is involved. The specimen to be mounted is placed into a cup of plastic or rubber with detachable bottom. Polymeric resin along with a hardener is mixed and poured around the specimen. Depending upon the resin and hardener, curing may take place in a few minutes to several hours. Sometimes a filler material like alumina is added alongside the resin and hardener for uniform solidification and provide more abrasion-resistant surface, which helps in edge retention.

### 8.2.4 Grinding and Polishing

The mount prepared is taken through various grinding and polishing steps to produce a suitable surface for microstructural observation.

In most of the equipment for automatic and semi-automatic grinding and polishing, specimen is moved around a rotating wheel covered with an abrasive, following an epicycloidal path. The scratch pattern is formed on the surface consisting of random arcs. These scratches should be uniform throughout the entire specimen surface before moving to the next step in the preparation sequence. Planar grinding rapidly removes large amounts of material from the specimen surface, which brings the group of specimens in the holder into coplanarity. This step prevents the occurrence of a situation where some specimens in the holder remain uncontacted with the abrasive while excessive pressure is applied on specimens in contact. Planar grinding also should remove the deformed material on the surface resulting from the previous sectioning operation. The disturbed material introduced during planar grinding is removed by fine grinding while creating a layer of disturbed material acceptable for the next step in the preparation sequence. Fine grinding is followed by polishing. Rough polishing is accomplished with a 3- to 6-μm diamond abrasive. Evaluating the surface of the specimen after rough polishing provides an idea whether the surface is acceptable, in which case the final polishing step is not needed. Otherwise, final polishing is performed using colloidal silica or aluminum oxide abrasive. The microstructure revealed should be true, undisturbed and free from any abrasive or polishing artifact.

## 8.3 NON-DESTRUCTIVE TESTS

Non-destructive testing (NDT) or non-destructive examination is performed without damaging the appearance or functionality of the finished part. Due to the complex structure of TSCs, interpretation through signals of NDTs becomes difficult. As a result, just a few techniques of NDTs have been implemented in the coating field. These techniques include:

*Visual inspection*: Through visual inspection, different surface flaws caused due to delamination, corrosion or contamination, and discontinuities are detected and examined. Image sensors are employed for visual records. For improved observation of cracks/defects, magnifying lenses can be used. External agents like dyes, fluorescent penetrants, and magnetic particles are also applied for easy detection of the flaws, based on which different techniques are developed.

*Laser inspection*: These methods use laser for surface inspection and dimensional measurements of coatings. Surface inspection involves detection of surface flaws and roughness. Dimensional measurements are carried out using a scanning laser gauge. Dimensional measurements can also be performed using laser triangulation sensors providing measurements of deviations based on surface alterations. Two sensors can be used for thickness measurement or measuring diameter of bores.

*Coordinate measuring machines*: These are used for 3D inspection, providing accurate and flexible measurements during and before the coating process, and for finished parts like coating after machining or rectification.

*Acoustic emission*: In this method, a sensitive piezoelectric transducer placed judicially is used to detect the stress wave radiated into the material, produced by the rapid release of energy because of crack formation, crack growth and/or deformation from any localized source within the coating. Generally, it is hard to separate the acoustic emission signals associated with elastic and plastic deformation. It is a difficult task to directly correlate the nature of cracking with the level of acoustic emission energy. Nonetheless, the technique is successfully implemented to check cracking in coatings, say during mechanical/wear test (e.g., bend, scratching).

*Laser ultrasonic technique*: This technique uses lasers for the generation and the detection of ultrasound. To gather information from the surface, waves are made to propagate through the surface of a material. The concentration of wave motion at the surface makes it useful for characterizing coatings and thin films. In this technique, waves are produced and detected at lengths beyond the coating dimensions, making the ratio of coating thickness to wavelength an important parameter for depth of penetration. The laser and its wave detection are performed at different places but on the same surface. Though many wave modes can be generated simultaneously by the laser pulse, surface longitudinal and Rayleigh wave velocities are the two types of ultrasonic waves which permit measurement of elastic modulus and Poisson's ratio.

*Thermography*: This technique determines flaws in the coating by finding anomalous hot spots after thermal excitation. Thermographic tests are of two types: active and passive. In the active or dynamic method, a controllable thermal source which reduces the effect of environment such as emissivity variations and ambient conditions is used to excite the material. It can determine the sub-surface spalling event by observing changes in the coating temperature during application of a transient thermal pulse. On the other hand, passive thermography directly estimates the alteration in surface temperature to identify the interested areas, which will have an abnormal hot spot with respect to the surroundings.

*Coating thickness*: Though the term "coating thickness" sounds simple, its measurement may not necessarily be straightforward. The method used to detect coating thickness may also be questioned. If the substrate shape and size permit, a simple caliper or micrometer can be used. For more complex geometries, the eddy current method, magnetic induction method, ultrasonic, capacitance, X-ray fluorescence and optical coating thickness measuring devices can be utilized.

## 8.4 DETERMINATION OF POROSITY

Major porosity in TSCs results from the incoming molten powder particles not being able to seal the protuberances of the surface formed by the formerly deposited splats. These irregularities could be un-melted particles, fragmented particles, oxide inclusions or surface topography of the substrate. Insufficient pressure inside the droplet not being able to break its surface tension will prevent the penetration into smaller cavities. Other than the pressure distribution inside the droplets, this phenomenon is controlled by other parameters like substrate topography, as well as the rate of solidification. In addition, due to the lack of fusion between particles, porosity also results due to the gases' expansion, produced during thermal spraying process. Other reasons for porosity include the shadow effect, splat-curling effect and roof effect. It becomes important to determine the role of varying spray parameters and porosity content in TSC in order to monitor its suitability for the intended purpose. The amount of porosity tolerable in a coating depends up on its application. Based on the wide size range of the porosity and the complex nature of the coating morphology, it may become difficult to quantify the porosity in coatings [3, 4].

The porosity testing methods include Test Method A and Test Method B to quantify percentage porosity from a metallographically polished specimen.

### 8.4.1 Test Method A or Direct Comparison

This method compares the images to microscopic fields of view on a prepared specimen, for comparison. A value is assigned to each figure representing varying degrees of porosity.

1. Properly position the prepared specimen under a microscope, diverting the image to a screen or video monitor. A hard copy print can also be recorded.
2. Adjust the magnification resolving all voids while filling the screen with entire coating thickness in the best way. In case it is not possible to cover the entire thickness of the coating without affecting the resolution of voids, it is recommended to prioritize the resolution of voids significantly contributing to the total porosity. The operator during analysis should be able to differentiate the oxides from epoxy infiltrated voids.
3. Next, the image adjusted on the screen needs to be compared with the figures, which should approximately be of the same size.
4. Note the figure value resembling the image. In the case of no close match, a rounded value to the nearest whole number may be considered; for example, if porosity in the present field lies between a figure representing values of 6.0% and 9.0% respectively, a 7.0 or 8.0 may be taken as an appropriate porosity value.
5. For less than 0.5% porosity in a field of view, it shall be reported as <0.5; while calculating the average area percentage of porosity, such fields should be considered zero.
6. If a field reveals more porosity than depicted in figure with the highest value, that field shall be as recorded outside range (OR). An estimated numerical value of the area percentage porosity shall be assigned to it; for example, a field estimated to possess 30.0% porosity shall be recorded as OR-30.

7. Under the same magnification, the aforementioned procedure shall be repeated and the value for at least ten random fields needs to be recorded. Make sure not to overlap or re-measure the fields.
8. A minimum of 10 prints corresponding to distinct fields of view are required for comparison using photo-micrographs. Again, no overlapping or re-photograph fields of view needs to be assured.

### 8.4.2 Test Method B or Image Analysis

1. Properly place the prepared specimen under the microscope and project the image onto the viewing screen.
2. Adjust a proper magnification allowing proper resolution of the voids while trying to fit the entre thickness on the screen.
3. After the determination of the best magnification, adjust the microscope aperture of the microscope and field diaphragms are adjusted for the best resolution and contrast.
4. Next step involves thresholding or image segmentation of porosity in the image. In segmentation, a binary image is created by appropriately choosing the gray values without detecting any features other than pores like oxides close to the threshold limits of porosity.
5. There are ways to exclude unwanted features which may be adopted for better segmentation; for example, opening involves removal of minute and thin objects without affecting the dimensions of large objects in the image.
6. Area percentage porosity should not be significantly altered while applying any binary image processing functions.
7. Microscopy techniques like darkfield, polarized light or fluorescence can alternately be used to improve the thresholding of porosity filled with a dye or treated epoxy.
8. Developing thresholding and image processing routine should follow with the confirmation that detection of porosity is precise under several fields of view.
9. Analyze a minimum of 20 separate fields of view while being careful not to overlap any previous field.
10. Make sure not to eliminate any coating features at the border of an image by any technique.
11. For comparing different specimens, use same instrument settings and objective lens.

Image analysis can determine various aspects of the porosity; for example, open and closed pores, distribution of pore size and spatial distribution of pores can be determined. Its simplicity and convenience have made it the most common technique for porosity determination in TSCs. Both optical and SEM images can be utilized for image analysis, but higher magnifying power and depth of field of SEM (other than that of the optical microscope) can help see fine cracks, interlamellar features, globular pores and pits.

### 8.4.3 Statistical Analysis

Since many specimens vary considerably in porosity in different fields of view, making the process of porosity determination not an exact measurement. A major uncertainty occurs with this variation. Hence, porosity measurement is incomplete if its precision within normal confidence is not calculated. It is assumed here that the normal confidence represents the expectation that 95% of the time, actual error lies within the specified uncertainty. Since Test Method A results are strictly based on direct comparison, these are excluded from statistical determinations beyond certain values including mean, maximum and minimum porosity.

For results generated via Test Method B, statistical determinations are presented below:

1. After the measurement of a desired number of fields, mean area percentage porosity is calculated according to:

$$\bar{x} = \sum \frac{x_i}{n} \tag{8.3}$$

where $\bar{x}$ is the mean, $x_i$ is the $i$th value and $n$ is the total number of measurements.

2. Standard deviation (s) is calculated according to the equation:

$$\mathrm{s} = \left( \frac{\Sigma \left( x_i - \bar{x} \right)^2}{n-1} \right)^{\frac{1}{2}} \tag{8.4}$$

3. The 95% confidence interval of each measurement is calculated as:

$$95\ \%CI = \frac{t.s}{\sqrt{n}} \tag{8.5}$$

The $t$ values (95% confidence interval multipliers) as a function of $n$ are presented in Table 8.1.

4. The relative percentage accuracy is calculated as:

$$\%RA = \frac{95\%CI}{\bar{x}} \times 100 \tag{8.6}$$

5. If % RA is considered as too high for the intended application, measurement of more fields to repeat the calculations of mean, standard deviation and % RA is done. For most of the purposes, the generally accepted precision is 10% RA (or lower).

**TABLE 8.1**
**95% Confidence Interval Multipliers [11]**

| No. of Fields (*n*) | *t* | No. of Fields (*n*) | *t* |
|---|---|---|---|
| 20 | 2.093 | 27 | 2.056 |
| 21 | 2.086 | 28 | 2.052 |
| 22 | 2.080 | 29 | 2.048 |
| 23 | 2.074 | 30 | 2.045 |
| 24 | 2.069 | 40 | 2.020 |
| 25 | 2.064 | 60 | 2.000 |
| 26 | 2.060 | ∞ | 1.960 |

### 8.4.4 X-Ray Computed Tomographic Investigation

Zhang et al. [4] investigated the porosity in a coating prepared by plasma electrolytic oxidation on a titanium substrate. A sub-volume was identified, representing the entire volume of the coating. Two-way validation between the X-ray CT and SEM images was explored to determine the segmentation of the pores. Top-down views of the surface of the sub-volume through X-ray CT and SEM are shown in Figure 8.1a and 8.1b, respectively. To distinguish the coating from the substrate and other background, an overall threshold based on grayscale values was used. The pores were identified using black top-hat transform on the virtual slices since they appeared darker. Smaller regions of volume <6 μm$^3$ were ignored for the purposes of avoiding errors during the quantification. The coating volumes and pores within the coating were calculated from voxel counting of the respective regions. Porosity was calculated as the ratio of the pore volume to the coating volume. To determine the coating thickness, voxels were counted perpendicular to the coating surface.

The X-ray CT results shown in Figure 8.2 have pores that are labeled in grey. The emergence of pores at the coating surface is evident from the image. The pores emerged around the edge of each nodule in a continuous band, as is revealed in several of the nodules, which were almost devoid of any large central pore. Comparing Figure 8.2a and 8.2b, connection can be drawn between the emergent pores and the underlying porosity. After segmentation, 373 well-resolved, isolated pores constituting a porosity volume fraction of 5.7% were revealed.

Through this study, the authors have shown the capability of X-ray CT in resolving and quantifying pores of volumes at least 6 μm$^3$.

## 8.5 DETERMINATION OF PHASES

Phases in TSCs are mostly analyzed by X-ray diffraction. Three primary methods are employed for X-ray diffraction quantitative analysis: (1) direct comparison, (2) internal standard, and (3) external standard. Direct comparison requires

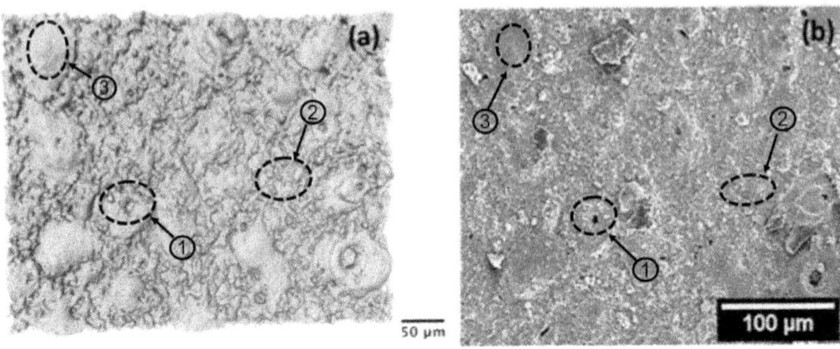


**FIGURE 8.1** (a) X-ray CT data of the selected 3D sub-volume with distinctive regions encircled: (1) nodule with a pore in center; (2) finely porous region; (3) nodule with no central pore. (b) SEM image in SE mode showing same region and same features.

*Source*: Reprinted with permission from [4]. Copyright 2016 American Chemical Society.

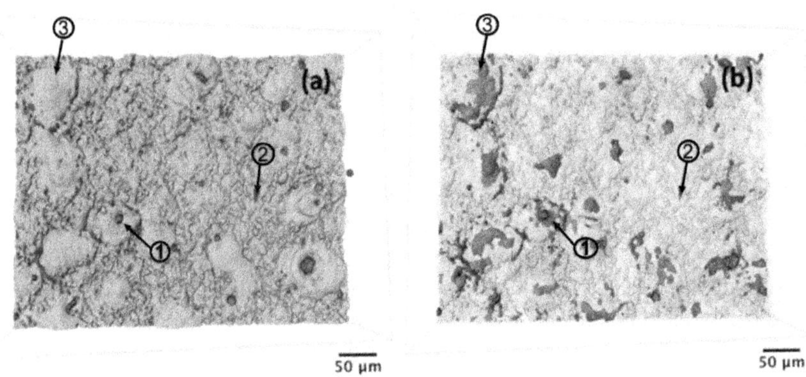


**FIGURE 8.2** X-ray CT images: (a) coating surface with highlighted pores: (1) nodule with a pore in center, (2) finely porous region, (3) nodule with no central pore; (b) pores within the coating.

*Source*: Reprinted with permission from [4]. Copyright 2016 American Chemical Society.

advance knowledge of the sample crystallinity. The internal standard method is only applicable to powdered samples to which a measured amount of a standard can be added. This method can also be applied to partially amorphous samples. The external standard method obtains the diffraction pattern from the coating surface which is then compared to an external standard [5]. The details of each method are presented herewith.

### 8.5.1 Direct Comparison

Quantification of phases using Rietveld analysis is done by calculating a diffraction pattern using an appropriate structural model taking into consideration all the phases identified during X-ray diffraction. The best least square fit between the calculated and the observed pattern is obtained by refining the detail of the model. Key model parameters (which include sample properties and measuring conditions) can be varied at each cycle to improve the match. Since the technique quantifies the crystalline phases in a routine way but can be used to make a relative comparison of the amorphous component present by making use of the index of crystallinity ($I_C$), which is the ratio of the integrated intensity (area under the peak) of Bragg's peaks of crystalline material and the total intensity of the spectrum for 2Θ comprising an amorphous hump [6].

### 8.5.2 Internal Standard

In TSCs (especially in plasma spraying coatings), rapid solidification of molten particles may result in the formation of amorphous component. Due to the quenching of the material during deposition, mobility of the molecules is retarded before it crystallizes, resulting in a non-ordered state. If all the phases in the mixture are known, then compared to any peak-intensity based method, Rietveld provides a higher accuracy. The following normalization equation holds:

$$\sum_i X_i = 1 \tag{8.7}$$

where $X_i$ is the weight fraction of the $i$th component of the mixture. To satisfy the above equation, refined weight fractions are overestimated in presence of an amorphous fraction. To resolve this problem, a combination of the Rietveld and internal standard methods can be employed. A known amount of standard is added and is refined using internal standards that are rescaled to obtain accurate calibrations [7]. The amorphous fraction $X_a$ can then be calculated as:

$$X_a = 1 - \sum_i X_i \tag{8.8}$$

$X_a$ can also be calculated as:

$$X_a = \frac{1}{1 - X_s}\left[1 - \left(\frac{X_s}{X_{sc}}\right)\right] \tag{8.9}$$

where $X_{sc}$ is the refined weight fraction of the internal standard and $X_s$ is the weight fraction of the original internal standard.

Similarly, weight fraction of a crystalline component $i$ can be estimated as:

$$X_i = \frac{1}{1-X_s}\left[\left(\frac{X_s}{X_{sc}}\right)X_{ic}\right] \tag{8.10}$$

where $X_{ic}$ is the refined weight fraction.

### 8.5.3 External Standard

When a powder sample is placed in an unpolarized X-ray beam, the integrated intensity of the (hkl) peak of the $i_{\text{th}}$ phase is proportional to the power per unit length of the diffraction circle [5].

$$I_i^{hkl} \propto \left[\frac{I_O}{16\pi R}\left(\frac{e^4}{m_e^2 c^4}\right)\frac{\lambda^2 M_{hkl} F_T^2}{v_a^2}(LP)\right]V_i \tag{8.11}$$

where $I_o$ is the intensity of the X-ray beam, $R$ is the radius of the goniometer, $e$ is the electron charge, $m_e$ is the mass of electron, $\lambda$ is the wavelength of radiation use, $M_{hkl}$ is the multiplicity factor, $F_T$ is the structure factor, $V_a$ is the unit cell volume, $LP$ is the Lorentz polarization factor and $V_i$ is the effective diffraction volume.

The expression for the integrated diffraction intensity for a given reflection from a given phase $I$ becomes:

$$I_i^{hkl} = K_i^{hkl}.V_i \tag{8.12}$$

where $K_i^{hkl}$ is a constant for a given phase $i$ and ($hkl$) reflection and $V_i$ is the effective volume of phase $i$ in the mixture.

For an infinitely thick sample, the effective volume can be expressed in terms of the volume fraction of the $i$th component:

$$V_i = \frac{A_O P C_i}{\mu_m} \tag{8.13}$$

Assuming that the sample is homogenous and the packing factor $P$ (mechanical density of the mixture to the weighted average X-ray density) and the volume fraction, $C_i$ are constant with depth.

$$\text{then, } I_i^{hkl} = K_i^{hkl}\frac{A_O P C_i}{\mu_m} \tag{8.14}$$

Intensity in terms of the weight fraction $W_i$ is then given by:

$$I_i = \frac{K_i A_O P C_i}{X_m P_i} W_i \tag{8.15}$$

where $X_m$ is the mass absorption coefficient of the mixture and $P_i$ is the density of the $i$th phase.

The ratio of the intensity for an unknown to the pure sample under identical diffraction conditions is:

$$\frac{I_i}{I_i^P} = W_i \frac{X_i}{X_m} \frac{P}{P^P} \quad (8.16)$$

$I_i^P$ is the diffracted intensity and $P^P$ is the packing factor of the pure phase.

Solving for the weight fraction $W_i$:

$$W_i = \frac{I_i}{I_i^P} \frac{X_m}{X_i} \frac{P^P}{P} \quad (8.17)$$

Above equation is the working equation for the external standard method.

## 8.6 MECHANICAL PROPERTIES

For coatings in service, determination of mechanical properties, including residual stress determination, adhesion, hardness, Young's modulus and toughness, is important. However, necessity for the coating to adhere to the base material throughout its application makes adhesion a property of major concern for TSCs.

### 8.6.1 Stress Determination

Residual stresses are the stresses present in a material even before the application of any external load. These are generated during thermal spraying and have an important role to play in the life and adhesion of the coating. The distribution of these stresses and their magnitude are affected by the processing parameters. Residual stresses are usually detrimental for the performance of any structure and have led to failures of structures, upon superimposition of the working loads, in many cases. Excess residual stresses may cause the separation of coating from the substrate or even cracking within the coating. These residual stresses are usually tensile in the coatings and that in the substrate are compressive. Their magnitude diminishes as we move away from the coating towards the substrate and becomes almost zero at the interface of coating and substrate [8].

Residual stress determination using diffraction involves determining the angles at which maximum intensity of peaks is obtained. Using Bragg's law, the interplanar spacing of the diffracting planes can be determined from these angles. In presence of stresses, the lattice spacing differs from that of the unstressed one, where this difference becomes proportional to the stress magnitude. For compressive stresses the spacing is larger, and for tensile stresses it is smaller. Upon tilting the specimen with respect to incoming X-ray beam, new grains diffract but via the same planes, the spacing will decrease as $2\Theta$ increases. Thus, the d-spacing can act as an indicator of strain (i.e., served as a strain gauge), and the sum of all the stresses can be measured.

### 8.6.2 Adhesion Test

Quality of TSCs and their performance strongly depend on the adhesion of the coating with the substrate. Adhesion is a measure of force required to cause separation between two adhering materials and is commonly termed as *adhesion strength*. Adhesion strength of coating is influenced by many intrinsic factors like powder characteristics, spray parameters and nature of substrate, and also by extrinsic factors including post treatment and service conditions. Since coating cannot be directly gripped, other approaches need to be made to ascertain its adhesive and cohesive strengths. Correspondingly, an adhesion test involves the removal of coating mechanically from the substrate and can be carried out in two ways: first, by detaching the coating normally with respect to substrate; and second, by its lateral detachment from the substrate. The test evaluates the strength/adhesion of a coating from the substrate to assess whether the coating is fit for service in any new application or for repairs to an already existing application. Many tests are available to check the adhesion strength of a coating, including the Scotch-tape method, ultra-centrifugal method, scrape adhesion tests, peel or knife tests, cathodic disbondment tests and pull-off adhesion tests. The pull-off test is the best-known test in terms of being most accurate and least subjective way of determining the coating adhesion. Low adhesion values are often attained because of some deficiency in the preparation of substrate surface and provide an indication of the premature failure during the service. In addition to mechanical test methods (destructive), some non-destructive methods are also available for evaluating the adhesion strength of coatings like the X-ray method, thermal method, electron beam method and so on [9–11].

#### 8.6.2.1 Pull-Off Adhesion Test

The methods and procedures for carrying out the adhesion test are described by ASTM D4541 and ISO 4624. The basic procedure is same in both cases and involves the fixing of a loading fixture/dolly to the coated substrate by means of an adhesive (Figure 8.3). Tensile force perpendicular to the surface is applied to remove the dolly along with the coating of substrate. Force is continuously increased until a specific value of the force is reached or a plug of the coating gets detached from the substrate. Both the force at which this detachment occurs and the type of resulting failure of the coating are recorded as the characteristic properties of the coating. The mode of failure of a coating under this tensile type test can be either interfacial occurring along the interface of the coating and substrate interface or cohesive failure in the coating. Further, mixed-mode failure can also occur combining these (i.e., interface and cohesive). Failure is assessed by examining the extent of adhesive and cohesive failure and by determining the actual interfaces involved with respect to total interfaces, while as pull-off strength is measured by computing the maximum load and the actual area stressed after proper calibration [9, 12]. For thermal barrier coatings (TBCs), which are basically duplex coating systems, cohesive failure can occur either in bond coat, at top coat or at its interface, while adhesive failure denotes the failure at bond coat/substrate interface [13].

**Apparatus and requirements**

1. *Tensile tester*: A tensile tester is capable of applying load perpendicular to the coated surface.
2. *Adhesion tester*: It includes:
    a. Loading fixtures/test dollies: it has a flat surface on one end to be attached to the coating and the other end to be attached to the tester.
    b. Detaching assembly: It contains a central grip for holding the fixture.
    c. Some base on the detaching assembly may be required to uniformly press against the coating surface.
    d. Provision to move the grip away smoothly and continuously from the base.
    e. Timer: To limit the rate of stress to less than 1 MPa/s in order to obtain the maximum stress in less than 100 s.
    f. Indicator and calibration system: To determine the actual force applied on the loading fixture.
3. *Solvent*: To remove primary contaminants of the loading fixture surface.
4. *Fine sandpaper*: To clean the coating without altering its integrity.
5. *Adhesive*: Holds the fixture to the coating without affecting the coating properties. To fail the coating, it is necessary that the cohesive and bonding characteristics of the adhesive are superior to that of coating.
6. *Clamps*: To hold the fixture in place during the curing of adhesive.
7. *Cotton swabs*: To remove excess adhesive.
8. *Circular hole cutter*: May be needed to drill through to the substrate around the fixture.

**Test procedure**

1. Clean the loading fixture and the substrate as recommended by the manufacturer.
2. Prepare the adhesive and apply to the fixture, the coating substrate or both using a suitable method. Preliminary screening should be carried out in order to determine the suitability of the adhesive for use. It should not cause any visible change to the coating.

3. Remove the excess adhesive from the surface around the fixture after pressing the dolly with a firm pressure.
4. Allow the adhesive to cure for a proper length of time while maintaining constant contact pressure between fixture and substrate surface.
5. Set up the pull-off equipment so that the load is applied to the center of the dolly and perpendicular to the surface. Set the force indicator to zero. Make sure that the equipment does not move during testing, as any misalignment reduces the pull-off strength.
6. Apply the load and increase it continuously at an even rate until failure.
7. Record the force at failure and the mode of failure.

Cai et al. [13] prepared two types of TBCs on Ni-based superalloy GH4169 substrate. Coating A was prepared by depositing a bond coat of CoCrAlY followed by an 8YSZ top coat, whereas coating B was exposed to high-current pulsed electron beam (HCPEB) irradiation prior to the deposition of the top coat. It was observed that failure in coating A was a mixed type that originated in an interface region between the bond coat and the top coat which then extended to others (epoxy failure), whereas the failure in coating B occurred in the glue line at the ceramic (top coat)/loading fixture interface. The mean value of bond strength increased from 40 MPa in coating A to 50 MPa in coating B after HCPEB irradiation.

The adhesion strength value for hydroxyapatite (HA)-coated Ti has been obtained in the range of 5–80 MPa. However, values around 25 MPa have been obtained for plasma-sprayed HA coatings, whereas a minimum value of ~50 MPa is required for HA coatings [14].

Brief descriptions of some other adhesion tests follow.

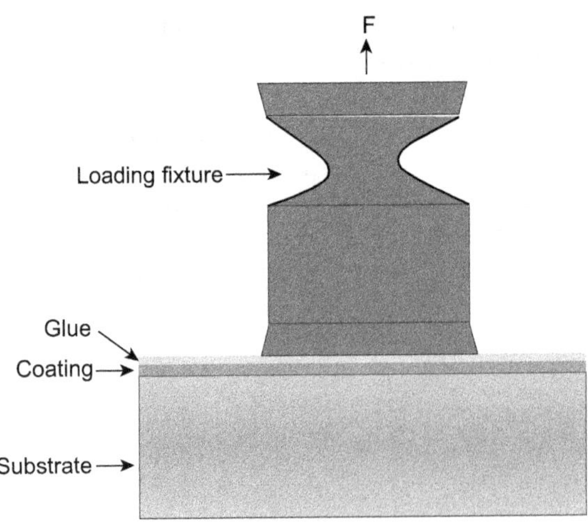


**FIGURE 8.3** Schematic showing the pull-off adhesion assembly for assessing adhesion strength of the coating.

#### 8.6.2.2 Scotch-Tape Method

A tape (pressure sensitive) is spread on to the surface of coating followed by the rapid stripping of the tape. Here an X-shape cut or cross-hatch cut pattern is made on the coating and tape is applied at the cut site. After the removal of tape, the cut area is analyzed for removal of coating from the substrate. This test is highly qualitative and is usually used for screening of the coatings with poor adhesion from those with stronger adhesion.

#### 8.6.2.3 Peel or Knife Tests

These tests involve directly holding onto the film and applying some sort of backing material to the film and then holding onto the backing. In another variant, a knife is used to pick at the coating establishing whether the adhesion at the coating/substrate interface is at an adequate level. The extent of difficulty in coating removal, as well as the area/region of the removed coating together, decides the coating's performance. Making two cuts down into the coating with a 30- to 45-degree angle between the legs, which intersect to form an X, followed by making an attempt to lift the coating with the point of the knife at the vertex of the cut, makes it a subjective test.

#### 8.6.2.4 Ultra-Centrifugal Method

Here the specimen is taken in the form of a rotor. The rotor is made to spin at very high speeds to generate the centrifugal force such that at some critical value of the centrifugal speed, the coating is no more able to sustain the centrifugal stresses and becomes detached.

#### 8.6.2.5 Scratch Method

A rounded tip (made of WC or diamond) is made to scratch against the coated surface by applying normal force. The normal force is increased till the tip makes a clear track over the coated surface after the complete removal of film along its path. The force at which this happens gives the measure of the adhesion strength.

#### 8.6.2.6 Cathode Disbondment Method

A coated substrate is made a cathode in the electrochemical cell with hydrogen gas as the fluid/electrolyte. Hydrogen enters the substrate-coating interface and is collected there, causing the blistering.

#### 8.6.2.7 X-Ray Method

This method works on the observation that poor adhesion at the film-substrate interface modifies the diffraction pattern and can be used to obtain qualitative information about the adhesion.

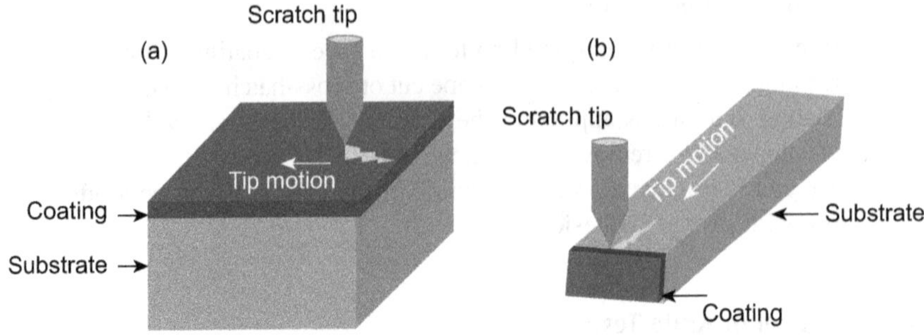


**FIGURE 8.4** (a) Scratch test carried out over the coating; (b) scratch test carried out along the cross section of coating.

#### 8.6.2.8 Adhesion Test Using Micro-Scratch

The scratch test is an easy method that has been used for a long time to measure coating/substrate adhesion. Figure 8.4 shows the schematics of the scratch test done using two different routes. This test makes it easy to determine the type of bond strength (adhesion/cohesion) that is critical for the coating under observation. During this test, two types of failures can occur. If the cone-shaped fracture occurs at the interface, it provides an indication about the problem in the coating adhesion, whereas its appearance within the coatings points at the issue of cohesion [15].

The adhesion strength can be determined using [16]:

$$L_c = \frac{A}{\eta\mu}\left[\frac{2EW}{t}\right]^{\frac{1}{2}} \tag{8.18}$$

where $L_c$ is the critical load, $A$ is the projected area of the indenter, $\eta$ is Poisson's ratio, $\mu$ is the coefficient of friction, $E$ is Young's modulus, $W$ is the work of adhesion and $t$ is the thickness of the coating.

#### 8.6.2.9 Weibull Distribution of Adhesion Data

Weibull distribution is primarily used for analysis of fracture data of brittle materials like glass. Weibull defined a function $P(x)$ as the probability of choosing a sample at random, where $P(x)$ is the probability of the element under test to support a load $x$ or may also be defined as the fraction of failures that have occurred up to load $x$. The probability of failure can be mathematically presented as [17]:

$$P(x) = \frac{n_x}{N+1} \tag{8.19}$$

where $n_x$ is the ranking of a sample failing at strength $x$ when the data set is arranged in an increasing order of $x$ and $N$ represents the total number of tests. As probability, $P(x)$ or any distribution function can also be written as [4]:

$$P(x) = 1 - e^{-\Phi(x)} \tag{8.20}$$

where $\Phi(x)$ is some suitable function of $x$. Weibull formulated the below function so that it satisfies the requirements in most of the systems as:

$$\Phi(x) = \frac{(x - x_u)^m}{x_o} \tag{8.21}$$

where $x_u$ is the minimum stress required for failure to occur (zero for brittle materials) and has significant value for metals. $x_o$ is the characteristic stress and $m$ is the Weibull modulus. Using equation (8.21) in equation (8.20), we get:

$$P(x) = 1 - \exp\left(-\frac{(x - x_u)^m}{x_o}\right) \tag{8.22}$$

or

$$\ln\left(\frac{1}{(1 - P(X))}\right) = \frac{(x - x_u)^m}{x_o}$$

or

$$\ln\left[\ln\left(\frac{1}{(1 - P(x))}\right)\right] = mln(x - x_u) - lnx_o$$

Assuming $x_u = 0$, the graph of ln($x$) versus ln(ln(1/[1 – $P(x)$]) yields a straight line whose slope and y-intercept are $m$ and –ln($x_o$), respectively. The Weibull modulus ($m$) provides a measure of the spread of data and gives information about the range of flaw sizes present. In general, higher Weibull modulus indicates more consistency of the parameter in question. The pull-off test or any other suitable test can be used to get adhesion-strength data, and Weibull distribution of failure strengths can be obtained by plotting of ln(ln(1/[1 – P($x$)]) as a function of ln($x$). The values of $m$ and $x_o$ can be extracted from the least-squares fit as a straight line. Equation (8.22) has also been written as [15]:

$$P(x) = 1 - \exp\left[-\left(\frac{x - x_u}{x_o}\right)^{\mathrm{m}}\right] \tag{8.23}$$

or

$$\ln\left[\ln\left(\frac{1}{\left(1-P(x)\right)}\right)\right] = m\ln\left(x-x_u\right) - m\ln x_o$$

where $x_0$ is the stress at 63.2% probability failure.

To calculate the Weibull modulus, values are sorted in increasing order based on the parameter $x$, and each value is assigned a probability of occurrence $P$ as per equation (8.19).

The Weibull modulus can qualitatively be observed by plotting the probability along the $y$-axis and the values of $x$ along the $x$-axis. The resulting plot is termed as the probability distribution of $x$. To determine the value of Weibull modulus, the plot of $\ln(\ln[1/(1 - P)])$ vs. $\ln(x - x_u)$ is followed by carrying out linear least-squares regression analysis. The slope of the resulting best-fit line is the Weibull modulus. Typical values of the Weibull modulus for some typical materials are provided in Table 8.2.

Erck et al. [17] carried out the pull-off tests of smooth and roughened zirconia samples coated with Ag up to a thickness of 1 μm (Figure 8.5). The Weibull moduli $m$ of the smooth Ag/zirconia specimens were close to 2.0. For rough specimens, $m$ was found to reduce from 2.0 to 1.7 or 1.6. This reduction in $m$ for rough surfaces indicates the broader size distribution of flaws. This indicates better mechanical interlocking of the rough coating making it exhibit a higher actual strength, but a lower Weibull modulus due to wide range of flaws.

Bull et al. [20] applied the Weibull analysis to scratch testing for an adhesion test of TiN coating (2–3 μm thick) on steel substrate by considering the importance of flaw distribution at coating-substrate interface change in this distribution that might occur by making the alterations in processing or roughness of the substrate. The Weibull modulus $m$, as well as the characteristic stress $x_o$, were found to decrease with the surface roughness [8].

**TABLE 8.2**
**Weibull Modulus of Some Materials**

| Material | $m$ |
|---|---|
| Brick, pottery, chalk | <3 |
| SiC, $Al_2O_3$, $Si_3N_4$ | 5–10 |
| Metals: Aluminum, steel | 90–100 |
| Extruded magnesium alloy | 92.6 |
| Magnesium-based glass alloy | 5–41 |
| HVOF_$TiO_2$ coating | 12–14 |

*Source:* Adapted from [18, 19].

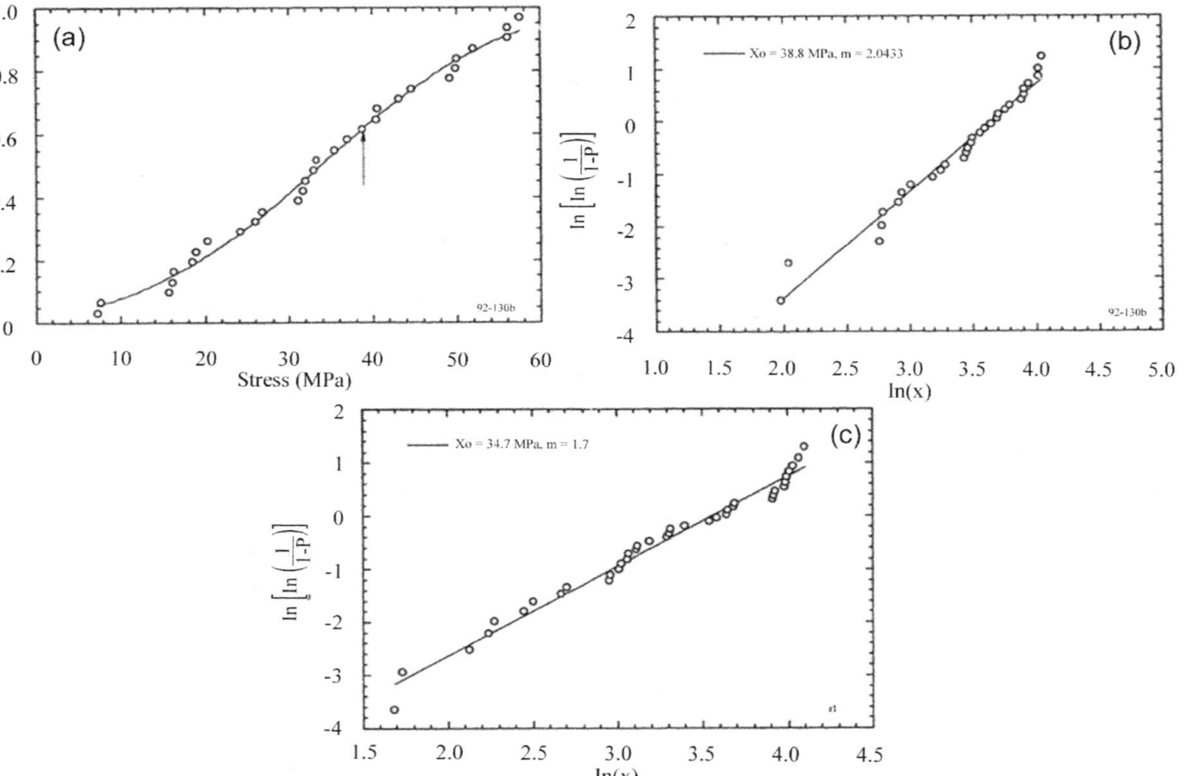


**FIGURE 8.5** (a) Probability distribution curve for smooth specimen; (b) Weibull distribution curve for smooth specimen; and (c) Weibull distribution curve for rough specimen.

*Source*: Reprinted with permission from [17].

## EXAMPLE

The fracture strength values of ceramic tiles obtained are 210 MPa, 345 MPa, 178 MPa, 276 MPa, 248 MPa, 262 MPa, 318 MPa, 235 MPa and 296 MPa. Calculate the Weibull modulus and comment on the reliability of this data.

## SOLUTION

Calculating the data for the *x*- and *y*-axes from the given data, as shown in Table 8.3.

Now plot ln(strength) versus ln(ln[1/(1 – F)]).

The value of *m* obtained from the slope of the plot in Figure 8.6 is 4.756, hence the data is quite reliable.

**TABLE 8.3**
**X and Y Values Calculated From the Given Data**

| Strength ($\sigma$, MPa) | Ranking, Ascending | *n* | *F* | 1/(1 – *F*) | ln(ln[1/(1 – *F*)], *Y* | ln(strength), *X* |
|---|---|---|---|---|---|---|
| 210 | 178 | 1 | 0.1 | 1.11 | –2.25 | 5.182 |
| 345 | 210 | 2 | 0.2 | 1.25 | –1.50 | 5.347 |
| 178 | 235 | 3 | 0.3 | 1.43 | –1.03 | 5.460 |
| 276 | 248 | 4 | 0.4 | 1.67 | –0.67 | 5.513 |
| 248 | 262 | 5 | 0.5 | 2.00 | –0.37 | 5.568 |
| 262 | 276 | 6 | 0.6 | 2.50 | –0.09 | 5.620 |
| 318 | 296 | 7 | 0.7 | 3.33 | 0.19 | 5.690 |
| 235 | 318 | 8 | 0.8 | 5.00 | 0.48 | 5.762 |
| 296 | 345 | 9 | 0.9 | 10.0 | 0.83 | 5.844 |

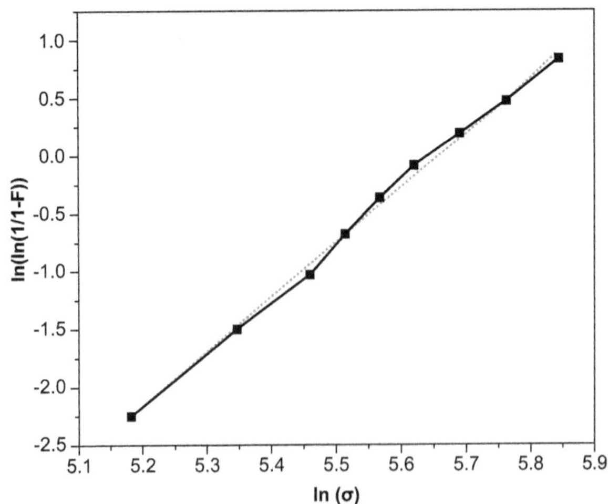


**FIGURE 8.6** Plot of ln($\sigma$) versus ln(ln[1/(1 – F)]) for the data provided in order to obtain the Weibull modulus.

### 8.6.3 Hardness Testing

The Vickers hardness test is the standard micro-hardness test adopted for TSCs in which a square-based pyramid made of diamond is used as an indenter. It works on the principle of identifying the depth of penetration and measuring hardness from the load and indentation area. Both the testing method and the direction of testing affect the hardness value. Different hardness values may be obtained depending on whether the tests are performed on the coating surface, orthogonal to the splats or along the cross section parallel to splats. Hardness measurements are performed using multiple indents. The specimen has to be prepared metallographically well (e.g., the preparation route tending to smear the coating can lead the underestimation of the hardness value). For macro-hardness measurement, the most commonly used test is Rockwell superficial hardness test. Here, initially a small load is applied at the start which tends to set the indenter into the specimen material without rising or sinking. Only after a minor load is a major load applied, and the depth of indentation is recorded in terms of hardness number. The thinner the coating, the more the chances of substrate influence on the results. For thickness greater than 250 μm, the substrate is considered to be of no problem.

Instrumented indentation gives information about elastic modulus, elastic/plastic work of indentation, and creep or cyclic behavior in addition to hardness information. Its force-displacement recording capability is sensitive to voids by providing a sudden increase in depth without an increase in the load, at a void. The automation of measurements is the most significant advantage of instrumented indentation.

### 8.6.4 Toughness

The indentation fracture toughness measurement technique is the most popular technique for measuring the fracture toughness of coatings using the Vickers indenter. There is no test specialized specimen required and can be performed on even small laboratory-based specimens. At low load regime, the cracks generated are Palmqvist cracks, whereas at a higher load regime, median cracks appear. When load is increased, formation of Palmqvist cracks is followed by the formation of median cracks. The two types of cracks are illustrated in Figure 8.7. Palmqvist cracks, unlike median or half-penny cracks, tend to penetrate no deeper than the indenter penetration. These are independent of each other, as shown in Figure 8.7a. The general equation to determine the fracture toughness from half-penny and Palmqvist-shaped cracks are provided in equations (8.24) and (8.25), respectively.

$$K_{1c} = k\frac{P}{a_c c^{\frac{1}{2}}} \quad \ldots \text{Half-penny crack } (c/a \geq 2.5,\ k = 0.129) \tag{8.24}$$

$$K_{1c} = k\left(\frac{E}{H_v}\right)^{\frac{2}{5}} \frac{P}{a_c l_c^{\frac{1}{2}}} \quad \ldots \text{Palmqvist crack } (0.25 \leq l/a \leq 2.5,\ k = 0.035) \tag{8.25}$$

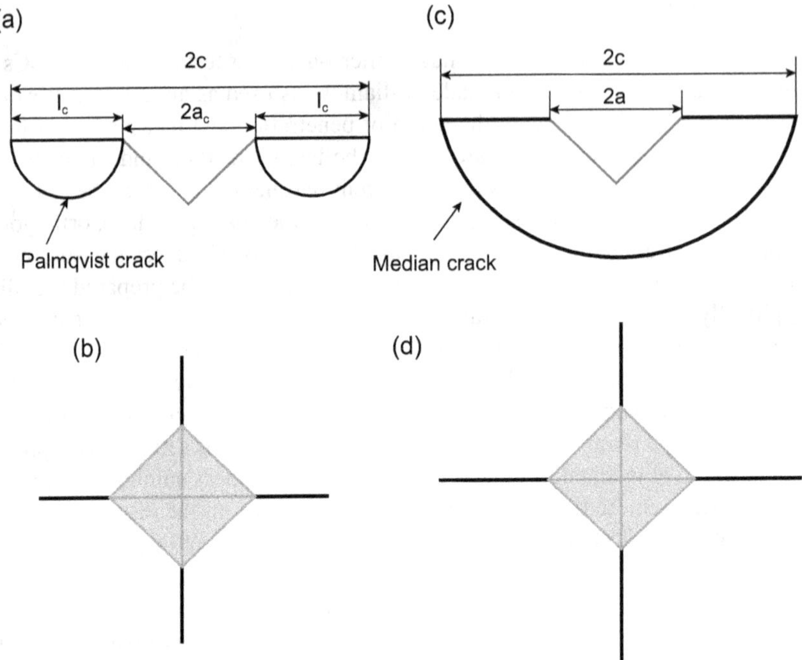


**FIGURE 8.7** Cross-sectional view of (a) Palmqvist cracks and (b) median cracks; top view of (c) Palmqvist cracks and (d) median cracks.

where $k$ is a constant (0.001–0.5), $P$ is the maximum load applied during indentation, $H_v$ is the Vickers hardness (GPa), and $E$ is Young's modulus (GPa). $a_c$, $c$ and $l_c$ are crack-related dimensions ($m$) and are defined as half the diagonal length of the indent, crack length (radial) and Palmqvist crack length, respectively (see Figure 8.7).

## 8.7 THERMAL PROPERTIES

For engineering applications, the thermal properties of coatings are important. The linear expansion coefficient ($\alpha$) plays an important role in residual stresses generated in the coating as a result of its different coefficient of thermal expansion that that of the substrate. Thereby, temperature during service condition should be such that it does not generate residual stresses enough to cause failure of the coating-substrate interface.

### 8.7.1 Coefficient of Thermal Expansion

Mechanical dilatometry is mostly used for the determination of the coefficient of thermal expansion (CTE) of TSCs, and the quantity determined is expressed in equation (8.26).

$$\alpha = \frac{1}{L}\frac{dL}{dT} \tag{8.26}$$

where $L$ is the length measured at temperature $T$.

Alternately, fractional change in volume with the temperature can also be determined, given as:

$$\alpha_v = \frac{1}{V}\left(\frac{dV}{dT}\right)_P \tag{8.27}$$

If the linear expansion in all the three directions is the same, then $\alpha_v = 3\alpha$.

### 8.7.2 Thermal Conductivity

Thermal conductivity ($k$) determines the temperature gradient created within the coating on application of heat. For coatings, it strongly depends up on the nature of coating, its roughness, defects, impurity and contact between the splats. Thermal conductivity is mostly determined from thermal diffusivity measurements using the relation:

$$k = \alpha.\rho.C_P \tag{8.28}$$

where $\rho$ is the density of the coating and $C_P$ is the specific heat at constant pressure. Heat capacity can be determined with a calorimeter.

Density at room temperature (R.T) is determined from the Archimedean porosimetry. To determine its value at any temperature $T$ other than room temperature, the thermal expansion coefficient needs to be known and is given by following expression:

$$\rho(T) = \frac{\rho_{R.T}}{1 + 3\alpha(T - 300)} \tag{8.29}$$

$\alpha$ can be determined using equation (8.30):

$$\alpha = 0.1388 \frac{L^2}{t^{\frac{1}{2}}} \tag{8.30}$$

where $t^{\frac{1}{2}}$ is the time taken to raise the temperature by half of the maximum, and $L$ is the thickness of the specimen. The coating diffusivity can be measured using laser flash apparatus. If it is measured with coating attached to the substrate, the relative method is applicable [21]. Suppose, substrate and coating have thermal conductivity of $k_1$ and $k_c$, respectively. The equivalent thermal conductivity of the coated sample is $k_2$. Then the laser flash method can be used to obtain values of $k_1$ and $k_2$. Knowing a theoretical relationship among $k_1$, $k_2$ and $k_c$, $k_c$ can be calculated. Thus, thermal conductivity can be calculated as equation (8.31):

$$k = \frac{Q}{\frac{At(T_2 - T_1)}{L}} \tag{8.31}$$

where $A$ is the area of thermal conduction, $Q$ is the heat flux, $t$ is time and $\left(\frac{(T_2 - T_1)}{L}\right)$ is the thermal gradient between the sample thickness $L$.

The coated sample whose thermal conductivity is to be measured is shown in Figure 8.8a. If substrate thickness is $H$ and coating thickness is $h$, then heat flow across the three thermally conductive areas are presented in Figure 8.8b. The substrate's thermal conductivity is provided as equation (8.32):

$$k_1 = \frac{Q_1}{\frac{At(T_2 - T_1)}{H}} \tag{8.32}$$

For the coated sample, equivalent thermal conductivity is given as (equation (8.33):

$$k_2 = \frac{Q_2}{\frac{At(T_2 - T_1)}{(H + h)}} \tag{8.33}$$

For coating, thermal conductivity can be provided by equation (8.34):

$$k_c = \frac{Q_3}{\frac{At(T_2 - T_1)}{h}} \tag{8.34}$$

where $Q_l$, $Q_2$, and $Q_3$ depict heat flow through the substrate material, composite and the coating, respectively. For steady-state heat conduction, heat entering the specimen equals the heat coming out of it, with constant temperature at each area, as presented in Figure 8.8b. In other words, during time $t$, heat flowing in and out at these surfaces should be the same; that is:

$$Q_1 = Q_2 = Q_3 \tag{8.35}$$

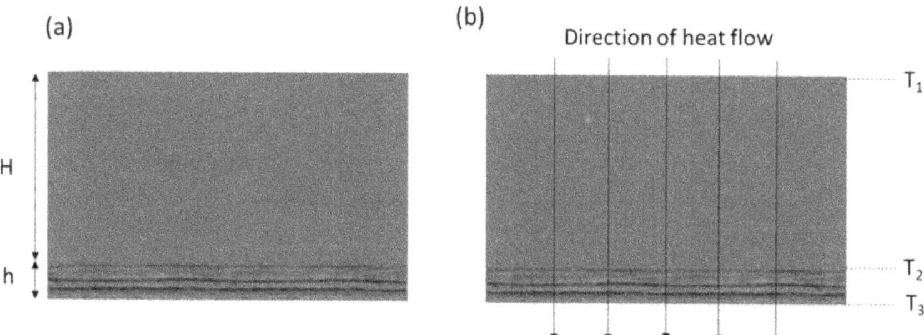


**FIGURE 8.8** (a) Schematic representation of a coated sample and (b) steady state heat flow through different areas (coating and the substrate) during directional heat flow exhibiting different temperatures.

*Source*: Adapted and redrawn from [21].

Substituting equations (8.32), (8.33) and (8.35) into equation (8.34), the resultant equation (8.36) should represent the thermal conductivity of the coating as:

$$k_c = \frac{H}{\frac{(H+h)}{k_2} - \frac{h}{k_1}} \tag{8.36}$$

It is observed that for same values of $k_1$ and $k_2$, $k_c$ would be the same.

Based on equation (8.42), $k_c$ is obtained knowing the specimen size and thermal conductivity of both the substrate and after the coating is deposited on the substrate.

### 8.7.3 Thermal Shock Resistance

The failure mechanism in a material as a result of significant temperature gradient under subjection of a sudden temperature change is called thermal shock; this mode of failure is prominently found in ceramics. Thus, thermal shock resistance (TSR) defines a material's resistance to failure under rapid temperature variations. Methods to evaluate TSR involve heating of a material and then quenching it in air or any other cooling medium like water, oil and so on. Repeated quench tests required to cause a certain extent of damage can be used as a measure of TSR. For an infinite slab heated symmetrically under steady heat flow, TSR parameter can be defined mathematically as:

$$TSR = \frac{\sigma_f (1-\upsilon)}{\alpha_l E} \tag{8.37}$$

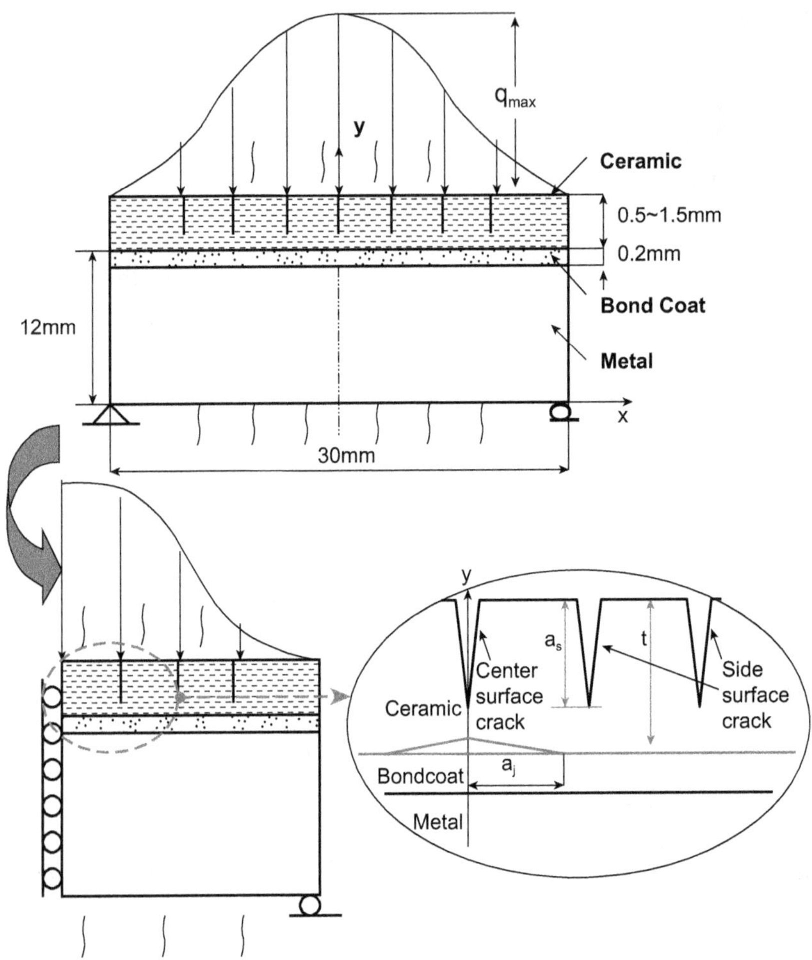


**FIGURE 8.9** Dimensions boundary conditions of the crack-growth FE model.

*Source*: Reprinted with permission from [24].

where $\sigma f$ is the fracture stress, $\upsilon$ is Poisson's ratio, $E$ is Young's modulus and $\alpha_1$ is the mean thermal expansion coefficient.

There are other shock resistance parameters which have not been discussed here.

Tests for thermal shock resistance measurement in thermal barrier coatings are specified ISO 14188:2012.

In thermal protection coatings, thermal expansion difference between the coating and the substrate may cause cracking and spallation of coating under thermal and mechanical loading. The presence of pre-existing cracks effects the shock behavior of the coating. Presence of multiple through-thickness cracks enhanced

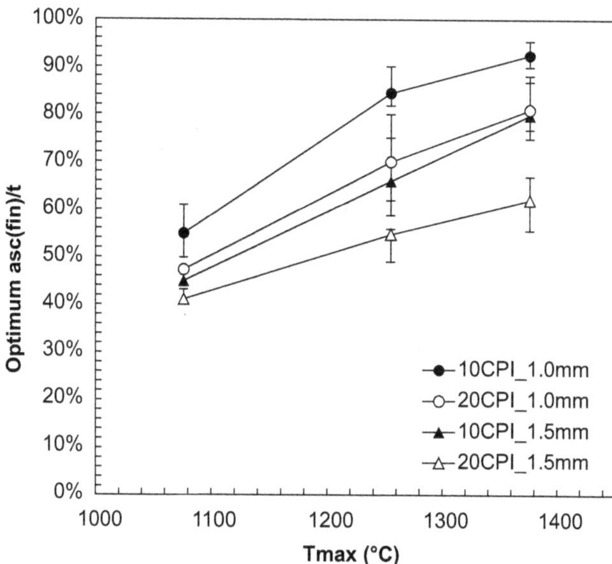


**FIGURE 8.10** Dimensions and boundary conditions of FEM, $a_{sc}$(fin)/$t$ is the center surface crack length, 10CPI_1.0 mm is 10 cracks per inch in a coating of thickness 1.0 mm, and so on.

*Source*: Reprinted with permission from [24].

the interface fracture resistance, increasing with increase in the density of the pre-existing cracks [22]. Zhou and Kokini [23] made an observation that it is not just the large density of cracks, but short surface crack length also decreased the driving force required for the crack propagation along the coating interface. This effect was more pronounced in coatings of higher thickness. Zhou and Kokini [24] further conducted laser thermal shock tests and developed analytical simulations. Figure 8.9 provides the dimensions and boundary conditions utilized in the computational finite element-FE analysis. It was observed that with the rise of maximum surface temperature, optimum pre-crack length increases (see Figure 8.10). It was essentially due to the coating requiring more strain tolerance to suppress the potential thermal fracture, which is provided by the larger pre-cracks. Moreover, small pre-crack density in the case of thin coatings required longer pre-crack length in order to inhibit the center surface crack growth.

## 8.8 WEAR RESISTANCE

Wear is the loss of material from a component surface as a result of its relative movement against another component or particles. Wear is a problem in almost all the industries where TSCs are used. In many machine parts, wear resistance dictates suitability of the utilized engineering part in order to provide satisfactory performance. Thermal spray technology can be used to improve surface properties by creating hard, wear-resistant coatings on these surfaces and is considered as one of the most versatile techniques ever utilized for coating material application to protect components from different types of wear. Hardness is usually considered as an indicator for the suitability of coatings to provide high wear resistance. Practically, a coating may undergo wear without having any effect on the substrate. It may also happen that the substrate undergoes deformation without any noticeable damage to the deposited coating. It has been observed that application of hard coatings on cutting tools may improve its life by a few to ten times. However, hard coatings exhibit limitations including porosity, poor bonding with the substrate and limited thickness. Still, coatings may undergo failure because of tensile, shear and compressive stresses causing cracking and spallation. While conducting the wear test of a material, it is important that the conditions used be appropriate and simulate the real application conditions. Owing to the limitation in the test system and difficulty in defining the conditions that occur in applications, it generally becomes impossible to exactly simulate the application conditions. Nonetheless, it is usually desirable to simplify the conditions to have more control of the testing, which in turn leads to a better understanding of the results and their relationship to the test conditions. During designing of a laboratory test, it is thus useful to consider aspects of the different test parameters. In this regard, the following points need to be considered while selecting a suitable wear test [25]:

1. The test measures the desirable properties of a material;
2. Forces and stresses involved are suitable;
3. Size, form and velocity are considerations for abrasive if it is present;
4. The type of contact that is involved between components: sliding, rolling, impact, erosion or a combination;
5. Test environment: temperature and humidity consideration;
6. Duration of the test;
7. Nature of the test testing materials.

It was already mentioned that hardness acts as an initial guide for the sustainability of coatings in applications necessitating higher wear resistance. However, different mechanisms of wear can prevail in a wearing material in service, making the assessment of the role of hardness complicated. Wear can be generally classified into six types briefly discussed below. All these mechanisms cause removal of material, and thus wear is often measured in terms of weight loss or volume loss of either of the two or both components of the tribo-couple. Parameters that are characteristics of the type of wear are then derived. Any type of wear produced, adopting any testing method is measured using optical or electronic microscopes providing wear track profile and debris distribution.

*Adhesive wear*: Loss of material occurs as a result of adhesive bond formation and its subsequent fracture during relative motion. Adhesive wear is evaluated using a pin-on-disc tribometer. In this method, a pin remains stationary in a lever arm and a disc is rotated beneath this pin. The horizontal and vertical displacements of the loading arm are recorded and utilized to determine the frictional force and wear of the pin, respectively. Wear occurs in the form of a groove or wear track on the disc, and the pin wears flat. The wear track is then evaluated using a profilometer providing the depth profile. The wear is provided by the wear coefficient $K_w$ as:

$$K_w = \frac{3 \times W_v \times H}{F \times d} \tag{8.38}$$

*Abrasive wear*: Sharp abrasive particles cut chips from the softer body. In abrasive wear testing, a counter body material in the form of a pin is rotated on an abrasive disc/cloth/paper which is mounted over a flat disc. One more approach is to interpose loose abrasive particles between these two surfaces, one of which slides or rotates on the other one which is stationary.

*Erosive wear*: Material wear occurs as a result of impingement of solid particles on the surface carried in a fluid. The wear caused by the liquid impingement through the collapse of bubbles is known as cavitation wear. At room temperature, an erosion test is conducted using an air blast tester. The eroding particles are fed from a vibratory hopper into a stream of gas. A known weight of erosive particles is directed onto the known weighted test discs, and then weight loss is measured. Erosion rate is then computed as:

$$Erosion\,rate = \frac{mass\,change\,in\,sample}{total\,mass\,of\,erosive\,particles} \tag{8.39}$$

*Surface fatigue*: Here, a material is exposed to fluctuating stresses and strains resulting in progressive localized permanent structural change. For coatings, for example, it occurs during rolling motion. Fatigue wear resistance is tested by using a rolling contact fatigue tester. The test subjects a cylindrical test specimen to rolling contact against a pair of grooved or crowned wheels to simulate ball and roller bearing loads, respectively.

*Corrosive wear*: This type of wear occurs due to the combined effect of corrosion and wear, which may result in the mechanical weakening of the surface and its wear at an increased rate. Corrosive wear test is same as that of the pin-on-disc test but performed in presence of lubricant or corrosive medium. The system acquires the potential current in addition to friction-force data.

*Fretting*: This occurs when two bodies are in intimate contact in presence of an oscillatory motion of little amplitude. Fretting wear test apparatus allows conducting fretting tests at an applied stress using a prescribed value of relative displacement (displacement amplitude).

*Pin on disc*: The pin-on-disc wear test (ASTM G99), which is an effective approach of estimating the coefficient of friction, frictional force and wear rate between two materials, is performed by rotating a pin against a stationary disc under a constant applied load, as shown in Figure 8.11.

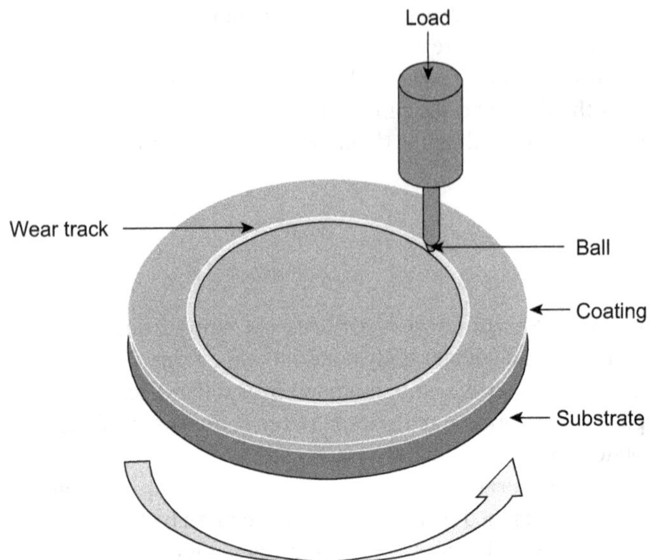


**FIGURE 8.11** Pin-on-disc test illustration to assess damage tolerance of a coating surface.

## 8.9 SUMMARY

Sprayed coating characterization is important, and the measurements acquired impact the reliability and reproducibility in thermally sprayed coatings' production. Thus, it is essential to carefully control the measurement uncertainty and statistics utilized properly. However, the provision to estimate and accommodate the human factor should remain indispensable in coating quality control.

### Questions for Self-Analysis

1. Why is metallographic preparation required? Briefly discuss the steps involved in it.
2. For $\delta = 0$, what would be the reliability of the coating after 1000 hours of service if scale parameter and shape factor of the Weibull distribution are 3000 hours and 2, respectively? (Answer: 0.895.)
3. Briefly discuss different types of wear.
4. How is thermal conductivity of a coating measured?
5. What is adhesion strength? Discuss the most common method of determining the adhesion strength of a coating.
6. What are residual stresses, and what is their effect on the thermally sprayed coating? What is the principle of residual stress determination in XRD?

## REFERENCES


[1] Davis, J.R., *Handbook of thermal spray technology*. 2004: ASM International.

[2] Fauchais, P.L., J.V. Heberlein, and M.I. Boulos, *Thermal spray fundamentals: From powder to part*. 2014: Springer Science & Business Media.

[3] Gan, J.A. and C.C.J.M. Berndt, Quantification and taxonomy of pores in thermal spray coatings by image analysis and stereology approach. *Metallurgical and Materials Transactions A*, 2013. **44**(10): p. 4844–4858.

[4] Zhang, X., et al., X-ray computed tomographic investigation of the porosity and morphology of plasma electrolytic oxidation coatings. *ACS Applied Materials & Interfaces*, 2016. **8**(13): p. 8801–8810.

[5] *Quantitative phase analysis of ceramic coatings*. https://www.lambdatechs.com/wp-content/uploads/dif12.pdf (wfcstaging.com). 1993.

[6] Welding, I.I.O. *Advances in technology and application: Proceedings of the international thermal spray conference*. in Thermal Spray 2004. 2004: Osaka Japan.

[7] Gualtieri, M.L., M. Prudenziati, and A.F. Gualtieri, *Quantitative determination of the amorphous phase in plasma sprayed alumina coatings using the Rietveld method. Surface Coatings Technology*, 2006. **201**(6): p. 2984–2989.

[8] Yi, J., et al., Determination of residual stresses within plasma spray coating using Moiré interferometry method. *Applied Surface Science*, 2011. **257**(6): p. 2332–2336.

[9] Fletcher, J. and D. Barnes, *Pull-off adhesion testing of coatings—Improve your technique*. 2015: Elcometer Limited.

[10] Mittal, K.L., Adhesion measurement of thin films. *Electrocomponent Science and Technology*, 1976. **3**(1): p. 21–42.

[11] Vaca-Cortés, E., et al., Adhesion testing of epoxy coating. Research Report No. 1265–6. Center for Transportation Research, 1998: p. 1–129.

[12] *ASTM D4541–09e1 standard test method for pull-off strength of coatings using portable adhesion testers*. 2010: ASTM International.

[13] Cai, J., et al., Adhesion strength of thermal barrier coatings with thermal-sprayed bondcoat treated by compound method of high-current pulsed electron beam and grit blasting. *Journal of Thermal Spray Technology*, 2015. **24**(5): p. 798–806.

[14] Remache, D., et al., *Delamination study of hydroxyapatite coatings for bone orthopedic implants. 24e me Congres Français de Mécanique*. Brest, 26 au 30 Août 2019.

[15] Vencl, A., et al., Evaluation of adhesion/cohesion bond strength of the thick plasma spray coatings by scratch testing on coatings cross-sections. *Tribology International*, 2011. **44**(11): p. 1281–1288.

[16] Jambagi, S.C., Scratch adhesion strength of plasma sprayed carbon nanotube reinforced ceramic coatings. *Journal of Alloys and Compounds*, 2017. **728**: p. 126–137.

[17] Erck, R., F. Nichols, and D. Schult, Weibull analysis applied to the pull adhesion test and fracture of a metal-ceramic interface. *Tribology Transactions*, 1994. **37**(2): p. 299–304.

[18] Guo, S., et al., Statistical analysis on the mechanical properties of magnesium alloys. *Materials*, 2017. **10**(11): p. 1271.

[19] Neilson, H.J., et al., Weibull modulus of hardness, bend strength, and tensile strength of Ni– Ta– Co– X metallic glass ribbons. *Materials Science and Engineering A*, 2015. **634**: p. 176–182.

[20] Bull, S., D.J.S. Rickerby, and C. Technology, New developments in the modelling of the hardness and scratch adhesion of thin films. *Surface and Coating Technology*, 1990. **42**(2): p. 149–164.

[21] Zheng, D., et al., Determining the thermal conductivity of ceramic coatings by relative method. *International Journal of Applied Ceramic Technology*, 2019. **16**(6): p. 2299–2305.

[22] Kokini, K., et al., Thermal fracture of interfaces in precracked thermal barrier coatings. *Materials Science and Engineering A*, 2002. **323**(1–2): p. 70–82.

[23] Zhou, B. and K. Kokini, Effect of pre-existing surface crack morphology on the interfacial thermal fracture of thermal barrier coatings: A numerical study. *Materials Science and Engineering A*, 2003. **348**(1–2): p. 271–279.

[24] Zhou, B. and K. Kokini, Effect of surface pre-crack morphology on the fracture of thermal barrier coatings under thermal shock. *Acta Materialia*, 2004. **52**(14): p. 4189–4197.

[25] Kennedy, D. and M.J. Hashmi, Methods of wear testing for advanced surface coatings and bulk materials. *Journal of Materials Processing Technology*, 1998. **77**(1–3): p. 246–253.

# 9 Processing Diagnostics and Spray Consistency

*Roopal Singh, K. Vijay Kumar and Anup Kumar Keshri*

**CONTENTS**



DOI: 10.1201/9781003321965-9

There is growing global activity in thermal spray coatings, due to the fact that there is a fast extension in research and development, and many functions are being commercialized. Even so, it has challenges with RRRs: repeatability, reproducibility and reliability for the production of quality and the dense coating. To achieve these RRRs, first we need to understand them.

Dwivedi et al. [1] attempted to define these RRRs as *repeatability*, which refers to the ability to repeat a process in order to achieve the same result. Checking repeatability entails measuring the same measurement with the same instrument and person or doing several measurements on the same parts or products with the same instruments and techniques. This definition can be applied to any coating. *Reproducibility* can be defined as the percentage change in average measurement acquired by more than one individual reviewing that very same component or thing using the similar measuring process. Many people are striving to reproduce the similar coating using the same processes, and reproducibility is usually used to illustrate dimensional variations. The chance that a coating will effectively serve its given function for a stated amount of time under a stated environmental setting is defined as *reliability*. As a result, it encompasses all facets of repeatability and reproducibility while also establishing basic benchmarks at every stage of the process. Many researchers have attempted to obtain these RRRs by better understanding and monitoring the parameters of the thermal spray process.

To improve the above-mentioned RRRs, one must optimize the thermal spray process parameters. Akbarnozari et al. discovered that using in-flight particle diagnostic systems for monitoring and controlling parameters like temperature and velocity, improved the reproducibility and repeatability of suspension plasma spray (SPS) coatings [2].

The main focus of this chapter is to give the knowledge to the reader about the in-flight particle parameters and their effect on splat-type distribution and coating microstructure. Also, this chapter gives knowledge about the different diagnostic systems for monitoring and measuring of in-flight particle temperature, velocity distributions. By measuring these process parameters, it will be useful to optimize the parameters using machine learning methods in order to achieve the above specified RRRs. Additionally, this chapter gives knowledge on the limitations in the monitoring and measuring of in-flight particle.

## 9.1 SPRAY PROCESSING PARAMETERS

There are many processing parameters of the plasma-spraying process in terms of starting powder characteristics, substrate condition, spray equipment and process variables, which all will affect the microstructural and mechanical properties of coatings [3]. All these parameters are depicted in Figure 9.1.

From the above figure, it is clear that the density and porosity of coatings mainly depends on splats (created when the accelerated, molten particles impact a

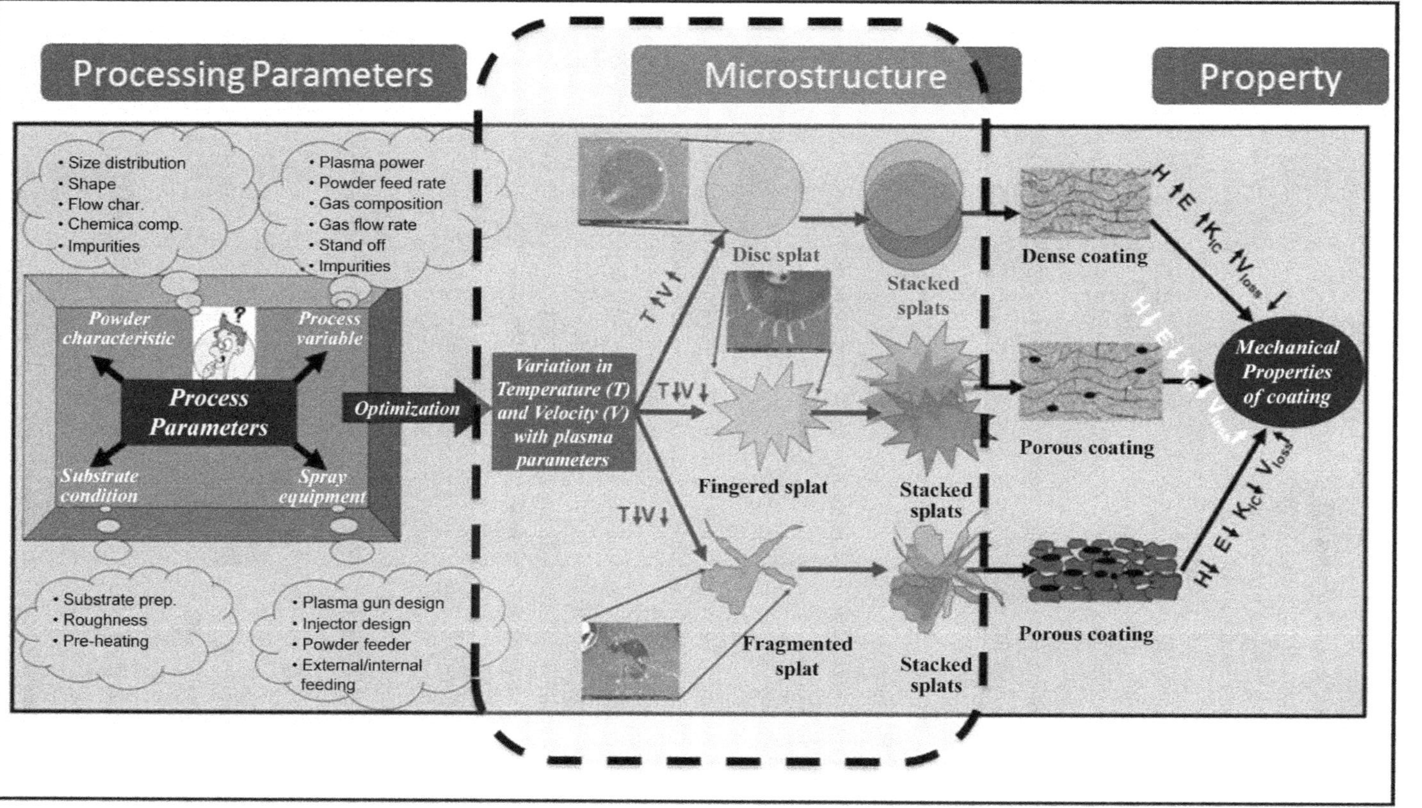


**FIGURE 9.1** Schematic of plasma-sprayed coating process parameters and correlation of mechanical properties with the evolved microstructure of deposited coating.

*Source*: Courtesy Anup Kumar Keshri, PhD thesis [3].

substrate surface) morphology, size of splats, and their stacking. Here, splats morphology mainly depends on starting powder characteristics, substrate condition and in-flight particles parameters [3]. This chapter mainly focuses on monitoring and controlling particle temperature and velocity during in-flight by using different diagnostic sensors. It is observed from Figure 9.1 that disc splats will give dense coatings and better mechanical properties like Young's modulus ($E$), fracture toughness ($K_{1c}$) with increasing temperature and velocity. Here, temperature and velocity of in-flight particles depend on input power, primary gas flow, size and shape of particles, and the distance (stand-off distance) at which the sensor is placed to measure the temperature. Temperature of in-flight particles increases by increasing input power, secondary gas (hydrogen) flow, and by reducing stand-off distance. In-flight particle velocity increases by increasing primary gas (argon) flow and by reducing stand-off distance.

### 9.1.1 Influence of Process Parameters on Coating Microstructure

The coating microstructure basically depends on the splats (splat-type distribution, inter-splat contact, size of splats, and their stacking). Different splat types are affected by in-flight particle parameters, substrate condition and spray angle. Figure 9.2 illustrates different splat shapes based on their molten states; that is, at higher temperature, a fully melted splat is formed; at medium temperature, a splat is partially melted; and at low temperature, a splat is less melted.

At low-impact particle temperatures, the microstructure of the coating appears porous with loose architecture, fragile cohesion; with lower consistency (Figure 9.3a–c); particles in semi-melted and un-melted states are ingrained within the coating, and inter-splat cracks are common and appear as horizontal fissures. At medium temperature, the microstructure pop is less porous with a dense structure (Figure 9.3d–f); we can see a better inter-splat bond because the number of inter-splat cracks is reduced since the coating contains fewer semi-melted and un-melted particles. At high impact temperatures, we can see a high, thick, fully flattened lamellae structure with high cohesiveness due to a higher percentage of completely melted particles, and we can observe inter-lamellar cracks and vertical cracks reproducing crosswise to the lamellae (Figure 9.3g–i) [4].

The effect of impact velocity of the particle on splat-type dispersion and microstructure of the coating can only be noticed at high particle impact temperatures because at low temperatures there is no effect on velocity. As the temperature rises, the velocity rises as well, resulting in disk-like and deformed splats [4].

## 9.2 PARTICLE IN-FLIGHT VELOCITY AND TEMPERATURE DISTRIBUTIONS

In-flight properties of powder particles are determined using altogether or regional area measurements.

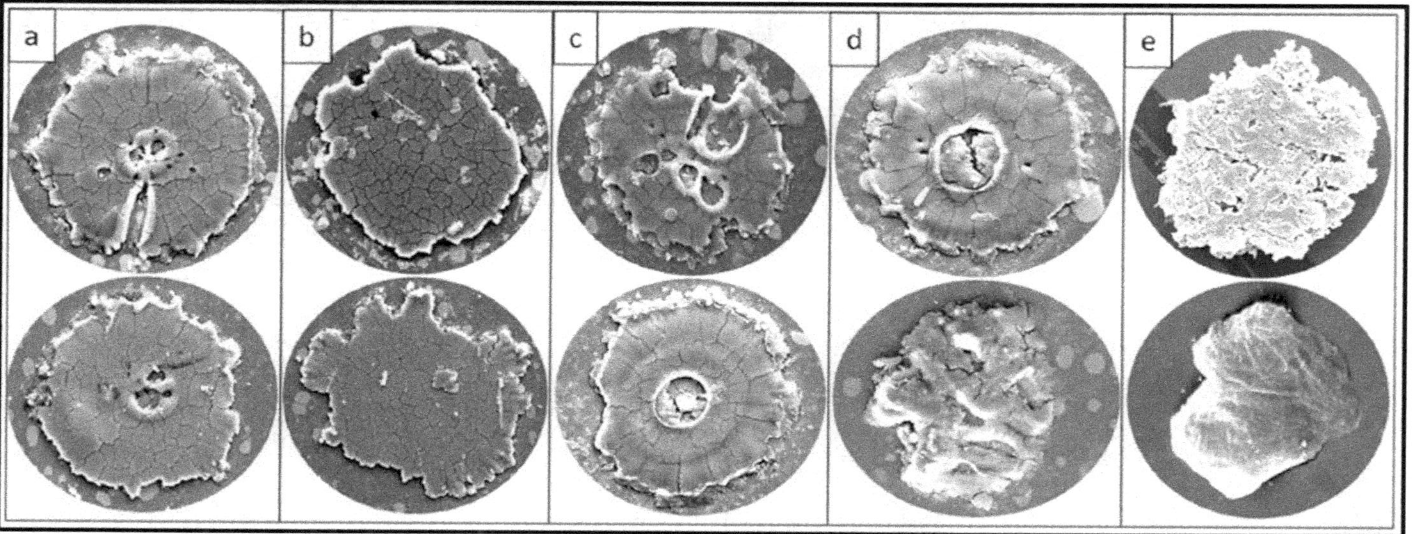


**FIGURE 9.2** Splats with various shapes and features: (a) a completely molten splat having a few irregularities, quenching micro-cracks and bubbles at the center; (b) completely molten splat with some irregularities and quenching micro-cracks; (c) completely molten splat with irregularities and bubbles; (d) partially molten splat having molten rim outside with solid core inside, and widely spaced quenching micro-cracks; (e) solid particle in a minor melting state, with no quenching micro-cracks.

*Source*: Reprinted with permission from [4].

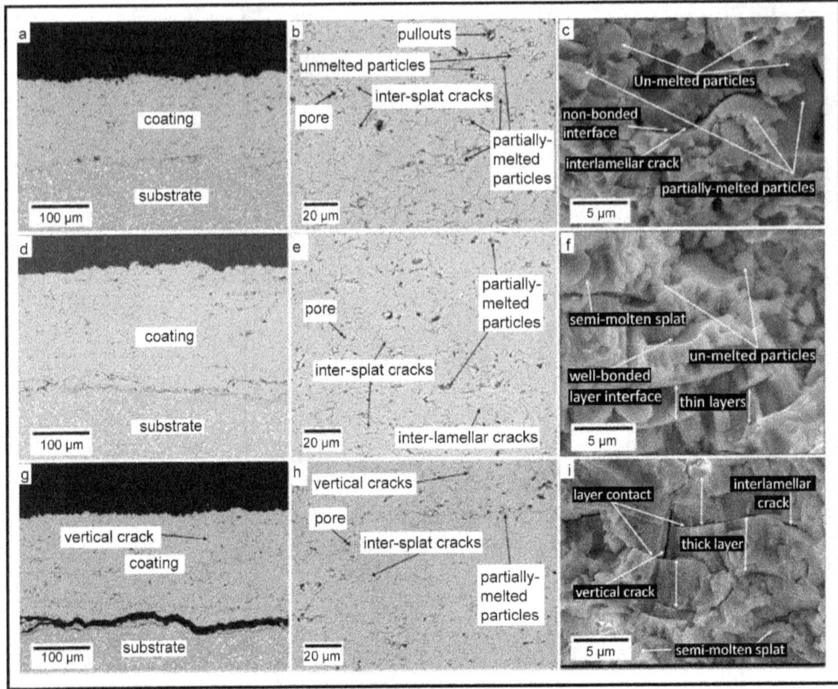


**FIGURE 9.3** Effect of particle impact temperature on microstructure (cross-section) of coating: (a–c) low temperature, (d–f) medium temperature and (g–i) high impact temperature.

*Source*: Reprinted with permission from [4].

1. *Local measurements*: Measurement is taken from a small volume which is a tenth of a mm$^3$. Individual particles can be measured. It necessitates observation of a proportionate statistical sample to offer a fair representation of mean temperature, velocity and particle diameter. Thermal spray procedures with particle mass flow rates ranging from 0.5 to 1 kg/h are required. High-speed detectors are used in the system, and the complete recording procedure takes only a few minutes. The most popular setup in this sector is the Tecnar DPV 2000, which is discussed in the following sub-section.
2. *Large measurement*: The calculations are made in a measuring volume of cylindrical chord, approximately tens of orders of mm$^3$ to contain a specified number of particles at any given point in time across the spray pattern. The derived particle parameters are not unique; rather, they are mean values. The image processing protocol involves (1) the original image, (2) the image acquired after filtering and (3) the total of several treated images. Measuring takes only a few seconds. Accuraspray and

Spray Watch, which are covered in the next section, are the most well-known configurations for these metrics.

## 9.3 MEASUREMENT OF IN-FLIGHT PARTICLE PARAMETERS: TEMPERATURE AND VELOCITY

### 9.3.1 Measurement of Temperature

Upon impact on the substrate surface, particle temperature is a crucial element in the thermal spray process, as it determines the quality and properties of the deposit created. The assessment of temperature of the particles in-flight is necessary in order to assure the repeatability of desired coating characteristics and to regulate the process. The plasma plume has two temperatures: plasma temperature and in-flight particle temperature [5]. Other than in-flight particle monitoring techniques, there are several ways to determine the temperature of a plasma plume. Optical emission spectroscopy, Enthalpy probe, computer tomography, and Boltzmann plot are the four methods [6]. In the case of in-flight particle temperature measurement, it is entirely dependent on two-color pyrometry, which is the measurement of radiance emitted in two or more wavelengths or shade bands by hot, incandescent particles. Planck's blackbody radiation law [6] is used to connect the temperature to intensities in a minimum of two wavelength bands.

#### 9.3.1.1 Temperature Measurement by Two-Color or Dual Wavelength Pyrometers

Two optical and electrical measurement channels (see Figure 9.4) are equal in structure in special pyrometers (two-color or dual wavelength pyrometers). Two wavelength ranges are kept close to each and adjusted to be highly narrow-banded, such that the effect of material specific characteristics like reflectance, emissivity from the target, is nearly identical to the two wavelengths.

The working principle of two-color pyrometer is based on Planck's law. An ideal black body emits from its surface element a thermal flux of spectral distribution:

$$E(\lambda) = \frac{K_1}{\lambda^5} \frac{1}{e^{\frac{K_2}{\lambda}T} - 1}, \text{W/Sr} \tag{9.1}$$

where $\lambda$ is the wavelength, $T$ is the temperature, and $K_1$ and $K_2$ are radiation constants.

$$K_1 = 2\pi \text{ h c}^2 = 3.741771 \times 10^{-16}, \text{Wm}^2$$

$$K_2 = \frac{hc}{k_B} = 1.438775 \times 10 - 2, \text{mK}$$

where $c$ is the velocity of light, $h$ is Plank's constant, and $k_B$ the Boltzmann constant [6].

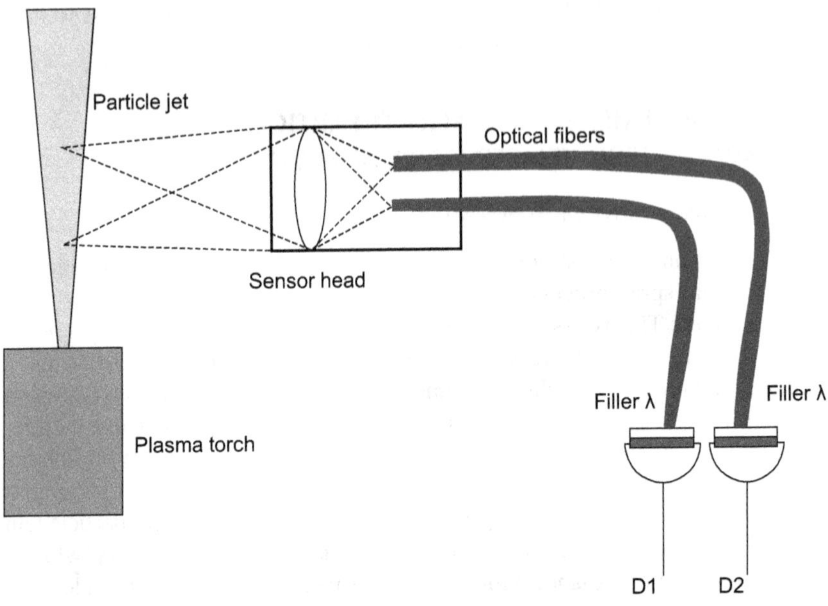


**FIGURE 9.4** Schematic illustration for measuring the temperature and velocity of particle using an optical arrangement.

*Source*: Adapted and redrawn from [9].

Two-color pyrometry measures in-flight particle temperature by filtering the thermal radiance emitted from a particle at two wavelength bands, whose ratio is given by:

$$\frac{E(\lambda_1)}{E(\lambda_2)} = \left(\frac{\lambda_2}{\lambda_1}\right)^5 \frac{\varepsilon(\lambda_1)e^{K_2/\lambda_2 T} - 1}{\varepsilon(\lambda_2)e^{K_2/\lambda_2 T} - 1} \tag{9.2}$$

where $\varepsilon$ is the emissivity providing the ratio of the radiation emission from a real body in comparison to an ideal black body.

For many applications, emissivities are unknown. Hence, gray body assumption, $\frac{\varepsilon(\lambda_1)}{\varepsilon(\lambda_2)} = 1$, is used, where it is assumed that the emissivity is not dependent on the wavelength.

The desired temperature of the particle is obtained by

$$T = \frac{K_2\left(\frac{1}{\lambda_2} - \frac{1}{\lambda_1}\right)}{\ln\left\{\frac{E(\lambda_1)}{E(\lambda_2)}\left[\frac{\lambda_1}{\lambda_2}\right]^5\right\}} \tag{9.3}$$

The limitation of two-color pyrometry measurements is that the temperature measurements can differ from the actual temperature of the particles due to non-thermal radiation sources, such as radiation emission from the plasma surrounding the particles, vaporized material and impurities, and that coming from the plasma source scattered by the particles.

**EXAMPLE PROBLEM 1**

The in-flight particle temperature measurements are made based on two-color pyrometry at two wavelengths 680 nm and 837 nm, the pyrometer measures the spectral intensities ratio of 0.85. If the emissivity does not change with the wavelength, find the temperature of the particle.

**SOLUTION**

If the emissivity does not change with the wavelength (i.e., $\frac{\varepsilon(\lambda_1)}{\varepsilon(\lambda_2)} = 1$):

Then the temperature of the particle $T = \dfrac{K_2\left(\dfrac{1}{\lambda_2} - \dfrac{1}{\lambda_1}\right)}{\ln\left\{\dfrac{E(\lambda_1)}{E(\lambda_2)}\left[\dfrac{\lambda_1}{\lambda_2}\right]^5\right\}}$

$K_2 = \dfrac{hc}{k_B} = 1.438775 \times 10^{-2}$ mK; $\lambda_1 = 837$ nm; $\lambda_2 = 680$ nm; $\dfrac{E(\lambda_1)}{E(\lambda_2)} = 0.85$

$$T = \frac{1.438775 \times 10^{-2}\,\text{mK} \times \left(\dfrac{837 - 680}{837 \times 680}\right)\dfrac{1}{\text{nm}}}{\ln\left\{0.85 \times \left[\dfrac{837}{680}\right]^5\right\}}$$

$$T = \frac{3.97 \times 10^3}{0.8761374}\,\text{K}$$

$$T = 4531\ \text{K}$$

The measured in-flight particle temperature is 4531 K.

### 9.3.2 Velocity Measurements

Particles' temperature and velocity are analyzed jointly for a more thorough characterization of in-flight particles. A time-of-flight technique is used to compute the velocity. Every particle in the measuring volume emits a signal that travels through two adjacent measurement slits positioned parallel to the overall velocity direction of the plume. The particle speed is then estimated from the distance, and time difference between the radiation peaks detected in each detector, knowing magnification of the optics [2].

## 9.4 IN-FLIGHT PARTICLE PARAMETERS DIAGNOSTIC SYSTEMS

The detecting elements of particle diagnostic systems can be identified. If the detecting element is based on a sensor system, it can increase the reproducibility and reliability of coatings. The most popular sensors are divided into two groups: (1) fiber-optic sensors, which use a pair of photo detectors for projection of measurement volume; and (2) CCD arrays (either 1D or 2D), which picture the plume in a single line at a set spray distance or in both vertical and spray directions [7].

Single-particle and ensemble methods are two types of measurement techniques. Single-particle methods assess the temperature of individual particles using high-speed pyrometry. From observations of a large enough number of individual particles, temperature distribution with its mean and standard deviation can be calculated. Ensemble methods observe particles in large numbers at the same time and provide a direct estimate of the mean temperature, but no information is provided either on shape or on width of the temperature distribution of the particles. In the thermal spray sector, both approaches are used. Single-particle approaches are ideal for scientific research, but they are time-consuming and difficult to implement. Ensemble approaches are appealing for control applications because they are insensitive to spray pattern position and provide fast response [8].

There are many techniques for the diagnosis of process parameters. Few of them are in-flight particle pyrometer (IPP), ThermaViz, Flux Sentinel, near infrared (NIR), spray watch, DPV 2000 and Accuraspray. All these will be discussed here.

### 9.4.1 In-Flight Particle Diagnostic System 1: In-Flight Particle Pyrometer (IPP)

Inflight Ltd. Co., Idaho Falls, Idaho, USA, created this two-color radiation pyrometer. The electronics, the optical measurement head, and the fiber cable that connects the two are the three primary components of the IPP. The objective lens can be swapped out for a regular 200 mm focal length lens. This lens creates a pencil-shaped measurement volume of 5 mm diameter with roughly 5 cm length [5]. It has a lengthy cylindrical measurement container (Ø 5 × 50 mm) providing average temperature measurement for a particle ensemble, at 10 Hz [6].

### 9.4.2 In-Flight Particle Diagnostic System 2: ThermaViz

This is a two-wavelength image pyrometer system entrenching a CCD camera by exemplary exposure times ranging from 5 to 20 seconds, developed by Stratonics

Inc., Laguna Hills, California, USA. Wavelengths at 625 and 800 nm are utilized in this two-color pyrometry to measure the temperatures, sizes and velocities of ensemble particles [6]. It shows live, real-time thermal imagery data, as well as thermal and dimensional data profiles, time histories and histograms.

### 9.4.3 In-Flight Particle Diagnostic System 3: Flux Sentinel

Cyber Materials LLC, Boston, Massachusetts, USA, created this CCD-based sensor capable of monitoring temperature, diameter and velocity of individual particles. In places where a considerable intensity is emitted above 2000 K, multiple-color pyrometry is employed to take temperature measurement of each particle using 125 nm wide color bands. The Stefan-Boltzmann law is used to compute particle velocity by adding the intensities of all pixels in a single particle's picture, which is roughly proportional to the particle's surface area, velocity and to the fourth power of temperature. Furthermore, particle diameter is measured photometrically [6].

### 9.4.4 In-Flight Particle Diagnostic System 4: Near Infrared (NIR)

This sensor was created by GTV Verschleißschutz GmbH in Luckenbach, Germany. It detects thermal radiation released by a single particle and employs the two-color pyrometry method to measure temperature using a two-emission signal ratio. The in-flight time and particle velocity are inversely related. Despite operating in the infrared range and using pulsed laser light, cold and slow particles can be spotted [6].

### 9.4.5 In-Flight Particle Diagnostic System 5: Spray Watch

Oseir Ltd., Tampere, Finland, created the spray watch diagnostic system. Using a high-speed camera, it is utilized to perceive and count luminous fast particles. The measurement occurs when the particles collide with a substrate kept at a distance in a pre-defined location within the plasma jet [4]. The digital photographs of the spray are taken using a quick-shutter CCD camera. The CCD array can cover measuring volume which can be adjusted between $18 \times 14 \times 5$ and $36 \times 28 \times 30$ mm$^3$. Using a dichromatic double mirror, the heated particles are imaged onto the CCD camera sensor. These digital images are then processed in an algorithm that recognizes each individual particle captured in the photos and determines their position, direction and velocity.

The two surfaces (front and back) of the mirror provide unique spectral bands and two-color double images of separate particles on the CCD sensor. The velocity and temperature of particles are averaged across the measurement volume, which is a disadvantage of utilizing spray watch providing lower temperature measurements than that obtained with DPV 2000 [6].

### 9.4.6 In-Flight Particle Diagnostic System 6: DPV-2000

TECNAR Automation Ltd., Saint-Bruno, Quebec, Canada, developed the DPV-2000 based on two-color pyrometry at two separate spectral ranges (790 nm and 990 nm) in the near infrared wavelength region. Particle velocity, diameters and temperatures can all be measured using it.

The sensor head (shown in Figure 9.5), the detection module (which includes the optical components and photo detectors), and the control module are the three primary components. When a particle passes across the view field of a sensor head, the scattering light at the plane containing two-slit mask forms a corresponding picture spot. The movement of such an image spot in the mask is translated into a two-peak signal and sent across the optical fiber to the detection module. A dichroic mirror spectrally separates the gathered optical signals, which are then filtered by two band pass filters centered at 995 nm and 787 nm, respectively. Signals obtained from the two detectors are amplified, digitized and routed to a computer.

Time lapse between the two signals (see Figure 9.6) provides the velocity measurement while knowing the distance between the slits and the lens magnification.

The particle velocity V from this two peak signal is provided by:

$$V = \frac{S}{TOF} \times \text{Lens optical magnification}$$

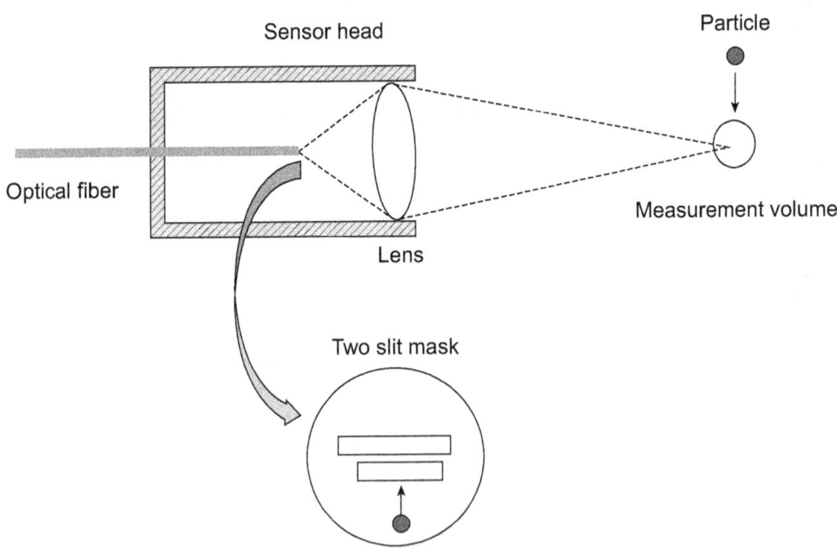


**FIGURE 9.5** Schematic demonstration of DPV-2000 sensor head.

*Source*: Adapted and redrawn from [9].

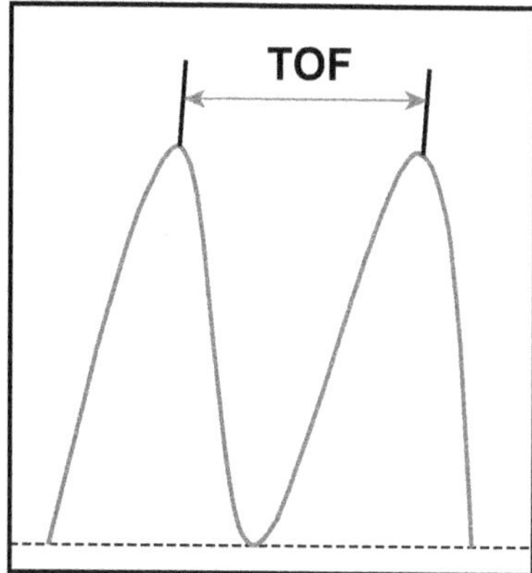


**FIGURE 9.6** Two-peak signal for measuring the velocity of the particle.

where TOF means time of flight between two light pluses collected as particle image spot moves from one slit to the second, and S is the separation distance between the slits.

### EXAMPLE PROBLEM 2

A DPV-2000 diagnostic system with 5x magnification lens is used to measure the velocity of a radiating particle passing the two-slit mask (distance between slits is 90 μm) and the time of flight is 100 ns. Calculate the measured velocity of the particle by DPV-2000.

### SOLUTION

The particle velocity V can be calculated simply as:

$$V = \frac{S}{TOF} \times \text{Optical magnification of the lens}$$

Therefore, S = 90 μm; TOF = 100 ns; and magnification of the lens = 5×

$$V = \frac{90\ \mu m}{100\ ns} \times 5$$

$$V = 4500\ m/s$$

The measured velocity of the particle is 4500 m/s.

### 9.4.7 In-Flight Particle Diagnostic System 7: Accuraspray

TECNAR Automation Ltd., Saint-Bruno, QC, Canada, created the Accuraspray diagnostic system (shown in Figure 9.7). It is based on similar concepts as that of the DPV-2000 but offers intermediate data depicting particle properties in a significantly wider assessment volume of approximately 3.25 mm. Velocities are

(a)

(b)

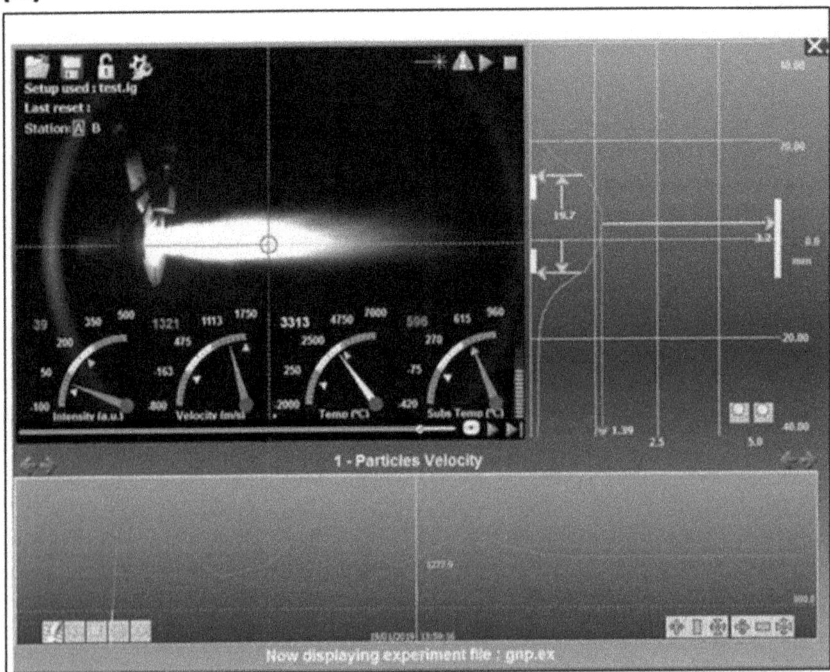


**FIGURE 9.7** (a) Digital image of Accuraspray sensor; (b) the screen is showing the live temperature and velocity of in-flight particle during the experiment.

*Source*: Open access from [10].

calculated using cross-correlation of data documented at two places close to each other [5]. It also uses a two-color pyrometer to monitor temperature. Furthermore, the temperature is measured by detecting electromagnetic radiation generated by the particles. Sensor delivers an average of data from ensemble particles passing through its 200 mm$^3$ measurement chamber. The signal obtained at a wavelength is the outcome of combined radiation from particles in the measuring volume, hence represents the ensemble temperature. This apparatus's minimum detectable temperature is roughly 900 °C [2].

There are a few Accuraspray versions based on single-point measurement (Accuraspray 4.0) and two-point measurement (Accuraspray 5.0, Accuraspray G3C). In the first one, the signal obtained from collected radiation captured at a point in the measurement is split using a dichroic mirror into two parts. On the other hand, two-point measurement system measures E ($\lambda_1$) and E ($\lambda_2$) at two separate points almost 3 mm apart along the spray axis. The signals are subsequently routed via band pass filters, allowing two unique wavelengths to reach each detector. The double-point measuring system offers some practical advantages in terms of measurement robustness, configuration ease, and cost-effectiveness. There is no requirement of an adichroic mirror in the configuration, making it easier to set up while maintaining consistent results. Also, from a business standpoint, it is less expensive to produce, which benefits both customers and developers [2].

## 9.5 LIMITATIONS OF IN-FLIGHT PARTICLE PARAMETERS MEASUREMENT

The temperature measurement of the in-flight particle is sometimes not precise, and in some circumstances inaccuracies might reach 20%. Furthermore, the value of average temperature of the particle is uninformative; if the measured average temperature value is greater than *Tm*, it doesn't mean all the particles are completely melted on impact. And, if the measured average temperature value is lower than *Tm*, it doesn't mean all the particles are melted on impact. With particle temperatures distribution, interpreting coating characteristics is significantly easier than with mean values or ensemble measurements.

The coating qualities will be substantially influenced by splat generation, solidification, splat flattening and layering; these are dependent on the spray gun speed, distance and the angle. To avoid temperature differences in coatings, we must manage the temperature of the coating surface and substrate, as well as the production of oxide during the coating's creation. Due to the temperature differential, residual tension may arise during the solidification process after coating. The coating properties can be considerably influenced by oxidation, which occurs through the diffusion or convection of molten particles and is not detectable by in-flight particle diagnostics. All of these variables play a role in regulating the coating temperature [11–21].

## 9.6 ONLINE CONTROL AND MONITORING

Until now, in-flight sensors have been used to assess particle attributes and then correlate them to calculated values and coating qualities. The main goal was to increase coating dependability and reproducibility by optimizing spray settings. In addition, a relationship between particle parameters and coating qualities must be modeled. For example, despite the little diversity in particle state, coating characteristics show a great deal of variation. Other coating data, like coating average temperature, stress distribution, and spray pattern, must be documented. Robot trajectory planning designated for high-quality spray coating processes should not only include accurate kinematics definition and control but should also ensure appropriate thermal guidance during deposition [22].

### 9.6.1 Spray Patterns through Robot Trajectory Planning

Robot trajectory planning can increase the coating's durability and functional qualities by establishing consistent spray patterns by programming the robot trajectory either online or offline. In comparison to online programming, offline programming yields superior results. Floristan et al. [22] and Deng et al. [23] are two studies worth reading for a better understanding.

## 9.7 DIAGNOSTICS OF COATING UNDER FORMATION

### 9.7.1 Coating Temperature

The temperature of the substrate at any stage of spray process (before, during or after) is critical for coating quality, consistency and repeatability. The following are some of the different methods for monitoring coating temperature during the spraying process:

1. Infrared (IR) pyrometers (wavelengths >6 μm). This reduces sensitivity to plasma, hot gases and particle radiation.
2. IR thermography.
3. Micro-thermocouples embedded in a metallic substrate with a quick reaction, which are used in conjunction with IR pyrometers to calibrate surface temperature data.

### 9.7.2 Stress Development

During the whole spray process, the stresses generated at the coating-substrate contacts are recorded in real time (preheating, coating, cooling). The temperature of the coated surface and the substrate linked to a water-cooled shield is monitored by the spray and deposit control (SDC) pyrometer (Figure 9.8). The substrate curvature is recorded by the linear variable differential transformer (LVDT). The SDC is also used to track particle trajectories in real time.

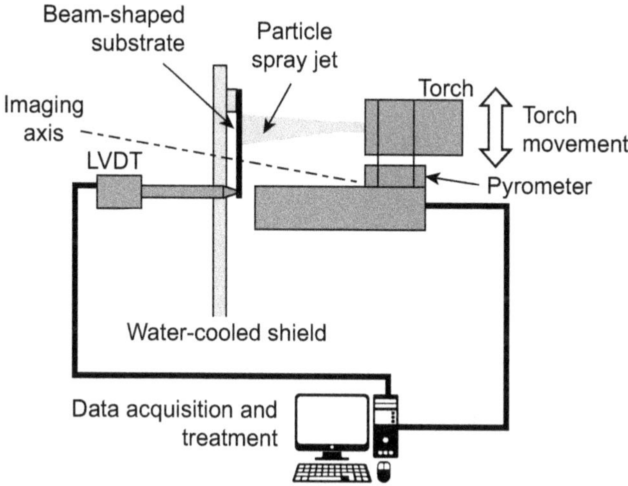


**FIGURE 9.8** Schematic representation of the set up used to measure the stresses developed during spray process.

*Source*: Open access from [24].

The most suggested method for measuring residual stresses that occur during the coating process in thermal spraying coatings is to determine the curvature of the substrate/coating combination. The specimen deflection photographed using a conventional camera can be converted to a $k$ value by assuming the uniform specimen curvature along its length. Figure 9.9, for example, anticipates the sample's curvature when sprayed at 400 °C and 600 °C. Stress distributions in the substrate and coating for the two different thermal histories are also predicted (Figure 9.4c) [25].

## 9.8 MACHINE LEARNING IN THERMAL SPRAY

The thermal Spraying coating phenomenon is linked to a number of complex physical processes. Because of various factors involved, and also due to nonlinear interdependence among these parameters, precision control and optimization of the thermal spraying techniques is a time-consuming and costly task. To quantify these complicated interactions and improve process repeatability, computer-aided approaches are required. To capture the associated complicated physical phenomena, simulation and modeling methodologies such as computational fluid dynamics (CFD) are frequently used. Although CFD has a lot of potential for understanding thermal spraying coating technology's sub-processes, the balance between model accuracy and computational cost has always been a difficulty in CFD [26]. Utilizing machine learning (ML) methods for the creation of a digital twin of the technique is a promising alternative for replacing the computationally

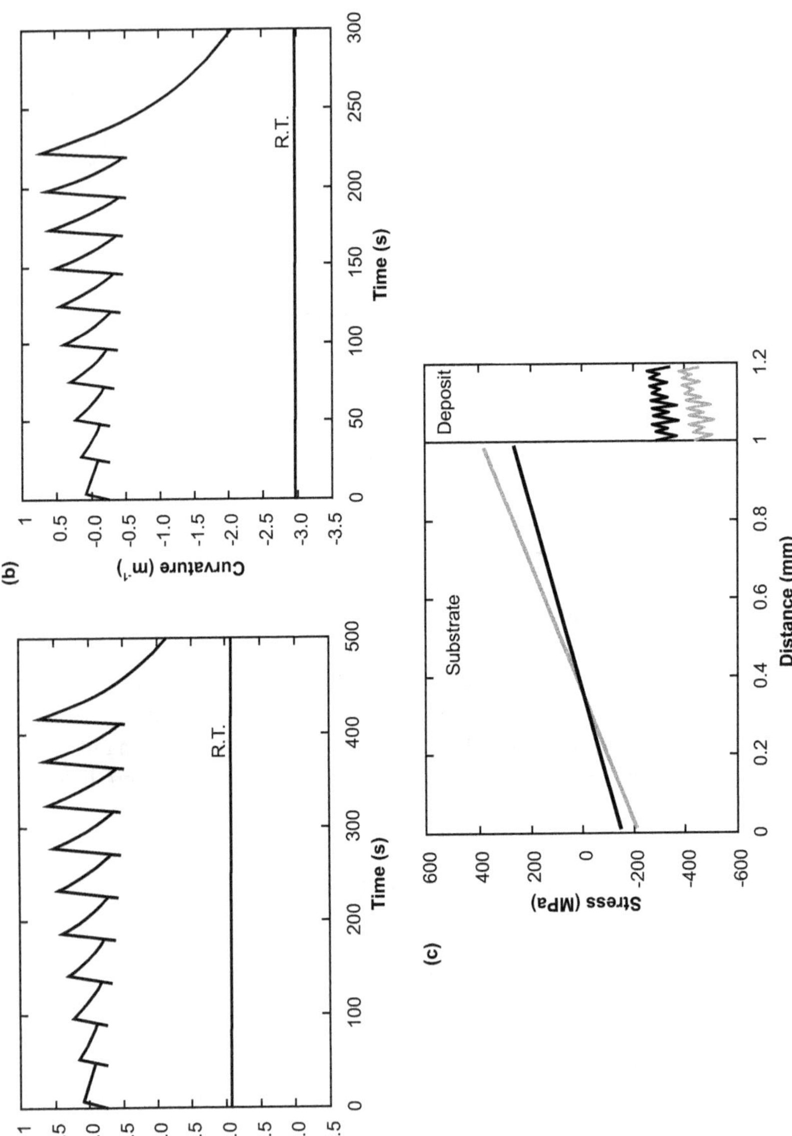


**FIGURE 9.9** (a–b) Curvature prediction of the specimen sprayed at 400 °C and 600 °C, respectively; (c) residual stress distributions for specimens, with thick lines for specimens sprayed at 400°C and thin lines for those sprayed at 600°C.

*Source:* Reprinted with permission from [25].

expensive CFD simulations in plasma spraying. Digital twin is basically a virtual and computerized replica of a physical system in the real space that includes information and data connecting the two systems, namely, virtual and physical [27]. This digital replication is mostly accomplished through the use of artificial intelligence approaches for system optimization, monitoring and prognostics. As a result, the considered system gains improved efficiency, accuracy and economic benefits [28]. Early research has primarily relied on experimental data to build digital twins for thermal spraying processes with the goal of predicting powder particle characteristics or managing the process parameters. In the literature, only a few studies have used simulated data sets to train ML models in thermal spraying [29–30].

### 9.8.1 Data Preparation and Structure of Simulated Data

The digital twin for the thermal spraying process is created using machine learning approaches to predict the properties of in-flight particles, which will vary depending on the input process parameters. This necessitates a large number of process parameter data sets, as well as particle properties obtained from spray jet CFD simulations. The data is prepared with two different design of experiments (DOE) methodologies: central composite design (CCD) and Latin hypercube sampling (LHS). After data is prepared, it is fed into a residual neural network (ResNet) and a support vector machine (SVM) to forecast particle properties [31]. Following that, the computed prediction accuracy of various ML models and DOE techniques is compared. Particle collisions and plasma flow turbulence cause randomness in particle behavior, because of which current ML techniques cannot provide a precise forecast of each particle's attributes. In plasma spraying, on the other hand, precisely predicting the average properties of particles is a critical performance indicator that can greatly benefit in analyzing the relationship of coating properties on process parameters, for example. To understand these relationships, the input variables are modified and changes in the system output are analyzed. CCD and LHS are employed to cover a set of typical input process parameters for the simulation. The DOE method takes into account the flow of primary gas and carrier gas, electric current, feed rate of the powder, size distribution of powder at the injection site, and stand-off distance as input process parameters. The DOE methodologies can be executed in MATLAB and then linked to the simulation runs' batch task scheduler to develop an automated data preparation pipeline.

DOE methods: On one hand, CCD is commonly employed in the development of second-order response surface models. It is based on a two-level design (full/factorial) with $2k$ points (where $k$ denotes independent variables) between the axes and a set of repeating points at the centroid ($N_0$). The computer experiments will give consistent results in multiple trials, unlike random errors in physical experiments. As a result, performing many runs at the centroid becomes useful in physical tests which correspond to CCD with six-factor fractional design ($2^{k-1} + 2k + N_0$) [29]. LHS design, on the other hand, is a space-filling strategy that divides the multidimensional experimental domain into *N* strata of equal marginal probability in order to reduce sample mean variance. To ensure that no point is too far away from a design point, the maximin LHS maximizes the minimal distance between every pair of experimental points inside the domain of experimentation. This will improve the developed model's prediction accuracy. As a result, LHS is an effective DOE method for computer experimentation [32]. Inputs and outputs of each CCD and LHS simulation are provided with indices in order to allocate them to the ML models.

### 9.8.2 Machine Learning Algorithms

ML models such as SVM and ResNet receive simulation data for training from DOE methods. The input parameters have already been mentioned, and the output particle properties are in-flight particle velocities, temperature, and particle $(x, z)$-coordinates on the virtual substrate at the specified stand-off distances. Different coordinates are obtained in the spray jet for particles of same size due to particle collisions and plume turbulence, thus varying in temperature and velocity temperature. As a result, while ML models are expected to accurately predict single particle properties, particle properties in average need to be reproduced with a significantly small error.

### 9.8.3 Support Vector Machine Algorithm

SVM theory is a supervised-learning approach that equally penalizes high and low errors using structural risk minimization and a symmetrical loss function too. It offers good generalization and prediction accuracy, and its computational complexity is independent of the dimension of the input space [33].

The linear SVM regression purposes to determine the true model *f* approximation in the form of:

$$g(x) = \left[w, \emptyset(x)\right] + b \tag{9.4}$$

where *w* is vector normal to *g*, and Ø and *g* are mapping function and bias parameter, respectively.

The values estimated from *g* show deviate from the true values *f*(*x*) by no more than ε; that is:

$$\left| g(x) — f(x) \right| \leq \varepsilon \tag{9.5}$$

The SVM algorithm will be implemented with the help of statistics and MATLAB's machine learning toolbox. A pre-processing step is performed to standardize the training data sets so that the inputs and outputs remain insensitive to the processing scales and magnitudes. Figure 9.10 presents a result set of the mean particle temperature from the CCD and LHS data.

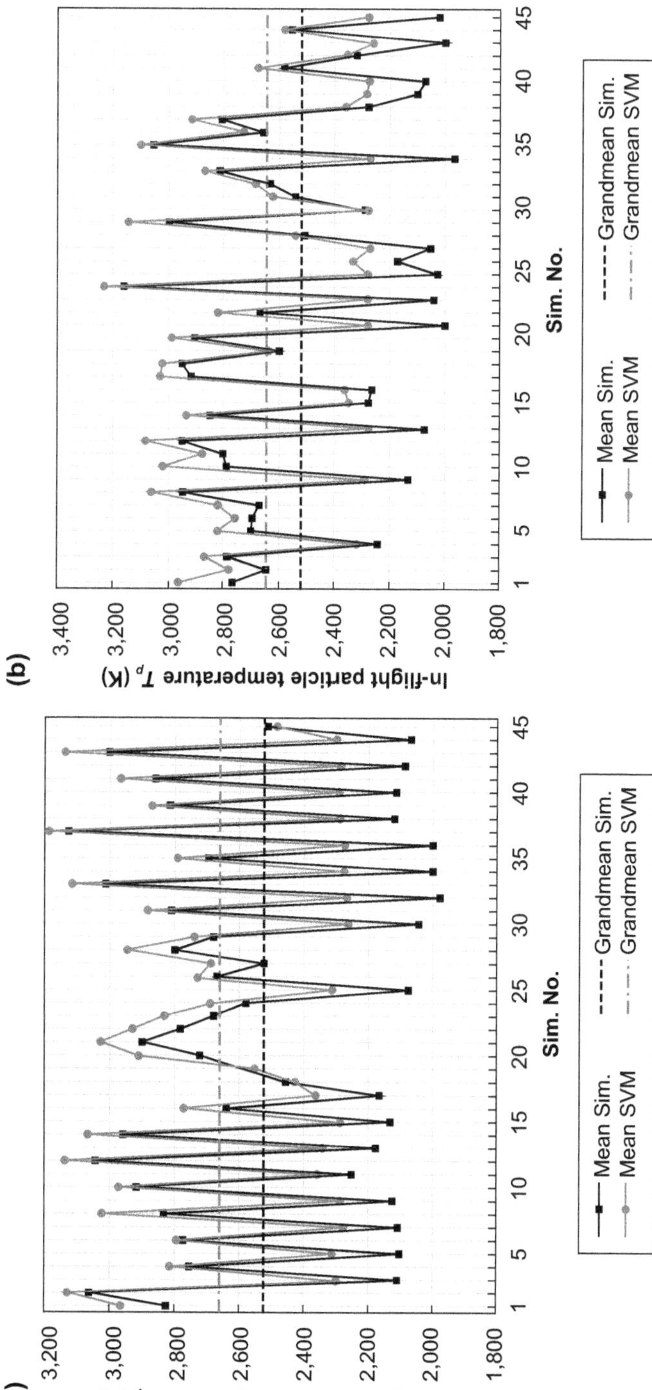


**FIGURE 9.10** Results showing mean temperature of particle per simulation for the SVM model using data sets from (a) CCD and (b) LHS [26].

*Source*: Open access from Springer Nature.

### 9.8.4 Residual Neural Network Algorithm

A multilayer perceptron characterized by a mathematical model that translates input values to output is known as an artificial neural network (ANN). The output vector $x(L)$ reflecting the ANN's prediction is determined with $L$ layers and a vector $x^{(0)}$ comprising input values by:

$$x^{(L)} = \sigma_2\left(W^{(L)}\sigma_1\left(W^{(L-1)}\sigma_1\left(\cdots\sigma_1\left(W^1x^{(0)} + b^{(1)}\right)+\cdots\right)+b^{(L-1)}\right)+b^{(L)}\right) \quad (9.6)$$

where $W^{(l)}$ is weight matrices and $b^{(l)}$ is bias vector ($l = 1, 2, \ldots, L$), $\sigma_1$ is a non-linear activation function, such as hyperbolic tangent or ReLU, and $\sigma_2$ is a linear activation function that differs from $\sigma_1$.

The purpose is to minimize the output vector $x^{(L)}$ deviation from the $y$ for a given target vector $y$. A loss function is frequently used to measure this deviation, with the square error being used for regression situations.

$$\mathrm{e} = \mathrm{y} - \mathrm{x}^{(\mathrm{L})2}_{\ \ 2} \quad (9.7)$$

It is worth noting that equation (9.7) only gives the error for a single output vector y. The average square error is obtained by summing together all errors and dividing by $N$ (number of entries in a training set). Appropriate $W(l)$ and $b(l)$, $l = 1, \ldots, L$, must be determined to minimize the error.

In reality, forward propagation is used to compute the prediction $x(L)$ of an ANN in equation (9.6), which predicts the output vector $x(l)$ of each layer $l = 1, \ldots, L$ of the network by:

$$x^{(l)} = \begin{cases} \sigma_1\left(W^{(l)^T}x^{(l-1)} + b^{(l)}\right), l = 1, \ldots L-1 \\ \sigma_2\left(W^{(l)^T}x^{(l-1)} + b^{(l)}\right), l = L \end{cases} \quad (9.8)$$

For deep networks training improvement, particular type of ANNs are designed, called ResNets. ResNet in comparison to a standard ANN has an added output and is used when equal number of neurons per hidden layer is set to the number of features. For the ResNet, the forward propagation formula can be written as:

$$x^{(l)} = \begin{cases} x^{(l-1)} + \sigma_1\left(W^{(l)^T}x^{(l-1)} + b^{(l)}\right), l = 1, \ldots L-1 \\ \sigma_2\left(W^{(l)^T}x^{(l-1)} + b^{(l)}\right), l = L \end{cases} \quad (9.9)$$

Figure 9.11 displays the structure of the applied ResNet, as well as the forward propagation technique of the ResNet in comparison to a standard ANN.

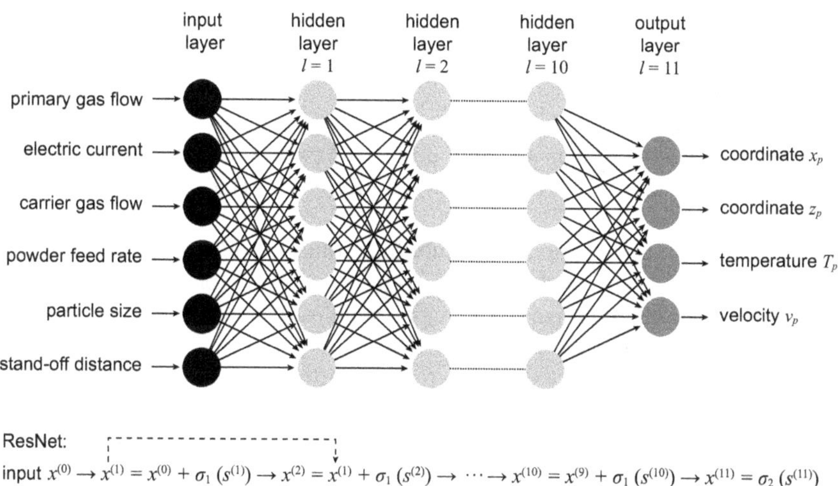


ResNet:

input $x^{(0)} \rightarrow x^{(1)} = x^{(0)} + \sigma_1 (s^{(1)}) \rightarrow x^{(2)} = x^{(1)} + \sigma_1 (s^{(2)}) \rightarrow \cdots \rightarrow x^{(10)} = x^{(9)} + \sigma_1 (s^{(10)}) \rightarrow x^{(11)} = \sigma_2 (s^{(11)})$

ANN:

input $x^{(0)} \rightarrow x^{(1)} = \sigma_1 (s^{(1)}) \rightarrow x^{(2)} = \sigma_1 (s^{(2)}) \rightarrow \cdots \rightarrow x^{(10)} = \sigma_1 (s^{(10)}) \rightarrow x^{(11)} = \sigma_2 (s^{(11)})$

with $s^{(0)} := W^{(l)^{\mathrm{T}}} x^{(l-1)} + b^{(l)}$, $l = 1,...,11$ for ResNet and ANN

**FIGURE 9.11** Comparison of the applied ResNet structure, its forward propagation procedure to compute the output vector $x^{(11)}$, to that of standard ANN [26].

*Source*: Open access from Springer Nature.

## 9.9 SUMMARY

The aim of this chapter is to discuss and reach a reliable system for measurement of in-flight particles parameters like temperature, velocity and diameter to achieve the best-quality thermal coatings to establish a thermal spray system with better repeatability, reproducibility and reliability. In this chapter, we discussed about several online process parameters like temperature, velocity and so on and their effect on the coating properties (i.e., microstructure, density, porosity). Additionally, the chapter discussed various measurement techniques to measure online in-flight particle parameters. Among all the techniques, DPV-2000 and Accuraspray provide better measurements of in-flight particle parameters. Two-color pyrometry is used in all the techniques to measure the particle temperature. The velocity measurement utilizes time between the two signals, separation between the slits, and the lens magnification. Also, we discussed the limitations of these diagnostic systems. In this chapter, we discussed online monitoring and control, and explained a feedback system (ANN optimization process) to optimize the process parameters to get the appropriate combination of parameters and limitations even after online monitoring. Additionally, we discussed the machine learning methods in thermal spraying to quantify the complicated interactions and improve process repeatability.

**Questions for Self-Analysis**

1. What are the most difficult aspects of producing a high-quality, dense coating?
2. What do (1) repeatability, (2) reproducibility and (3) reliability mean?
3. What are the plasma spraying processing parameters which will have an effect on the microstructure and mechanical characteristics of the coating?
4. Which procedures are utilized for the estimation of temperature of the in-flight particles and the temperature of the plasma plume?
5. What are the non-thermal radiation sources that will affect two-color pyrometry temperature measurements?
6. What are the various strategies for diagnosing in-flight particle process parameters?
7. What role does the substrate temperature play in coating quality?
8. How is the developed stress at the coating measured and monitored?
9. Why are ANN models used in thermal spray coatings?
10. What function does machine learning play in the thermal spraying process?
11. What is the purpose of DOE methods?
12. What are some of the most popular machine learning algorithms? How do they function?

## REFERENCES


[1] Dwivedi, G., Wentz, T., Sampath, S., Nakamura, T. (2010); Assessing process and coating reliability through monitoring of process and design relevant coating properties. *J Therm Spray Technol* 19(4):695–712.

[2] Akbarnozari, A., Ben-Ettouil, F., Amiri, S., Bamber, O., Grenon, J.-D., Choquet, M., Pouliot, L., Moreau, C. (2020); Online diagnostic system to monitor temperature of in-flight particles in suspension plasma spray. *J Therm Spray Tech* 29:908–920.

[3] Keshri, A. K. (2010); Comprehensive process maps for synthesizing high density aluminum oxide-carbon nanotube coatings by plasma spraying for improved mechanical and wear properties, Thesis, FIU.

[4] Tillmann, W., Khalil, O., Baumann, I. (2021); Influence of spray gun parameters on inflight particle's characteristics, the splat-type distribution, and Microstructure of plasma-sprayed YSZ coatings. *Surf Coat Technol* 406:126705.

[5] Swank, W. D., Fincke, J. R., Haggard, D. C. (1995); A particle temperature sensor for monitoring and control of the thermal spray process, *Advances in Thermal Spray Science & Technology*, C. C. Berndt and S. Sampath, Eds., ASM International, Houston, TX, p. 111–116.

[6] Mauer, G., et al. (2011); Plasma and particle temperature measurements in thermal spray: Approaches and applications. *J Therm Spray Technol* 20(3):391–406.

[7] Wroblewski, D., Reimann, G., Tuttle, M., Radgowski, D., Cannamela, M., Basu, S. N., Gevelber, M. (2010); Sensor issues and requirements for developing real-time control for plasma spray deposition. *J Therm Spray Technol* 19(4):723–735.

[8] Fincke, J. R., Haggard, D. C., Swank, W. D. (2001); Particle temperature measurement in the thermal spray process. *J Therm Spray Technol* 10(2):255–266.
[9] Moreau, C., Bisson, J. F., Lima, R. S., Marple, B. R. (2005); Diagnostics for advanced materials processing by plasma spraying. *Pure Appl Chem* 77(2):443–462.
[10] Islam, A., Mukherjee, B., Pandey, K. K., Keshri, A. K. (2021); Ultra-fast, chemical-free, mass production of high quality exfoliated graphene. *ACS Nano* 15(1):1775–1784.
[11] Li, L., Vaidya, A., Sampath, S., Xiong, H., Zheng, L. (2006); Particle characterization and splat formation of plasma sprayed zirconia. *J Therm Spray Technol* 15(1):97–105.
[12] Montavon, G., Berndt, C. C., Coddet, C., Sampath, S., Herman, H. (1997); Quality control of the intrinsic deposition efficiency from the controls of the splat morphologies and the deposit microstructure. *J Therm Spray Technol* 6(2):153–165.
[13] Sampath, S., Jiang, X. (2001); Splat formation and microstructure development during plasma spraying: Deposition temperature effects. *Mater Sci Eng A* 304–306: 144–150.
[14] Syed, A. A., Denoirjean, A., Hannoyer, B., Fauchais, P., Denoirjean, P., Khan, A. A., Labbe, J. C. (2005); Influence of substrate surface conditions on the plasma sprayed ceramic and metallic particles flattening. *Surf Coat Technol* 200(7):2317–2331.
[15] Fauchais, P., Fukumoto, M., Vardelle, A., Vardelle, M. (2004); Knowledge concerning splat formation: An invited review. *J Therm Spray Technol* 13(3):337–360.
[16] Chandra, S., Fauchais, P. (2009); Formation of solid splats during thermal spray deposition. *J Therm Spray Technol* 18(2):148–180.
[17] Gill, S. C., Clyne, T. W. (1990); Stress distribution and material response in thermal spraying of metallic and ceramic deposits. *Met Trans* 21B:377–385.
[18] Neiser, R. A., Smith, M. F., Dykhuisen, R. C. (1998); Oxidation in wire HVOF-sprayed steel. *J Therm Spray Technol* 7(4):537–545.
[19] Espie, G., Denoirjean, A., Fauchais, P., Labbe, J. C., Dubsky, J., Schneeweiss, O., Volenik, K. (2005); In-flight oxidation of iron particles sprayed using gas and water stabilized plasma torch. *Surf Coat Technol* 195:17–28.
[20] Syed, A. A., Denoirjean, A., Fauchais, P., Labbe, J. C. (2006); On the oxidation of stainless steel particles in the plasma jet. *Surf Coat Technol* 200:4368–4382.
[21] Trifa, F.-I., Montavon, G., Coddet, C. (2007); Model-based expert system for design an simulation of APS coatings. *J Therm Spray Technol* 16(1):128–139.
[22] Floristan, M., Montesinos, J. A., Garca-Marn, J. A., Killinger, A., Gadow, R. (2012); Robot trajectory planning for high quality thermal spray coating processes on complex shaped components. In: *ITSC 2012*. ASM International, Materials Park, OH, e-Proceedings
[23] Deng, S. H., Cai, Z. H., Fang, D. D., Liao, H. L., Montavon, G. (2012); Application of robot offline programming in thermal spraying. *J Therm Spray Technol* 206(19–20):3875–3882.
[24] Fauchais, P., Vardelle, M. (2010); Sensors in spray processes. *J Therm Spray Technol* 19(4):668–694.
[25] Gill, S. C., Clyne, T. W. (1994); Investigation of residual stress generation during thermal spraying by continuous curvature measurement. *Thin Solid Films* 250(1–2):172–180.
[26] Bobzin, K., Wietheger, W., Heinemann, H., Dokhanchi, S. R., Rom, M., Visconti, G. (2021); Prediction of particle properties in plasma spraying based on machine learning. *J Therm Spray Technol* 30(7):1751–1764.

[27] Kritzinger, W., Karner, M., Traar, G., Henjes, J., Sihn, W. (2018); Digital Twin in manufacturing: A categorical literature review and classification. *IFAC-PapersOnLine* 51(11):1016–1022.

[28] Negri, E., Fumagalli, L., Macchi, M. (2017); A review of the roles of digital twin in CPS-based production systems. *Procedia Manuf* 11:939–948.

[29] Choudhury, T. A., Berndt, C. C., Man, Z. (2013); An extreme learning machine algorithm to predict the in-flight particle characteristics of an atmospheric plasma spray process. *Plasma Chem Plasma Process* 33(5):993–1023.

[30] Zhu, J., Wang, X., Kou, L., Zheng, L., Zhang, H. (2020); Prediction of control parameters corresponding to in-flight particles in atmospheric plasma spray employing convolutional neural networks. *Surf Coat Technol* 394:125862.

[31] Myers, R. H., Montgomery, D. C., Anderson-Cook, C. M. (2016); *Response Surface Methodology: Process and Product Optimization Using Designed Experiments.* John Wiley & Sons, Schamberger.

[32] Joseph, V. R., Hung, Y. (2008); Orthogonal-maximin Latin hypercube designs. *Stat Sin*:171–186.

[33] Awad, M., Khanna, R. (2015); Support vector regression. In *Efficient Learning Machines* (pp. 67–80). Apress, Berkeley, CA.

# 10 Bulk Nanostructure and Near-Net Shape

*Ariharan S and Kantesh Balani*

## CONTENTS



According to the definition, “nanometer-sized,” “nanosized” or “nanostructured” materials are particles or an internal structure with one dimension less than 100 nm. In 1973, the first identification of nanostructured features was observed in the thermal spray coatings. One of the agenda items of the current advanced work is aimed at producing nanostructured coatings. Nanostructured bulk products are at the front line of materials engineering due to their outstanding properties [1, 2]. Its high surface-to-bulk ratio resulted in the high-volume fraction of the interfaces. Thus, it leads to relatively excellent utilization of nanostructured bulk materials compared to that of coarser materials. As an example, ceramics exhibit a significant improvement in dimensional stability, thermal insulation (high phonon scattering at the interfaces), mechanical properties (particularly hardness and fracture toughness) and tribology (the variation in the material fracture and removal mechanism of the ultrafine structure) [2, 3]. The interest of nanostructured bulk coatings in thermal spraying is briefly described, and the exploitation of the process to new insight is also described in the current chapter.

DOI: 10.1201/9781003321965-10

# 10.1 DESIGN AND CONTROL OF BULK NANOSTRUCTURE

## 10.1.1 Nanostructured Coatings

An important characteristic of thermal spraying is its capability in the retention and production of the nanostructured coatings of thickness in the range of a few tens of microns to several millimeters. Almost all classes of materials, such as metals, ceramics and composites, are used to form products of nanostructures using thermal spray technology. Historically, the development of thermal spray deposition is required to improve the efficiency of aircraft gas turbine engines by increasing the operating temperature. Later, the application domain quickly extended to several major areas, such as the automobile industry, energy and environment sector, surgical implants, diesel and gas turbine engines and more. In thermal spraying techniques for producing the coatings, the powders or wires for materials are exposed to a combustion flame or plasma plume, and the solid/semi-solid state materials are accelerated towards the parts or substrate using the high-velocity gas stream. The high-impact forces of the accelerated particles at the substrate encourage good adhesion between the particle and substrate. At the same time, densification will also be achieved due to successive impingement of accelerated particles. Nevertheless, the inability to prepare the reproducible coating is a major problem due to the fluctuations in the parameters associated with the instrumentation and others. It is difficult to control the characteristics of the coatings, such as composition, microstructure, residual porosity and technical problems associated with the feeding rate of the powder to the spray gun. In addition, it appears to be difficult to enhance the characteristics of the coatings with the direct route of enhancing the properties with the existing technology. Similarly, the problem of powder feed that flows into the spraying gun is also complex and cannot be solved easily. So several scientific approaches have been attempted, and the properties of the coatings have been improved to overcome these major issues. More achievements need to be innovated to appreciate the complete prospective of achieving consistent and uniform delivery of nanosized powders into the spraying gun. Thus, an improvement has been attained by optimized parameters of two major deciding aspects, such as starting materials and thermal spraying instrument–related factors. As part of starting materials, related characteristics (e.g., particle and/or agglomerate size, shape, density, heat capacity, thermal diffusivity) are major factors that affect the properties of the deposited coating. The nanosized powder fed through the slurry and in-situ generated nanostructures from a precursor are methods utilized to overcome the existing problems of obtaining bulk nanostructured components [4]. The process optimization with a control system and in-situ monitoring is required to synthesize a specific nanostructured coating.

The attention focused on the mechanisms of the fabrication of nanostructured coatings using thermal spraying as a step towards tuning the system capability. The comprehensive sensing and control system intended for thermal spraying is schematically shown in Figure 10.1. Further, analytic methods need to be established on the following points:

1. The velocity and spatial distributions of the powder particles in the plasma plume or combustion flame;
2. The type of the chemical reactions formed due to the temperature of the plasma plume or combustion flame;
3. The characteristics of semi-solid particles impingement and splat quenching phenomena on the substrate after being exposed to high temperature.

Thermal imaging and laser strobe illumination are some of the convenient methods to determine the spatial distribution of in-flight semi-solid particles and their velocity distributions. The chemical reactions that occur at the particles due to suitable thermodynamic conditions during the spraying are monitored using customary spectral sensors. The chemical species present in the flame or plasma environment can be explored with the help of laser-induced fluorescence. In addition, it could be an important factor in the thermal spraying to deposit different classes of materials with various powder properties. It can be readily valued with the help of the entire thermal history of the nanoparticles during the transit in the spraying environment, which mostly determines the microstructure of the deposited coating.

Experimental data collected from various in-situ imaging and flame analysis is used to validate simulations and theoretical modeling of the thermal spray

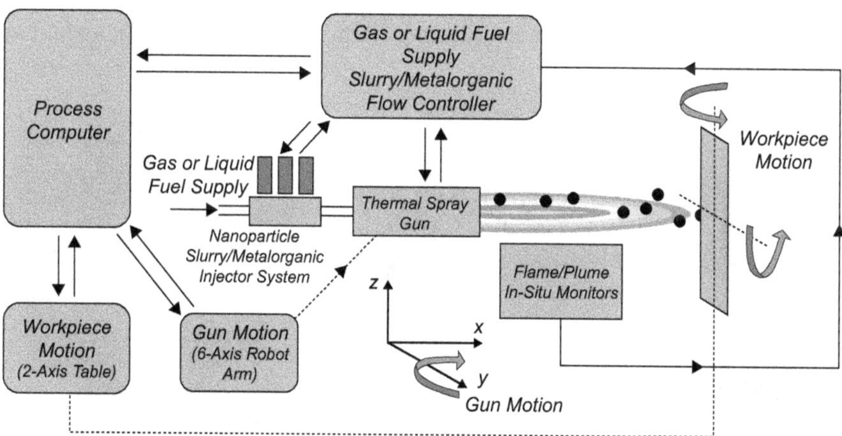


**FIGURE 10.1** Processing of nanostructured coatings with intelligent feedback system in the thermal spray system.

*Source*: Adapted and redrawn from [5].

process. General optimization of all variables are needed, as gas and particle flow rates, gas pressure and substrate (stand-off distance, temperatures) will be correlated with the microstructure and morphology of the coatings. It will be optimized with computer simulations and real-time data analysis that leads to a resultant process simulator with intelligent control algorithms and implement in the development of the system. It is expected that will provide a fundamental design and construction of the thermal spray assembly, which includes a real-time process control system. The control system attached with the thermal spray system includes multi-axis (six-axis) robotic control of the spraying gun as well. It will allow a uniform deposition as per the shape of the complex parts in a reproducible model. It is important for the production of a desired microstructure with a microsized or nanosized regime in a high productivity rate, where comprehensive examination of different components is not practically possible. Successful thermal spraying can be realized with the consideration of modeling, numerical simulation and diagnostics.

### 10.1.2 Technology of Nanostructured Coatings

#### 10.1.2.1 Synthesis of Coarse Powders with Nanostructures

*Spray conversion processing* (*SCP*): The synthesis of coarser powder that contains nanostructured particles and quenching of molten particles are two practical ways to prepare the agglomerated nanostructured particles, which are synthesized by various methods including the bottom-up (solution or precipitation) and top-down approach (ball milling). It can also be obtained employing different sources (electrons, ions or photon) bombardment on a base and produce vapors in a controlled atmosphere. Later, these vapors react with suitable species in the controlled environment that produce nanostructured particles of specific composition. It has been introduced to synthesize nanograined cermets [6], (Figure 10.2), and it has

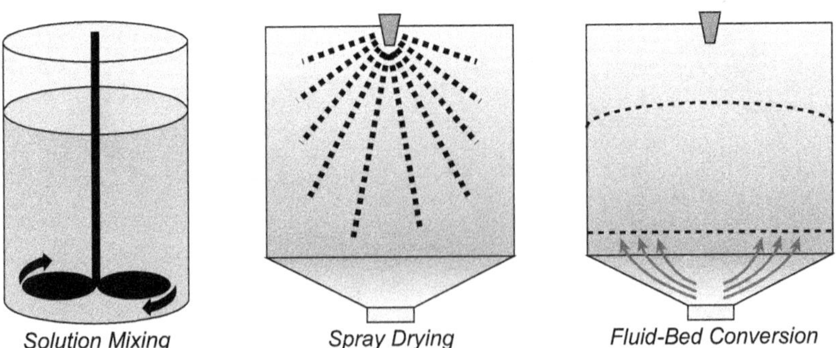


**FIGURE 10.2** Schematic representation of nanosized powder synthesis from solution precursors, spray conversion process.

*Source*: Adapted and redrawn from [5].

validated the viability of thermal spraying of nanosized powders. The SCP process contains three major steps:

1. Aqueous solution preparation with constituent salts;
2. Precursor (homogeneous) powder by spray;
3. Preparation of precursor powder of controlled structure by fluid bed conversion.

The SCP processing route is unique in some respects:

1. Semi-solid state slurry formation from the mixture of the particle in the aqueous solution;
2. No requirement of mechanical stirring to achieve a specific morphology;
3. Synthesis of fine structure (nanometer size) using the chemical precursor.

An optimized reprocess has been developed to as-synthesized nanosized powders into sprayable agglomerates that will be easily flowable through the standard powder feed systems. The major dissimilarities in the preparation of coating with nanosized and microsized powders in thermal spraying are schematically shown in Figure 10.3. Micro-sized particles will experience the melting of particle surface only due to the low residence time of powders in the combustion flame or

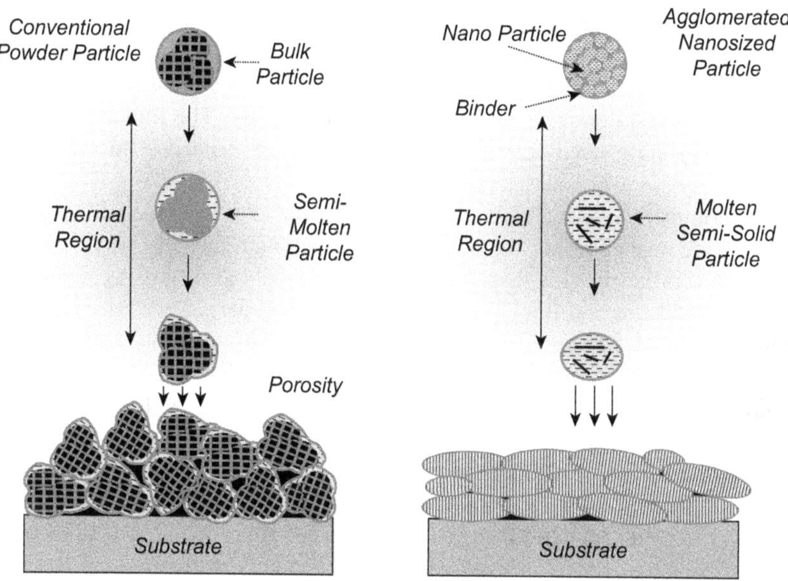


**FIGURE 10.3** Schematic of thermal spray deposition of relatively finer and coarser particles.

*Source*: Adapted and redrawn from [5].

plasma plume. But in the case of nanosized powder, it will be homogeneous or bulk melting. During impinging of semi-solid powder on the substrate, nanosized particle shows an easier flow to form a lamella than the micro-sized particles. Therefore, it forms a dense coating with a nearly uniform nanostructure.

The first scientific report on SCP [7] was on the reprocessing of as-synthesized nanosized WC-Co powder, which was effectively utilized for thermal spraying. The rapid kinetics of nanosized WC particles suspended in liquid Co leads to predict the relative quantities of individual WC and Co phases. Over the pseudo-binary eutectic in the WC-Co phase diagram, the degree of superheat is the controlling factor. The high superheat of powder leads to a decrease in its viscosity, and hence the spreading rate after impinging of the semi-solid particles with the substrate will be high. Under selected circumstances, these particles display a dynamic viscosity and it decreases with a high shear rate. In addition, the turbulent stream at the impact of semi-solid particles is useful to disintegrate the nanoparticle agglomerates, and that will promote homogeneity in the microstructure of the deposited coating.

The high hardness of the nanostructured composite coating is achieved by nullifying the decarburization during the exposure of the powders in the high-temperature flame/plasma. Low-pressure thermal spraying is one of the methods to avoid decarburization. The protection of the nanosized powder particles is assured from the severe oxidation in the low-pressure plasma plume. This modification in the process leads to an achievement of high-density thermal spray coatings and high hardness. In alloys with high Co content, the splat-quenched coating consists of uniform dispersion of nano-WC in the Co-rich phase [6, 7]. On contrary, the decarburization of WC-Co is not severe in HVOF coatings. It is due to relatively shorter residence time and low temperatures of powder particles, compared to that of plasma spraying. The important innovations of SCP have been introduced to all classes of materials during thermal spraying.

The commercially available fine nanosized particles or milled from coarser particles are suspended in the solution cooling media (usually, alcohol) called slurry. Generally, the size distribution of the powders after milling depends on the duration of the milling. The aggregation of fine particles is avoided by the addition of dispersing/stabilizing agents in the suspension. Commercially available dispersion agents, such as Trusan 471, KB4027, Dolapix CE64, oligomeric, sodium polyacrylate, Hydropalat, tetrasodium diphosphate and phosphate ester are used to formulate the suspension for $Al_2O_3$, $ZrO_2$, $Y_2O_3$, $TiO_2$, $Ca_5(PO_4)_3OH$ (hydroxyapatite) and $LaMnO_3$ (perovskite) [8]. The slurries are synthesized by dispersing the precursors in deionized water, ethanol, acetone and so forth. Then, the slurry can be supplied through a container followed by an injector using the compressed gases and vessel/peristaltic pump. The gas pressure in the pressurized vessel will control the slurry injection velocity. Two peristaltic pumps in the feedstock system enable the design of the setup for multi-layer or functionally graded coatings by controlling the pump's rotation. This dynamic arrangement uses the system to coat different materials with individual feed rates.

The injection of slurry feed regulates some of the key important issues, such as the trajectory, thermal history of the particles in plasma/flame. Presently, the solution is injected into (1) a nozzle (diameter: 0.05–1.2 mm), followed by (2) a spray atomization system. The continuous slurry injection is promoted and it can be disturbed due to a piezoelectric system [9]. It has the advantage to form uniform sized slurry droplets that are comparable with the spraying nozzle exit diameter. The injection velocity of the slurry, $v_1$ fed into the combustion flame is measured as per equation (10.1) [10]:

$$v_1 = \frac{q_l}{\rho_l \, A_i} \tag{10.1}$$

where $q_l$, $\rho$ and $A_i$ are the feed rate, density of the slurry and the area of the nozzle exit, respectively. Practically, the velocity is influenced by the pressure in the vessel and the nozzle orifice size. Also, the wide or narrow size distribution of the slurry droplet's size at the nozzle will be controlled with an atomization system. Among two modes of injection (radial and axial), the axial mode of injecting the slurry into the thermal spraying showed a rapid heat-up of the particles with more homogenous grains in the coatings. The phenomena of flight in the flame are dependent on the feedstock used for the spraying, such as:

1. Agglomerated nanosized powders;
2. Solution with a liquid precursor;
3. Slurry, suspension of nanosized powders.

After exposure of slurry droplets into the jet of plasma or flame, the kinetic phenomena of droplets are comparable with the conventional thermal spraying and it influences the microstructural features of the coatings. The liquid or slurry feedstock processing has exemption with size reduction of droplets by vaporization of the liquid and it is considered at the starting steps of in-flight particles. Thermal phenomena of different types of nanosized powder in flight influence the size of features or grains. The major thermal effects are shown in Figure 10.4.

#### 10.1.2.2 Agglomerated Nanostructured Solids

The agglomerated porous particles possess slow heat diffusion compared to that of the solid particles. Therefore, the melting initiates from the surface of the agglomeration (Figure 10.5), and it is anticipated that the nanostructure in the agglomerate could have vanished during melting. Accordingly the parameters need to be tuned in the thermal spraying system to retain the nanostructure of the powders. It can be achieved by increasing the in-flight particle velocity, which reduces the residence time in the jet or flame.

Figure 10.5b shows the retention of nanostructure in the deposited coating from the agglomerated powder particle of poor diffusion of heat. Further, the substrate temperatures have to be maintained in such a way that they should retain

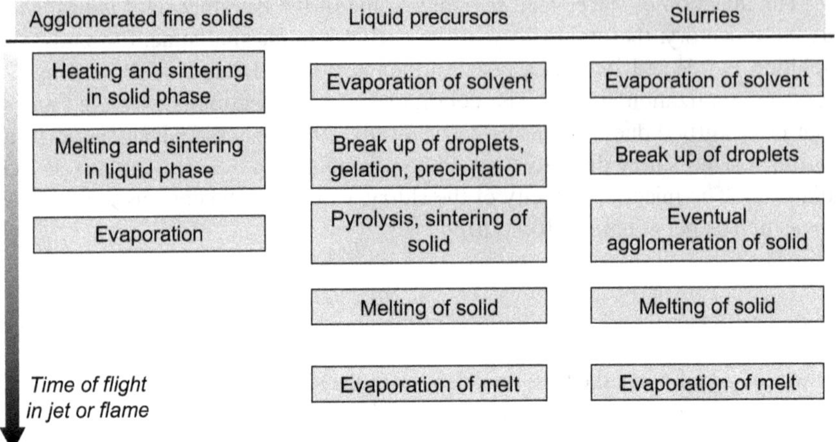


**FIGURE 10.4** Thermal phenomena during the flight of nanosized feedstock in flame or plasma.

*Source*: Adapted and redrawn from [8].

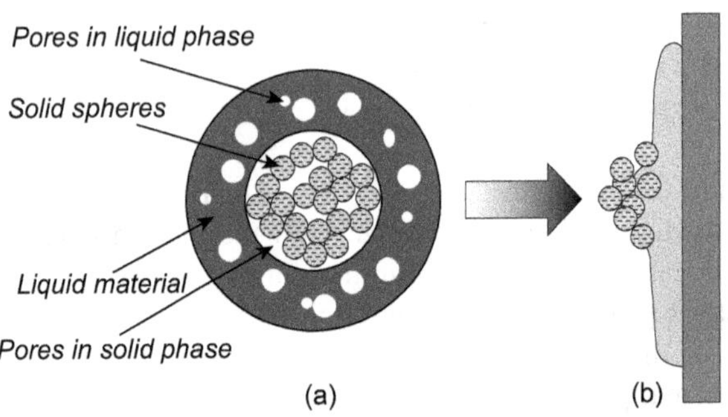


**FIGURE 10.5** (a) Melting of agglomerated porous particle and (b) lamella formed from the partially melted particle.

*Source*: Adapted and redrawn from [8].

the nanostructure from the grain coarsening. It can be achieved by cooling coated substrate at a high rate.

#### 10.1.2.3 Liquid Precursors

The behavior of liquid precursors in plasma or flame focused on the temperature inside liquid droplets and powder synthesis due to precipitation. The precipitation of the droplets fragmented at their primary interaction after their exposure into plasma, followed by the evaporation of the liquid phase. Subsequently, the precipitates occur as per the following four possible routes:

1. Formation of solid particles from the uniform solute concentration and volume precipitation;
2. Realization of super-saturation condition at the surface of droplet leads to the development of dense solid shell at the surface; later, these are fragmented and separated individually;
3. Liquid phase in the solid shell formed in the route will further form small solid spheres after shattering of solid shell;
4. The development of a deflated elastic shell due to the super-saturated surface of the droplet.

The follow-up stages include melting of the solid phase and evaporation of the liquid phase used to disperse the particles. It is also an important step to decide the shape and size of the powder end product.

#### 10.1.2.4 Suspension of Fine Particles

The thermal spraying of suspended fine particles in a liquid has significant differences for spray drying:

- Hot heating environment with combustion flame or plasma plume, etc.;
- Boiling of liquid used for the suspension of fine particles;
- Relatively short exposure time (milliseconds) in the flame or plasma;
- Binder-free solution.

The vaporization of liquid from the droplet happens at the initial stage of the droplet's entry into the high-temperature environment. On the vaporization, the fine particles accumulate together due to sintering. The beginning stage of the process is responsible for the droplet breakup and so on.

### 10.1.3 Microstructure of Coatings with Agglomerated Fine Powder

The impingement of sprayed powder particles on the substrate causes it to deform and solidify. Then, it forms the lamellae with two primary shapes [8]:

1. A pancake shape, due to semi-solid particles with low energy (both thermal and kinetic energy);
2. A flower shape, due to splashing of the semi-solid particles.

The microstructure of the coating mainly depends on spraying feedstock. Thermal spraying of agglomerated particles from nanosized powders ends up in any form. It depends on the state of the particles (molten or semi-molten form), as shown in Figure 10.5. The microstructure contains the molten particles/grains, as well as the sintered agglomerates (Figure 10.5b). The final microstructure will be of bimodal distribution of grains. The first maximum could have corresponded to the nanoparticle size in the starting agglomerated powder, and the second maximum will depend on the solidification rate of the molten or semi-molten deposit.

The lamellas formed from the liquid precursors depend on the form of a molten or semi-molten particle before impact. Pancake-shaped lamella will form due to volume precipitation and complete melting of nanosized solid particles. Here, the pancake-shaped lamella is preferable compared to that of the flower form. It is due to the forces that keep the liquid together being higher than the one that resulted from the dynamic force of liquid splashing during the impingement. However, there is no splashing in the case of semi-solid particle deformation. Nevertheless, the coating microstructure will be of deposits with hollow shells without sintering and melting of fine particles during pyrolysis of organometallic decomposition. Therefore, the un-altered particles during the poor thermal exposure are accommodated in the coatings as tiny crystals. Similarly, the sprayed slurries have pancake-shaped lamellae. In addition, most of the small particles do not deform to form an efficient lamella due to insufficient kinetic energy during spraying.

The solidification occurs simultaneously after the deformation of molten particles and the columnar structure established inside lamellae. Thus, the final coated product will be empowered with the surface of superior properties with the stability of nanostructures. It will help to develop near-net shaped products of complex shapes with materials that are known to be difficult to process.

## 10.2 NEAR-NET SHAPE PROCESSING

Processing of near-net shaped products using thermal spraying comprises powder feed, melting and acceleration of the solid/semi-solid for the deposition on a substrate or mandrel [11]. The coating deposited on the structure is called a mandrel, which will be in the shape of the anticipated product. Later, the deposited structure detached from the mandrel and is used as a stand-alone part. A wide range of simple and complex shapes are prepared using this near-net processing that contains several steps (Figure 10.6) [11]. The process provides refractory metals and ceramics specifically due to their highly brittle nature and difficulty to machine for the preparation of products. The big advantage of this process is the effective usage of materials with the least wastage in the raw materials. In addition, the prepared shape requires no or slight finishing for the final use. Mostly, the density of the prepared products is subject to in-flight particle velocity that controls the particle's exposure in the high-temperature region. Thus, it affects the rotating mandrel with a particular kinetic energy. Further, the computer-controlled spraying gun and the rotating mandrel are crucial, especially for the production of critical shapes. The combustion flame or plasma implies the melting of almost all the material is suitable for spraying of near-net shape with specific thickness and geometries. However, the essential key for the development of nanostructured near-net shape is the processing route. Therefore, the

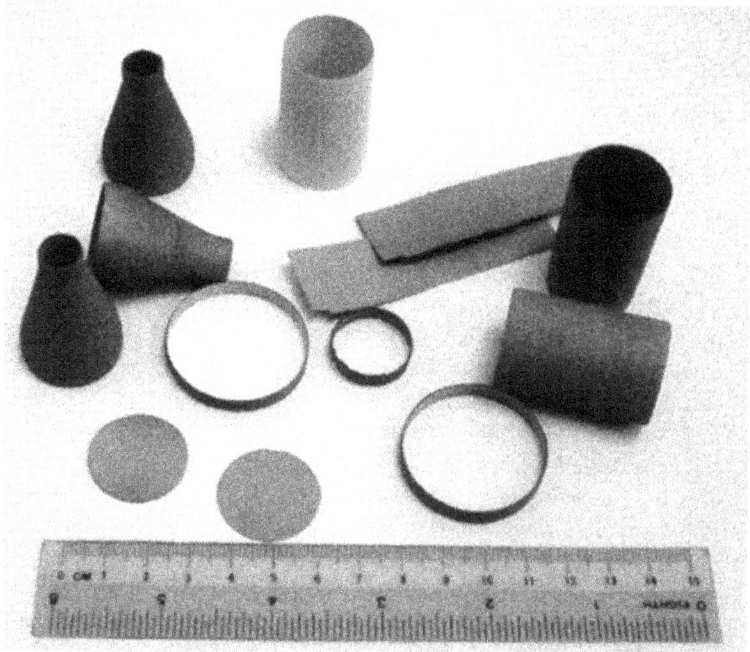

**FIGURE 10.6** Spray-formed nanostructured components of different dimensions.
*Source*: Reprinted with permission from [13].

processing route is designed in such a way that the nanostructured features are retained without any major changes in the final product. It is attained by optimal controlling of thermal spray parameters, powder feedstock, and the substrate or mandrel temperature. As an example, a tapered aluminum (6061 alloy) mandrel was used to synthesize a nanostructured $Al_2O_3$-dispersed composite as arc cylinders for the application in the energy and electrical industry. A mandrel with a height of 100 mm, diameter of 62 mm (bottom) and 57 mm (top) [11] was used for the preparation of the arc tube. The design condition for this is motivated with the help of the replication method to fabricate a lightweight X-ray mirror for the advanced telescope in astrophysics [11, 12]. Also, the criteria for the design of mandrel will depend on the application of the near-net shaped structure. Further, a random selection of mandrel materials will affect the final quality of the near-net shaped product. So, the choice of mandrel material will be restricted based on the processing route and complexity of the shape.

## 10.3 MANDREL CHOICE FOR NEAR-NET SHAPE PROCESSING

The choice of mandrel material for thermal spraying should possess at least high stability during thermal exposure, be chemically inert with the deposits and be easily removable after completion of the coating process. But the ideal mandrel material should possess the following characteristic features, as listed here:

1. *Made of coating precursor materials*
   The process to develop the results of the near-net shape in the effective use of precursor materials as a choice of the mandrel is preferred. It lends to the synthesis of the coating over the mandrel, and there is minimal wastage of sprayed parts during the removal of the mandrel by machining in a destructive way. Also, the final machining can be of rough finishing because precursor material itself can act as coating composition.
2. *Smooth surface finishing*
   The smooth surface finishing (with $R_a$ of a few nm) of the mandrel surface is preferred to avoid the damage of freestanding structure during mandrel removal. Comparatively, a rough mandrel surface (with $R_a$ of few tens of μm) leads to strong interlocking of coatings with the mandrel, hence coating separation will be difficult.
3. *Withstand high-temperature exposure*
   The mandrel should be stable during the thermal spraying to provide support for the coatings to form the desired shape. Therefore, the mandrel should be stable from burn-off, high-temperature oxidation. A small amount of non-uniform porosity induced due to burn-off can increase the coating interlocking with the rough mandrel surface, which should be avoided.
4. *Chemically inert with the spraying materials*
   A chemically inert mandrel surface avoids strong metallurgical bonding of coatings, which leads to easy removal of the mandrel that provide freestanding parts. It can be adjusted by making the mandrel surface with the composition of the coating itself. Later, it can be removed during the machining at an average finishing level.
5. *Shape should be the negative figure of the desired product*
   As a general rule, the mandrel should be the negative figure of the desired product. But, it is not necessary to build up a complex product. It is recommended to coat with a robotic arm (six-axis) controlled spraying gun to get the desired product with a uniform microstructure, because the properties (e.g., porosity, hardness) of the coatings highly depend on the direction of deposits of mandrel surface.
6. *Large difference in coefficient of thermal expansion (CTE) with coated materials*
   The large difference in the CTE increases the tensile stress at the interface, which leads to easy removal of the mandrel. The mandrel-coating system is rapidly cooled or heated to create a high mismatch in CTE and thus increases the stresses at the coat/mandrel interface. It is one of the major methods for successful removal of the mandrel.
7. *Ease of machinability*
   The low-density and ductile mandrel can be easily removed by machining.

8. *Minimal final machining*
   It is related to the ease of machining at the end of the mandrel removal. Though, if it is hard to remove the mandrel, final finishing should be easy to achieve to utilize the end product without any further processes.
9. *Design with sufficient heat transfer coefficient*
   The proper design of the mandrel is an important factor to control heat transfer during the spraying period. During the thermal spraying, the heat has to be dissipated quickly to retain nanostructures in the coating. Also, it avoids the hot spot, and thus it maintains uniform microstructure throughout the coatings.
10. *Withstand at the rapid temperature change*
    A high rate of heating/cooling leads to poor thermal shock resistance, and thus it induces failure of the coating. In addition, the rapid rise in the temperature cause cracking as the expansion coefficient rate of metal/alloy and non-metallic mandrel are dissimilar. It leads to catastrophic failure of the coatings, particularly in the case of ceramic coatings. So, it is desirable to choose the mandrel materials with an optimum CTE to avoid the peel-off at the pre-matured condition with insufficient coating thickness.

The choice of mandrel depends on several variables, as mentioned above. As a prerequisite, the mandrel should withstand high heating and cooling rates during the coatings processes. In addition, the maximum temperature of the mandrel should be less than its softening or melting point to avoid the stimulation of the failure on the coated structure. Thus, as the primary selection condition, graphite is one of the choices for the use of a mandrel. A dissimilar CTE between the mandrel and coating material resulted in high contraction after thermal spraying. Thus, it assists in the release of coating from the mandrel. Wide variation in the CTE mismatch is not recommended, as it is expected to promote the crack initiation at the interface during the preparation of the coatings. As an example, for mandrel choice, the design criterion for the next-generation space telescope is focused with the help of the replication method to fabricate lightweight X-ray mirrors (Figure 10.7) [11, 12].

As shown in Figure 10.7, solid cylinder-shaped first-generation mandrels have holes of specific dimensions located at strategic positions to place the thermocouples. The conventional mandrel offered trouble in the removal of the coating. Therefore, a low tapering angle (~2°) was introduced in the design of the second-generation mandrel. It predicted that the tapering angle would comfortably separate the coating. However, the heat dissipation issue is the main problem of the early generation mandrels and is attributed to the tolerance less shape of the mandrel design that causes the cracks on the surface. The localized melting (hot spot) due to high-temperature exposure is also one of the reasons for the damages in the coatings. To overcome the limitation of the first- and second-generation mandrels, the third-generation mandrel has evolved with the design to improve heat transfer

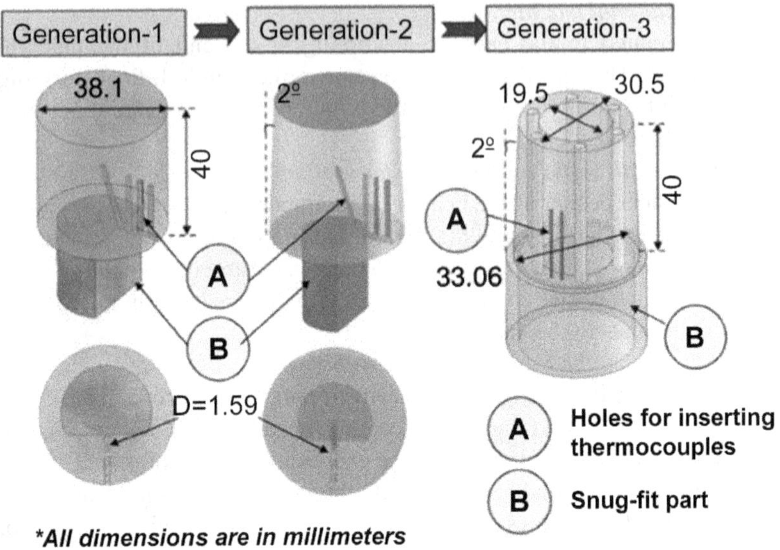


**FIGURE 10.7** Three generations of the mandrels for space telescope X-ray mirrors.

*Source*: Reprinted with permission from [14].

coefficient using a hollow-shaped mandrel with six holes as shown in Figure 10.7. The main advantage of this third-generation mandrel is that there is a provision to cool inner walls using a coolant. The lowest part of the mandrel is fitted snugly onto a hollow tube while spray coating. So, it permits the coolant to enter into the mandrel completely (hollow shaft as well).

## 10.4 MANDREL REMOVAL FOR NEAR-NET SHAPE

Based on the selection of suitable mandrel, the method of mandrel removal from the coated structure will be either in a destructive or non-destructive approach. The non-destructive method of mandrel removal will be preferable due to the ability to reuse the mandrel, which causes less contamination for the environment.

1. *Machining of the mandrel*
   Machining of the mandrel is one of the basic operations to remove the mandrel to achieve a simple freestanding product. The different types of machining will be utilized based on the complexity of the shape to remove from the deposited coating. Also, the destructive way of removal by machining leads to restricting the reusability of the mandrel and it will increase the cost of the end product.
2. *Selective degradation*
   This removes the mandrel by oxidizing it by exposure to an extremely corrosive or oxidizing environment. Thus, the mandrel will be selectively

degraded and the deposited coating will be separated. Also, the corrosive or oxidizing environment has to be chosen in such a way that it should not affect or diminish the actual properties of the near-net shape. So, the mandrel should be chosen with lower oxidation resistance than that of coating material. Usually, different grades of graphite can be preferred as a mandrel to coat ceramic materials by removal with selective degradation. It is due to oxidation of graphite that will occur at a relatively low temperature (~550–650 °C), and almost all ceramic coating will be stable at this temperature without any alteration in the initial property of the coating. But the oxidation graphite gives a serious concern to the environment and reusability.

3. *Rapid cooling or heating*
   Rapidly changing the temperature of the mandrel-coating system by cooled or heating will be utilized to remove the mandrel. The rapid cooling or heating increases the tensile stresses and thus increases the CTE mismatch at the mandrel-coating interface to assist in the relief of the parts. Also, the temperature change should be lesser than the critical temperature of the ceramics to withstand the thermal shock resistance (especially for ceramic coatings) of the deposit. The advantages of mandrel removal using the rapid cooling or heating method are reusability of the mandrel, environmental friendly and minimum final finishing.

The spray-deposited $Al_2O_3$ on several mandrels with varying thickness is one of the examples to prepare a freestanding $Al_2O_3$ for space application as X-ray mirrors [11]. After spray deposition of $Al_2O_3$ on aluminum (6061 alloy), the temperature of the mandrels is reduced using liquid nitrogen (–190 °C) to increases the CTE mismatch and assist the release of the $Al_2O_3$ shell. The CTE for aluminum mandrel is ~$25 \times 10^{-6}$ $K^{-1}$, whereas the $Al_2O_3$ shell has CTE $7.0 \times 10^{-6}$ $K^{-1}$. In addition, the cooling was executed critically to minimize the crack formation in the $Al_2O_3$ shell due to poor thermal shock resistance.

## 10.5 SUMMARY

The high demand for the practical advantages of an efficient coating rationalizes the economical effort in an effective nanosized feedstock. The deposited coatings using liquid precursor and nanoparticle suspended solutions are ongoing in the investigation and expansion stage. The usage of liquid feed in most of the spraying techniques is highly challenging. In addition, it demands alteration in the procedures associated with the instrumentation part of the spraying method. Other than the nanostructured coatings, the preparation of near-net shaped products is a highly demanding forming process using thermal spraying. A combination of experimental and theoretical approaches using computer simulations can be utilized for the efficient preparation of near-net shaped products. The effective computer algorithm combined with the mandrel and coating system will be the scope for the advanced development in the spray-forming processes. Thus, it leads to a procedure to handle the most common problem of heat flow in the mandrel-coating system and remove the experimental error to make near-net shaped parts effectively with reduced time and energy.

### Questions for Self-Analysis

1. What are the advantages of nanostructured coatings?
2. How is thermal spraying beneficial to produce a nanostructured coating?
3. What is the comprehensive sensing and control system used in thermal spraying for diagnosing and establishing effective coatings?
4. What are the different synthesis routes in thermal spraying to obtain nanosized features in the bulk coatings?
5. What are the unique steps involved in the spray conversion processing route to obtain nanostructured features in the materials?
6. What are the important characteristics of the powder feed to synthesize nanostructured bulk materials?
7. What are the advantages of thermal spray deposition of nanostructured powders instead of conventional powders?
8. Which two primary shapes can be easily formed in the thermal-sprayed coatings?
9. List the thermal phenomena that occur during the flight of nanosized feedstock of different types in jet or flame of the thermal spraying.
10. What are the main characteristics of the substrate materials to produce a near-net shaped product?
11. What are the ideal characteristics for selecting mandrel material?
12. What are the methods to remove the mandrel for obtaining the near-net shaped product?
13. How is the selective material degradation useful to remove the mandrel?

## REFERENCES

[1] McPherson R (1973) Formation of metastable phases in flame-and plasma-prepared alumina. *J Mater Sci* 8:851–858

[2] Schodek DL, Ferreira P, Ashby MF (2009) *Nanomaterials, nanotechnologies and design: an introduction for engineers and architects.* Butterworth-Heinemann

[3] Dahotre NB, Nayak S (2005) Nanocoatings for engine application. *Surf Coatings Technol* 194:58–67. https://doi.org/10.1016/j.surfcoat.2004.05.006

[4] Strutt PR, Kear BH, Boland RF (2003) Nanostructured feeds for thermal spray systems, method of manufacture, and coatings formed therefrom. U.S. Patent No. 6,579,573.

[5] Kear BH, Strutt PR (1995) Nanostructures: The next generation of high performance bulk materials and coatings. *KONA Powder Part J* 13:45–55. https://doi.org/10.14356/kona.1995009

[6] McCandlish LE, Kear BH, Kim BK (1990) Chemical processing of nanophase WC-Co composite powders. *Mater Sci Technol* 6:953–957

[7] Kear BH, McCandlish LE (1993) Chemical processing and properties of nanostructured WC-Co materials. *Nanostructured Mater* 3:19–30. https://doi.org/10.1016/0965-9773(93)90059-K

[8] Pawlowski L (2008) Finely grained nanometric and submicrometric coatings by thermal spraying: A review. *Surf Coatings Technol* 202:4318–4328

[9] Blazdell P, Kuroda S (2000) Plasma spraying of submicron ceramic suspensions using a continuous ink jet printer. *Surf Coatings Technol* 123:239–246. https://doi.org/10.1016/S0257-8972(99)00440-5

[10] Fauchais P, Etchart-Salas R, Delbos C, et al. (2007) Suspension and solution plasma spraying of finely structured layers: Potential application to SOFCs. *J Phys D Appl Phys* 40:2394–2406. https://doi.org/10.1088/0022-3727/40/8/s19

[11] Agarwal A, McKechnie T, Seal S (2003) Net shape nanostructured aluminum oxide structures fabricated by plasma spray forming. *J Therm Spray Technol* 12:350–359

[12] Agarwal A, McKechnie T, Seal S (2002) The spray forming of nanostructured Aluminum Oxide. *JOM* 54:42–44. https://doi.org/10.1007/BF02709093

[13] Devasenapathi A, Ng HW, Yu SCM, Indra AB (2002) Forming near net shape free-standing components by plasma spraying. *Mater Lett* 57:882–886

[14] Patel RR, Keshri AK, Dulikravich GS, Agarwal A (2010) An experimental and computational methodology for near net shape fabrication of thin walled ceramic structures by plasma spray forming. *J Mater Process Technol* 210:1260–1269. https://doi.org/10.1016/j.jmatprotec.2010.03.012

# 11 Case Studies

*Moumita Mistri, Shivani Gour, Alok Bhadauria, K. Vijay Kumar, Ariharan S, Rubia Hassan, Pooja Rani, Ashutosh Tiwari, Anup Kumar Keshri and Kantesh Balani*

## CONTENTS



DOI: 10.1201/9781003321965-11

Thermal spray coating technology comprises a plethora of techniques which operate at different temperature ranges and generate various coatings used to enhance or restore surface properties. This chapter aims at providing the reader with an overview of different types of coatings used for engineering applications, such as coatings used for biomedical aspects, thermal barrier coatings, Al-based coatings, ultra-high temperature coatings and so on deposited via commonly used thermal-spray techniques. It highlights the complexities associated with commercial application of thermal spray techniques and factors that determine the choice of a coating technique. It comprises eight case studies focusing on important areas of application. Each case study will comprehensively cover the principals and techniques used for coating application, material selection, coating types, quality control, recent advances in processing techniques and the effect of processing parameters on coating performance. This chapter can act as a comprehensive guide for identifying suitable thermal-spray techniques and materials for coating applications.

## 11.1 CASE STUDY 1: PLASMA-SPRAYED COMPOSITE COATINGS FOR JOINT ARTICULATION

Since the first successful surgical intervention of low friction arthroplasty (LFA) in the 1960s by Sir John Charnley, biomedical research has constantly been thriving to find balance in longevity and biological safety of total bone and joint implants. The increasing satisfactory rate of success and survivorship of primary hip arthroplasty have encouraged millions of patients of high global median age for replacement procedures to ameliorate degenerative and inflammatory diseases such as osteolysis, osteoarthritis (OA), avascular necrosis, rheumatoid arthritis (RA), fractured neck of femur and so on. There were 752,440 primary total hip replacement (THR) surgeries registered between 2012 and 2020 in the United States [1] and 1,251,164 and 2003–2020 in the UK [2]. However, periprosthetic osteolysis, aseptic loosening, dislocation, or infection due to wear from articulation of hip prosthesis limits the implant longevity, predominantly in young and physically active patients. Biomaterials research now focuses on extending implant survivorship beyond ten years of follow-up [3] to avoid catastrophic clinical complications and the adverse health consequences of revision surgery [4], which is of aggravated morbidity and 20 times higher cost [5] occurring over 15% primary THR cases [6] (~50% THR needing revision surgery within less than five years [7]).

### 11.1.1 Description of Concept

A hip joint is a ball-and-socket synovial joint formed by the articulating rounded ball or femoral head fitted into a shallow socket called an acetabulum which is located inside the pelvis surrounded by tendons and ligaments (Figure 11.1a). Hip joints are subjected to cyclic loading as high as $10^6$ cycles in a year at static (sitting, standing, or lying) postures [8]. Hip contact forces (HCFs) rise to four to five times the bodyweight at walking at 3 km/h [9], which can intensify up to ten times the bodyweight during strenuous dynamic activities such as running or jumping

**(a) Hip joint implant components**

Worn-out particles
Acetabular component
Polymer acetabular cup liner
Femoral head
Femoral neck
Femoral stem

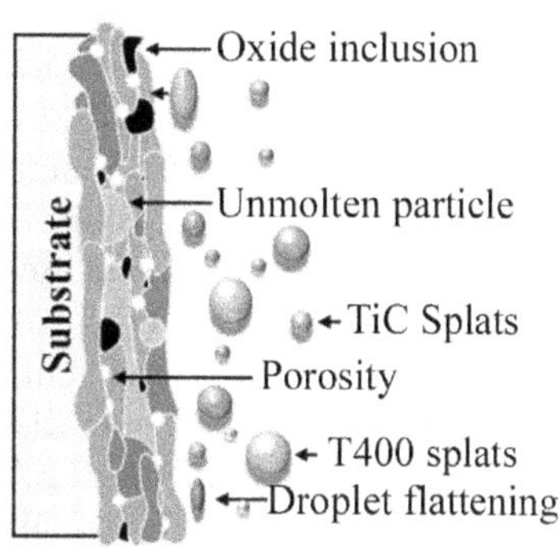


**FIGURE 11.1** Schematic representation of (a) hip joint components and (b) T400-TiC composite coating cross section in hybrid suspension-powder plasma spray.

*Source*: Moumita Mistri, Tribomechanical Study of Carbide-Laden Plasma-Sprayed Composite Coatings, PhD thesis, 2021.

[10]. Different bearing couples such as metal on polyethylene (MoP), metal on metal (MoM), ceramic on ceramic (CoC), and ceramic on polyethylene (CoP) are in practice. Excellent biocompatibility, hemocompatibility, low cost, good manufacturability, wear resistance, low friction torque, adequate tensile strength/fatigue strength/fracture toughness of metallic biomaterials (e.g., 316L stainless steel, Ti-6Al-6V, and Co-Cr-Mo alloys) compared to ceramics or polymers have envisioned them the longest enduring and most widely used joint replacement materials. However, metallosis induced by corrosion/wear/infection in a blood/tissue milieu in MoM couples may trigger immunological reactions such as allergic reactions, local anaphylaxis, and inflammation, which consequently demand heavy intake of immunosuppressants or other treatments, further instigating deleterious effects on patient's health [11]. Furthermore, higher (>100 GPa) Young's modulus of metallic implants than that of the natural bone (3–20 GPa), causing stress shielding during load-bearing [12], results in non-integration with host tissue, poor bone resorption, and thus increased vulnerability to post surgery microbial infections, and successive implant failure requiring revision surgery.

Various approaches such as plasma sputtering and etching, plasma deposition, laser ablation, sol-gel, hot isostatic compaction, ion beam coating and plasma spraying have been directed towards modifying the bulk properties of metallic biomaterials to render (1) osteointegration [13] at an interface of tissue and bioimplant with insignificant local tissue reactions, and antibacterial properties impeding biofilm formation; (2) high resistance to mechanical and chemical wear, and corrosion resistance; and (3) mechanical properties withstanding static and dynamic loads facilitating bone-to-implant load transfer to reduce the stress shielding [14]. Some of the industrially embraced surface engineering approaches are plasma spray by Exactech (USA), CP Ti plasma spray by OMNI (USA), porous Ti plasma spray by Stryker (USA), DuPuy (USA) and Zimmer (USA) [15]. For effective bio-integration to increase the longevity of orthopedic implants, (1) surface asperity produced by "overspray" feedstock particles and (ii) porosity in the order of 100–600 μm generated by thermal stresses, inter-splat oxidation, intra-splat cohesion, or cracking of trapped, un-melted, resolidified particles, are beneficial microstructural features in the biological scaffold [16].

Cast (ASTM F75, [17]), wrought ASTM F1537 and ASTM F799 [18] $Co_{28}Cr_6Mo$ find widespread commercial success in bearing articulation against UHMWPE due to excellent mechanical, corrosion and wear resistance. Although the major alloying elements of cobalt alloys such as Co, Cr (Co and Cr are bioactive), Mo and Ni constitute essential trace elements in the human body, the in-vivo leached out particles (of ~40 μm) due to wear/corrosion or tribocorrosion may impart local toxicity [19]. Although wear particle–induced higher risk of cancer associated with inflammatory cell-induced corrosion (ICIC) and mechanically assisted corrosion (MAC) has remained a lingering concern for CoCrMo alloys, the analytical concentration for Co ions seems acceptable up to 2 μg/L in blood [20]. Furthermore, the stress-shielding effect due to 10 times higher elastic modulus (200–230 Gpa) and ultimate tensile strength (430–1028 Mpa) of the Co-alloys compared to that of human bone expedite the requirement of surface engineering.

### 11.1.2 Tribaloy T400-Carbide Composite Plasma-Sprayed Biological Coatings

Tribaloy T400 ($Cr_{8.5}Mo_{28.5}Si_{2.6}Co_{balance}$ in weight) consists of ~50 vol% hard intermetallic amorphous Laves phase of FCC or HCP $Co_3Mo_2Si$ or CoMoSi (m.p.: 1560 °C) dispersed in lamellar eutectic matrix and Co solid solutions ($M_{23}C_6$, $M_7C_3$, and MC). Cr forms hypoeutectic $M_7C_3$ granting solid solution strengthening, or it forms hypereutectic $M_{23}C_6$ providing oxidation and corrosion resistance. Reduced stacking fault energy in presence of Mo and Cr prompts transformation of FCC ® HCP in the Co matrix in T400 at ~400 °C resisting the abrasive wear, enhancing hardness with reasonable workability. However, presence of 50 vol% of Laves phase (hardness ~1300 HV) and precipitation of secondary carbide phase at 873–1373 K causes compromise in ductility and poor crack propagation resistance, yielding low fracture toughness (20 Mpa $m^{1/2}$), which is addressed by reinforcement of TiC through a novel approach of hybrid suspension-powder plasma spray [21] (Figure 11.1b) in this study. Owing to high melting point (3160 °C), low atomic number, low density (4.93 $g/cm^3$), high hardness (2800–3200 HV), lower friction coefficient, good thermal fatigue resistance and so forth, TiC coatings have been of paramount research interest [22]. Suspension spray enables the carrying of sub-micrometer and nanometer-sized particles in fluid (without choking the nozzle) but with a limited deposition rate, whereas axial powder spray can provide a high deposition rate. Thus, hybrid suspension-powder plasma spray permits obtaining thick coatings with retained refined microstructure that typically exhibit superior tribomechanical properties compared to that of its individual deposition techniques.

Larger particle size of T400 (10–45 μm) constituting different Co/Mo phases of lower melting point (1,495–2,623 °C) than spraying flame at a temperature of 3,500 °C is ascribed to a high quenching rate (> ~$10^6$ K/s), giving rise to porosity. Partial vertically porous feathery microfeatures, arising from segmentation of TiC (of ~2.2 μm particle size) and momentum alteration, are seemingly attributed to the lattice micro-strain (of ~0.96 ± 0.22%) induced by the compressive stress and low Stokes number. Co-existing TiC phases (~24 wt%) dispersed in CoMoSi/$Co_3Mo_2Si$ intermetallic Laves, and $Co_7Mo_6$/$Co_2Mo_7$ eutectic phases of T400 in T400-TiC show reduced interlamellar porosity than that of powder-sprayed T400. It provides an indication that the powder feed rate merely affected the suspension feed rate and demonstrate the major mechanism for spraying the composite coating as splat formation and mechanical interlocking. The reduced lattice constant of a = 4.17 Å w.r.t. literature value of 4.38 Å [23] suggesting 13.7% lattice contraction for TiC does not allow B1 → B2 transition [24] under plasma flow of 300 SPLM. Thermodynamically stable rutile $TiO_2$ is present through diffusion from TiC at high temperature and pressure. The fast quenching of sub-stoichiometric $TiC_{1-x}$ caused high carbon vacancy % [25] and/or substitution of carbon lattice position by oxygen during T400-TiC deposition presumably resulting in ~24.9% lattice contraction with the lattice constant of 3.98 Å in TiC.

A synergistic enhancement in mechanical properties including elastic modulus (by 36%), and Vickers hardness (by 82.5%), in T400-TiC (E of 184.5 GPa, H of ~11.5 GPa) as compared to T400 (E = ~135 GPa, H = 6.3 GPa) is the result of carbide phase strengthening [26] through TiC reinforcement (E of ~160.3 GPa, H of ~7.7 GPa) along with the presence of $TiO_2$ [27]. The deviation of experimental mechanical properties from the theoretical elastic modulus (440–500 Gpa, [28]) and Vickers hardness (2800–3200 HV, first-principle method [28]) of TiC has resulted from different coating micro-features instigated from a distinct spraying technique adopted herein. Further, TiC reinforcement in the composite caused an augmentation in its elastic behavior which reflected through the reduction in the magnitude of wear resistance index, expressed as $E_r^2/H^3$, ($E_r^2/H^3$ decreased from 86.5 in T400 to 27.0 in T400-TiC). Hertzian contact theory estimates a higher magnitude of Hertzian contact diameter (HCD, equation (11.1)), resulted in an enhanced interaction in the tribo-couple causing more wear.

$$HCD = 2\sqrt[3]{\frac{3RF_n}{4E^*}} \tag{11.1}$$

where $R$ and $F_n$ are the radius of the counter-body and applied load in fretting wear experiments, respectively. $E^*$ is the reduced elastic modulus estimated from instrumented micro-indentation.

An enhancement in COF of T400-TiC by ~18.4% (~0.45) than that of T400 (0.39) occurred in fretting wear. Reduction of HCD by ~10.1% in T400-TiC (96.1 μm) compared to T400 (106.93 μm) suggests reduced interaction of the coating surface with the counter body which speaks of a significant improved in its damage tolerance under reciprocating wear. The grain refinement in presence

**TABLE 11.1**
**Deposition Techniques and Corresponding Physical and Tribomechanical Properties of Powder-Sprayed T400, Suspension-Sprayed TiC, and Suspension-Powder T400-TiC Coatings (Adapted from [21])**

| Sample | Coating Thickness (μm) | 2D % Porosity | Mechanical Properties[1] | | | Fretting Wear Properties[2] | | |
|---|---|---|---|---|---|---|---|---|
| | | | E (GPa) | Hv (GPa) | $E_r^2/H^3$ | μ (COF) | HCD (μm) | $\left\lvert F_n/H \right\rvert$ |
| T400 | 97.3 ± 8.8 | 3.22 ± 0.59 | 135.2 ± 19.5 | 6.3 ± 1.2 | 86.5 | 0.39 ± 0.01 | 106.9 | 1.587 |
| TiC | 67.8 ± 7.1 | 2.66 ± 0.05 | 160.3 ± 19.1 | 7.7 ± 0.9 | 66.5 | 0.44 ± 0.02 | 101.1 | 1.298 |
| T400-TiC | 95.2 ± 7.6 | 1.19 ± 0.11 | 184.5 ± 14.0 | 11.5 ± 1.2 | 27.0 | 0.45 ± 0.01 | 96.1 | 0.869 |

*Notes:*
1 Microindentation parameters: 500 mN load, 1000 mN/min loading rate and 10 s dwell time [21].
2 Fretting wear parameters: 10 N load, 10 Hz frequency, $10^5$ cycle and 100 μm reciprocating length.

of TiC due to T400 phases of hexagonal ($MgZn_2$) structures has helped [29] in augmented wear resistance in T400-TiC. The factor $F_N$/H, conceptually converging to P/$P_m$ (P or $F_N$ is the applied load, and $P_m$ refers to the hardness of softer counterpart) included in Archard's equation [30], also suggests enhanced tribomechanical response.

Upon TiC reinforcement (contact angle ~93.9°), T400-TiC shows a decrease in contact angle (~95.7°) by 21.4% in relation to T400 (~121.8°). Surface asperities ($R_a$ of ~17 μm in TiC, $R_a$ of ~14 μm in T400-TiC) may accentuate the extent of surface wicking causing deviation from the thermodynamically defined intrinsic Young-Laplace contact angle; however, it qualitatively envisages a predictive index of cytocompatibility [31]. Adequate cell viability of TiC along with concomitant $TiO_2$ passivating layer, demonstrates an efficient osteogenic integration free of toxicity through surface free energy alteration, and nanosized and microsized roughness retention [32]. TiC and T400-TiC display higher cell viability by 46% and 24%, respectively as compared to that of the positive control. The coexisting longer T400 and finer TiC splats influence the surface roughness and wettability, which in conjugation with the corresponding surface chemistry, further elicits a favorable cellular response in T400-TiC.

The effect of TiC synergistic reinforcement in eliciting the augmented mechanical and fretting wear behavior in T400-TiC through establishing sequential lamellar microstructure deployed via hybrid suspension powder spray is schematically illustrated in Figure 11.2. Also, the changes in surface wettability in the composite coating with respect to T400 or TiC coating alone is incorporated to correlate the suitable cellular response achieved herein. Overall, its enhanced tribomechanical properties, attributed to carbide-strengthening, and favorable fibroblast proliferation evinced in suspension-powder composite T400-TiC compared to that of powder-sprayed T400, corroborate its potential as femoral head articulation for hip joint prosthesis.

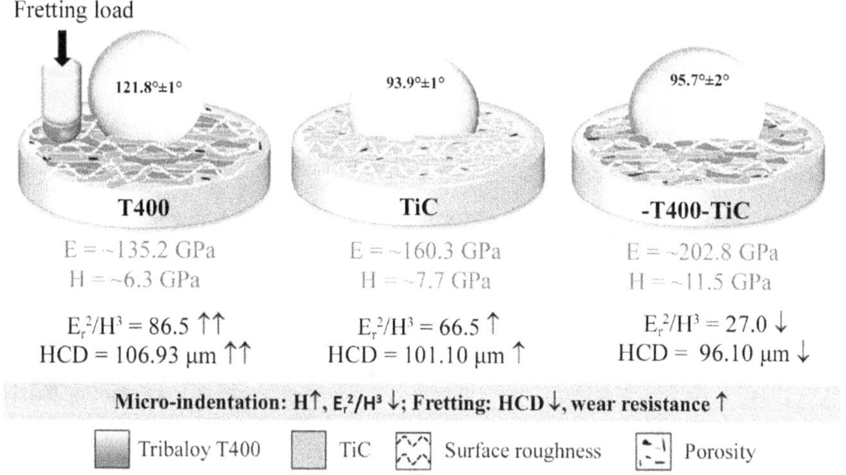


**FIGURE 11.2** Schematic representation of the effect of TiC reinforcement in tribomechanical performance and surface wettability of T400-carbide coatings.

Source: Moumita Mistri, Tribomechanical Study of Carbide-Laden Plasma-Sprayed Composite Coatings, PhD thesis, 2021 [21].

## REFERENCES


[1] American Joint Replacement Registry Releases 2021 Annual Report, Showing increase in number of hip and knee procedures despite pause due to COVID-19 (n.d.). www.aaos.org/aaos-home/newsroom/press-releases/merican-joint-replacement-registry-releases-2021-annual-report/ (accessed January 7, 2022).

[2] R. Brittain, P. Howard, S. Lawrence, J. Stonadge, M. Wilkinson, T. Wilton, S. Dawson-Bowling, A. Watts, C. Esler, A. Goldberg, S. Jameson, T. Jennison, A. Toms, E. Young, C. Boulton, D. Taylor, O. Espinoza, V. Mccormack, C. Newell, M. Royall, M. Swanson, Y. Ben-Shlomo, A. Blom, E. Clark, K. Deere, C. Gregson, A. Judge, E. Lenguerrand, A. Price, D. Prieto-Alhambra, J. Rees, A. Sayers, M. Whitehouse, NJR statistical analysis, support and associated services (n.d.). www.njrcentre.org.uk (accessed January 15, 2022).

[3] D.J. Berry, W.S. Harmsen, M.E. Cabanela, B.F. Morrey, Twenty-five-year survivorship of two thousand consecutive primary Charnley total hip replacements: Factors affecting survivorship of acetabular and femoral components, *J. Bone Joint Surg. Am.* 84 (2002) 171–177. https://doi.org/10.2106/00004623-200202000-00002.

[4] R. Pivec, A.J. Johnson, S.C. Mears, M.A. Mont, Hip arthroplasty, in: *Lancet*, Elsevier, 2012: pp. 1768–1777. https://doi.org/10.1016/S0140-6736(12)60607-2.

[5] J.N. Katz, B.E. Earp, A.H. Gomoll, Surgical management of osteoarthritis, *Arthritis Care Res. (Hoboken).* 62 (2010) 1220–1228. https://doi.org/10.1002/acr.20231.

[6] N.N. Mahomed, J.A. Barrett, J.N. Katz, C.B. Phillips, E. Losina, R.A. Lew, E. Guadagnoli, W.H. Harris, R. Poss, J.A. Baron, Rates and outcomes of primary and revision total hip replacement in the United States Medicare population, *J. Bone Jt. Surg.—Ser. A* 85 (2003) 27–32. https://doi.org/10.2106/00004623-200301000-00005.

[7] S.D. Ulrich, T.M. Seyler, D. Bennett, R.E. Delanois, K.J. Saleh, I. Thongtrangan, M. Kuskowski, E.Y. Cheng, P.F. Sharkey, J. Parvizi, J.B. Stiehl, M.A. Mont, Total hip arthroplasties: What are the reasons for revision? *Int. Orthop.* 32 (2008) 597–604. https://doi.org/10.1007/S00264-007-0364-3.

[8] G. Bergmann, F. Graichen, A. Rohlmann, Hip joint loading during walking and running, measured in two patients, *J. Biomech.* 26 (1993) 969–990. https://doi.org/10.1016/0021-9290(93)90058-M.

[9] G. Giarmatzis, I. Jonkers, M. Wesseling, S. Van Rossom, S. Verschueren, loading of hip measured by hip contact forces at different speeds of walking and running, *J. Bone Miner. Res.* 30 (2015) 1431–1440. https://doi.org/10.1002/jbmr.2483.

[10] D.J. Cleather, J.E. Goodwin, A.M.J. Bull, Hip and knee joint loading during vertical jumping and push jerking, *Clin. Biomech.* 28 (2013) 98–103. https://doi.org/10.1016/j.clinbiomech.2012.10.006.

[11] P.H. Wooley, S. Nasser, R.H. Fitzgerald, The immune response to implant materials in humans, *Clin. Orthop. Relat. Res.* 326 (1996) 63–70. https://doi.org/10.1097/00003086-199605000-00008.

[12] Q.H. Zhang, A. Cossey, J. Tong, Stress shielding in periprosthetic bone following a total knee replacement: Effects of implant material, design and alignment, *Med. Eng. Phys.* 38 (2016) 1481–1488. https://doi.org/10.1016/j.medengphy.2016.09.018.

[13] R.I.M. Asri, W.S.W. Harun, M. Samykano, N.A.C. Lah, S.A.C. Ghani, F. Tarlochan, M.R. Raza, Corrosion and surface modification on biocompatible metals: A review, *Mater. Sci. Eng. C* 77 (2017) 1261–1274. https://doi.org/10.1016/j.msec.2017.04.102.

[14] M.I.Z. Ridzwan, S. Shuib, A.Y. Hassan, A.A. Shokri, M.N. Mohamad Ib, Problem of stress shielding and improvement to the hip implant designs: A review, *J. Med. Sci.* 7 (2007) 460–467. https://doi.org/10.3923/jms.2007.460.467.

[15] M. Borroff Professor Paul Gregg Peter Howard Professor Alex MacGregor Keith Tucker, C. Esler Alun John Matthew Porteous, A. Goldberg, P. Baker Simon Jameson, R. Beaumont James Thornton Melissa Wright Elaine Young, O. Forsyth Anita Mistry Claire Newell Martin Pickford Martin Royall Mike Swanson, P. Yoav Ben-Shlomo Professor Ashley Blom Emma Clark Professor Paul Dieppe Alison Smith Professor Jon Tobias Kelly Vernon, National Joint Registry for England and Wales Healthcare Quality Improvement Partnership NJR Management team and NJR Communications (2012) 17. www.njrcentre.org.uk (accessed December 9, 2021).

[16] V.K. Balla, S. Bodhak, S. Bose, A. Bandyopadhyay, Porous tantalum structures for bone implants: Fabrication, mechanical and in vitro biological properties, *Acta Biomater.* 6 (2010) 3349–3359. https://doi.org/10.1016/j.actbio.2010.01.046.

[17] A. Standard F75–98, Standard specification for cobalt-28 chromium-6 molybdenum alloy castings and casting alloy for surgical implants (UNS R30075) 1, *Annu. B. ASTM Stand.* (2007) 7–10. https://doi.org/10.1520/F0075-98.

[18] ASTM Standard F1537–11, Standard specification for wrought cobalt-28-chromium-6-molybdenum alloys for surgical implants (UNS R31537, UNS R31538, and UNS R31539), *ASTM Int.* (2012). West Conshocken, PA. https://doi.org/10.1520/F1537-00.

[19] D.R. Boardman, F.R. Middleton, T.G. Kavanagh, A benign psoas mass following metal-on-metal resurfacing of the hip, *J. Bone Jt. Surg.—Ser. B* 88 (2006) 402–404. https://doi.org/10.1302/0301-620X.88B3.16748.

[20] C. Delaunay, I. Petit, I.D. Learmonth, P. Oger, P.A. Vendittoli, Metal-on-metal bearings total hip arthroplasty: The cobalt and chromium ions release concern, *Orthop. Traumatol. Surg. Res.* 96 (2010) 894–904. https://doi.org/10.1016/j.otsr.2010.05.008.

[21] M. Mistri, S. Joshi, K.K. Kar, K. Balani, Tribomechanical insight into carbide-laden hybrid suspension-powder plasma-sprayed Tribaloy T400 composite coatings, *Surf. Coatings Technol.* 396 (2020) 125957. https://doi.org/10.1016/j.surfcoat.2020.125957.
[22] Y.T. Pei, D. Galvan, J.T.M. De Hosson, A. Cavaleiro, Nanostructured TiC/a-C coatings for low friction and wear resistant applications, *Surf. Coatings Technol.* 198 (2005) 44–50. https://doi.org/10.1016/j.surfcoat.2004.10.106.
[23] J.H. Jang, C.-H. Lee, Y.-U. Heo, D.-W. Suh, Stability of (Ti,M)C (M=Nb, V, Mo and W) carbide in steels using first-principles calculations, *Acta Mater.* 60 (2012) 208–217. https://doi.org/10.1016/j.actamat.2011.09.051.
[24] B. Winkler, E.A. Juarez-Arellano, A. Friedrich, L. Bayarjargal, J. Yan, S.M. Clark, Reaction of titanium with carbon in a laser heated diamond anvil cell and reevaluation of a proposed pressure-induced structural phase transition of TiC, *J. Alloys Compd.* 478 (2009) 392–397. https://doi.org/10.1016/j.jallcom.2008.11.020.
[25] S. Mahade, K. Narayan, S. Govindarajan, S. Björklund, N. Curry, S. Joshi, Exploiting suspension plasma spraying to deposit wear-resistant carbide coatings, *Materials (Basel).* (2019). https://doi.org/10.3390/ma12152344.
[26] S.H. Whang, *Nanostructured metals and alloys: Processing, microstructure, mechanical properties and applications*, Elsevier, 2011. https://doi.org/10.1533/9780857091123.
[27] S. Kumaraguru, G.G. Kumar, S. Raghu, R. Gnanamuthu, Fabrication of ternary Ni-TiO2-TiC composite coatings and their enhanced microhardness for metal finishing application, *Appl. Surf. Sci.* 447 (2018) 463–470. https://doi.org/10.1016/j.apsusc.2018.03.216.
[28] Y. Yang, H. Lu, C. Yu, J.M. Chen, First-principles calculations of mechanical properties of TiC and TiN, *J. Alloys Compd.* 485 (2009) 542–547. https://doi.org/10.1016/j.jallcom.2009.06.023.
[29] K. Kubota, M. Mabuchi, K. Higashi, Processing and mechanical properties of fine-grained magnesium alloys, *J. Mater. Sci.* 34 (1999) 2255–2262. https://doi.org/10.1023/A:1004561205627.
[30] H.C. Meng, K.C. Ludema, Wear models and predictive equations: Their form and content, *Wear.* 181–183 (1995) 443–457. https://doi.org/10.1016/0043-1648(95)90158-2.
[31] T.P. Kunzler, T. Drobek, M. Schuler, N.D. Spencer, Systematic study of osteoblast and fibroblast response to roughness by means of surface-morphology gradients, *Biomaterials.* 28 (2007) 2175–2182. https://doi.org/10.1016/j.biomaterials.2007.01.019.
[32] N. Huang, P. Yang, Y.X. Leng, J.Y. Chen, H. Sun, J. Wang, G.J. Wang, P.D. Ding, T.F. Xi, Y. Leng, Hemocompatibility of titanium oxide films, *Biomaterials.* 24 (2003) 2177–2187. https://doi.org/10.1016/S0142-9612(03)00046-2.

## 11.2 CASE STUDY 2: THERMAL BARRIER COATINGS (THERMAL INSULATION)

Thermal barrier coatings (TBCs) are extensively applied to metallic components for providing the thermal insulation employed at elevated temperature environments such as gas turbine blades, combustion zone of the engines [1]. The typical TBCs system consists of a ceramic layer called top coat (generally made of stabilized $ZrO_2$) serves as a thermal barrier due to its lower thermal conductivity, and metallic bond coat (MCrAlY (M = Ni and Co) improves the hot corrosion oxidation resistance of metal substrate (generally made of Ni- or Co-based substrate). Generally, yttria (6–8 wt%) stabilized $ZrO_2$ is used as a top-coat ceramic material owing to its physical, chemical and thermal properties (thermal conductivity 2 $Wm^{-1}K^{-1}$ at room temperature) [2, 3]. At present, the improvement in the efficiency of the engine is envisioned via increasing the inlet temperatures of the turbine. Therefore, TBC-coated systems provide better thermal insulation through reduction in thermal conductivity. So, for reducing the thermal conductivity, several strategies are utilized such as incorporating the nanostructured coating, which offers significant advancement in thermal insulation performance and improvement in mechanical and physical properties via refinement of grain size in comparison to that of conventionally made TBCs. Also, with addition of more than one rare earth element, stabilized zirconia ($ZrO_2$) has shown more reduction in thermal conductivity and enhance the phase stability than that of YSZ.

Kai et al. [4] used disc- and tube-shaped Ni-based substrate to apply the atmospheric plasma sprayed (APS) thermal barrier coating. Nanometer sized YSZ powder with grain size of 20–30 nm was plasma sprayed under standard parameters using 7M APS equipment [4]. For comparison purposes, a conventional YSZ is also sprayed to almost similar thickness (~500 μm) with standard parameters from Sulzer Metco. Thermal diffusivity ($\alpha$) and specific heat ($C_p$) were determined via the laser flash method and differential scanning calorimetry, respectively. Thermal conductivity ($k$) of coating samples was calculated via the following equation (11.2):

$$k = \alpha C_p \rho \tag{11.2}$$

where $\rho$ is the density of the coating measured via the Archimedes principle.

Figure 11.3a represents the experimental value of thermal conductivity for conventional and heat-treated coatings. The values of thermal conductivity of plasma-sprayed coatings are in between 0.88 to 1.1 $Wm^{-1}K^{-1}$. In both cases (plasma sprayed and after heat-treated coatings) thermal conductivity values are decreasing with the increase in measuring temperature (25 °C to 1300 °C). Coating heat treated at 1100 °C for 10 h showed comparatively larger value of thermal conductivity as compared to that of plasma-sprayed coatings and lower thermal conductivity than 100 h heat-treated coating. Coating heat treated at 1100 °C for 100 h showed ~25% increment in thermal conductivity than plasma-sprayed coating, owing to coating sintering.

Nanostructured coating (NC) showed further reduction in thermal conductivity than conventional coatings owing to presence of large amount of porosity (~19% in as-sprayed nanostructured coating and ~16% in conventional coating). Figure 11.3b represents thermal conductivity values of plasma-sprayed and heat-treated nanostructured coating. The values of thermal conductivity are in between 0.71–0.79 $Wm^{-1}K^{-1}$ for plasma-sprayed NC.

Thermal conductivity is directly proportional to the heat treatment time. It increases ~60% for heat treatment time of 100 h in composition with plasma-sprayed NC. So, in conclusion, thermal conductivity of NC was reliant on its porosity content, and porosity changes with the heat treatment time as sintering effect comes into the picture.

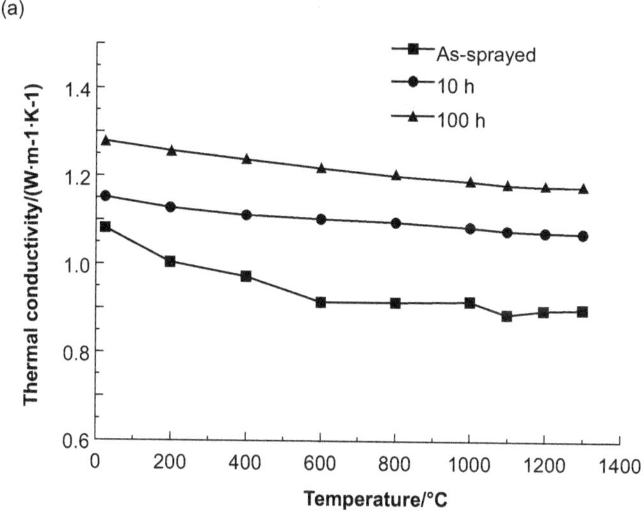


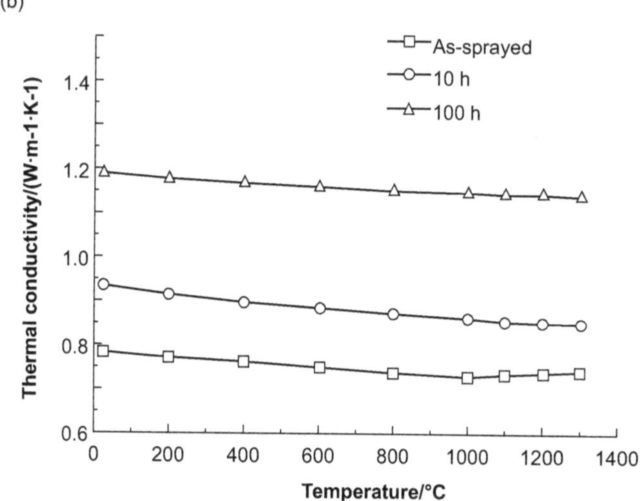


**FIGURE 11.3** Thermal conductivity versus temperature curves: (a) conventional coatings; (b) nanostructured coatings.

*Source*: Reprinted with permission from [4].

In another research work [5], they deposited bond coat (NiCoCrAlYHf) on the Ni-superalloy substrate via arc ion plating physical vapor deposition (AIP-PVD, A-1000). Both substrate and bond coat were heated at 870 ± 30 °C for 3 ± 1.0 h to promote the elemental diffusion for improving the interfacial bonding. The electron beam physical vapor deposition (EB-PVD) was used for top layer of Er-($ZrO_2$), Y-($ZrO_2$) and Y-Er-($ZrO_2$) with deposition rate of 4.0 ± 1.0

μm/min in TBCs. In coating characterization part, scanning electron microscopy (SEM) characterizes the cross section of the coatings. Figure 11.4a shows the representation of Y-Er-($ZrO_2$) coating comprising a ceramic layer (80–85 μm), metallic bond coat (24–28 μm) and substrate. The microstructure of ceramic layer is a characteristically columnar structure owing to EBPVD technology, though the microstructure of metallic bond coat is characteristically equiaxed owing to AIPPVD technology [6]. Figure 11.4b shows XRD pattern of as sprayed Er-($ZrO_2$), Y-($ZrO_2$), and Y-Er-($ZrO_2$), which represents whole tetragonal structure of $ZrO_2$. High intensity peaks of the Y-($ZrO_2$) indexed as (101), (110), and (211) at the angle of 30.28°, 35.18° and 59.98°, respectively, for $ZrO_2$ phase. For Er-($ZrO_2$) strong diffraction XRD peaks indexed as (101), (200) and (211) at angle of 30.38°, 50.56° and 60.02°, respectively, for $ZrO_2$ phase and for Y-Er-($ZrO_2$) the strong XRD peaks at 35.25°, 60.48° and 74.23° indexed as (110), (211) and (220), respectively, for $ZrO_2$ phase. Generally tetragonal structure of $ZrO_2$ in pure form in coating is stable, and during heating and cooling process it would not transform into other phases of $ZrO_2$, due to this TBCs shows good thermal insulation, hence good thermal properties [5]. So, stable tetragonal structure is quite helpful for better thermal shock life of TBCs. Coatings analysis showed lower thermal conductivity helping to improve thermal insulation. Generally, lower diffusivity and thermal conductivity is a measure of the thermal insulation ability of the TBCs. Figure 11.4c shows the values of thermal diffusivity in the different ceramic top-coating materials Er-($ZrO_2$), Y-($ZrO_2$), and Y-Er-($ZrO_2$) which was determined by Netzsch LFA 427. Further, thermal conductivity was calculated via equation (11.2). The Y-($ZrO_2$) top coat shows the thermal conductivity of 0.921 $Wm^{-1}K^{-1}$ and 0.927 $Wm^{-1}K^{-1}$ at 1000 °C and 1200 °C, respectively (Figure 11.4d). Both thermal conductivity and diffusivity of Er-($ZrO_2$), and Y-Er-($ZrO_2$) top coat are significantly lower than that of Y-($ZrO_2$) top coat. The thermal conductivity and diffusivity are decreasing in the temperature range 25–600 °C. Further, Y-Er-($ZrO_2$) and Er-($ZrO_2$) top coat showed an increase in thermal conductivity and diffusivity above 600 °C and 800 °C, respectively. In particular, Y-Er-($ZrO_2$) top coat showed thermal conductivity of 0.515 $Wm^{-1}K^{-1}$ and 0.596 $Wm^{-1}K^{-1}$ at 1000 °C and 1200 °C, respectively, which is ~40% less than the YSZ (Y-($ZrO_2$)) and 20% lower thermal conductivity than that of Er-($ZrO_2$) top coat. In all top coat (Er-($ZrO_2$), Y-($ZrO_2$) and (Y-Er-($ZrO_2$)), enhancing the number of the optical photons might be responsible for increasing the available phonon-to-phonon interactions. The comparatively complex elemental composition, which reduces the mean-free-paths of the phonons [7] leads to robust phonon-to-phonon scattering, which results in a decrease in the thermal conductivity. Also, point defects like vacancies or substitution atoms have a noteworthy effect on thermal conductivity of the top-coat materials used in this study.

XRD analysis shows that the crystal structure of the top-coat materials (Er-($ZrO_2$), Y-($ZrO_2$), and Y-Er-($ZrO_2$)) are pure tetragonal. Also, several point defects were presented in $ZrO_2$ tetragonal structure due to interatomic force, ionic radius and difference in the mass of Y, Er and Zr elements. So, lower thermal conductivity leads to an improvement in the insulation ability of TBCs.

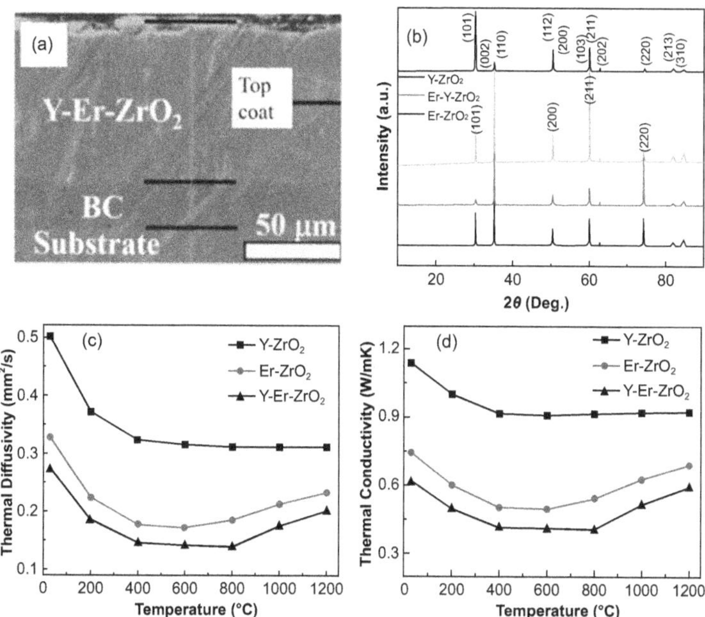


**FIGURE 11.4** (a) SEM image showing cross sectional microstructure of Y-Er-($ZrO_2$) TBC, (b) XRD, and (c) and (d) thermal diffusivity and conductivity of Er-($ZrO_2$), Y-($ZrO_2$), and Y-Er-($ZrO_2$) TBCs.

*Source*: Reprinted with permission from [5].

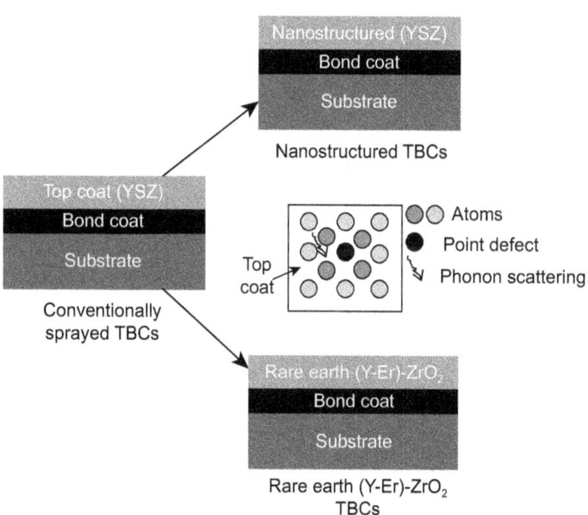


**FIGURE 11.5** Schematic representation of thermal barrier coatings with different top-coat ceramic layers.

It can be concluded from these case studies that the amount of porosity increases in nanostructured coating, and also with the introduction of more rare earth materials stabilized $ZrO_2$ in the top coat, causes decrease in the thermal conductivity of TBCs resulted in higher thermal insulation, aiding TBCs to withstand higher temperatures (schematics shown in Figure 11.5).

## REFERENCES

[1] Vassen R, Stuke A, Stöver D, Recent developments in the field of thermal barrier coatings, *Therm. Spray Technol.* 18 (2009) 181–186

[2] Leib EW, Vainio U, Pasquarelli RM, Kus J, Czaschke C, Walter N, et al., Synthesis and thermal stability of zirconia and yttria-stabilized zirconia microspheres. *J Colloid. Interface Sci.* 448 (2015) 582–592

[3] Patdure NP, Gell M, Jordan EH, Thermal barrier coatings for gas turbine engine applications, *Science* 296 (2002) 280–284

[4] Kai W, Hui P, Hongbo G, Shengkai G, Effect of sintering on thermal conductivity and thermal barrier effects of thermal barrier coatings. *Chinese J. Aeronaut* 25 (2012) 811–816

[5] Shen Z, Liu Z, Rende M, He L, Liu G, Y-Er-ZrO2, thermal barrier coatings by EB-PVD: Thermal conductivity, thermal shock life and failure mechanism, *Appl. Surf. Sci. Advances* 3 (2021) 100043

[6] Sampath S, Schulz U, Jarligo MO, Kuroda S, Processing science of advanced thermal-barrier systems, *MRS Bull.* 37 (2012) 903–910

[7] Pan W, Phillpot SR, Wan C, Chernatynskiy A, Qu Z, Low thermal conductivity oxides. *MRS Bulletin* 37 (2012) 917–922

## 11.3 CASE STUDY 3: NON-WETTABLE COATINGS

Wettability is defined as a liquid's ability to spread over a surface as measured by the contact angle (CA), which is the angle produced at the point of engagement between tangent of the liquid droplet surfaces and a solid planar surface. Generally, if the measured water CA >90°, the surface is considered non-wettable (hydrophobic); if the measured water CA <90°, the surface is considered wettable (hydrophilic). Furthermore, if the measured water CA >150° with a contact angle hysteresis (CAH) <10°, the surface is regarded as superhydrophobic and phenomenon is also referred to as the *lotus effect* [1]. Self-cleaning, anti-fogging, anti-fouling, anti-icing and corrosion resistance are just a few applications of these surfaces. Naturally, these surfaces are found on lotus leaves or rose petals [2].

The wettability can be greatly shaped by the surface free energy and its geometrical structure. And, the spreading coefficient is a factor that determines the wetting or anti-wetting of a liquid droplet. It is defined as equation (11.3):

$$S = \gamma_{SA} - \gamma_{SL} - \gamma_{LA} \tag{11.3}$$

where $\gamma$ is surface tension and S, A, L are solid, air and liquid, respectively. $S < 0$ denotes no-wetting, while $S = 0$ denotes full wetting.

The wettability of a smooth solid surface can be determined by equation (11.4):

$$\cos\theta = \frac{\gamma_{SA} - \gamma_{SL}}{\gamma_{LA}} \tag{11.4}$$

Wenzel, Cassie and Baxter models can determine the wettability of rough surfaces. The surface area of a rough surface is higher than that of a smooth surface, according to the Wenzel model, and the liquid fully enters cavities and comes into contact with the whole level material.

The CA is measured by the following equation in the Wenzel model, equation (11.5):

$$\cos\varnothing = r\cos\theta \tag{11.5}$$

where Ø is CA, $r$ is roughness, and $\theta$ is smooth surface equilibrium CA.

Instead of liquid droplets, the Cassie model argues that air is trapped in the cavities, resulting in a composite interface. For CA measurement, the Wenzel presented equation (11.6):

$$\cos\varnothing = f_{SL}\cos\theta_1 + f_{LA}\cos\theta_2 \tag{11.6}$$

where $f_{SL}$, $f_{LA}$ represent the area fraction of droplet in contact with solid and air respectively, and $f_{SL} + f_{LA} = 1$. Also, $\theta 1$ and $\theta 2$ are CAs of solid and air with liquid droplet respectively [2].

Previously, non-wettable surfaces were made of organic materials or polymers, which had lower surface energy, higher corrosion resistance, and were more readily available. But, owing to their weak thermal and mechanical stability and limited wear resistance, its hydrophobicity was destroyed by hostile conditions. Inorganic materials, on the other hand, offer greater mechanical stability than organic materials. Because of their excellent chemical stability, unique electrical structure, and inexpensive cost, rare earth oxides (REOs) such as cerium oxide ($CeO_2$) have a high hydrophobicity among inorganic materials. However, the hydrophobicity of $CeO_2$ may deteriorate due to adsorption of air contaminants when exposed to hostile environment for long period. Rolling the water droplets or dissolving the pollution under sunshine can solve the contamination problem. Even so, some contaminants, such as oleic acid, are not removed by rolling the water. As a result, a coating that both removes contamination and protects the hydrophobicity of the material is required. In this regard, $CeO_2/TiO_2$ composite coating is the best solution for adsorption of contaminations, owing to a small band gap (1.54 eV) along with photocatalytic act of $TiO_2$ [3].

The non-wettable surfaces have been fabricated by various methods such as electro-deposition method, dip coating, casting, electro-spinning, vapor deposition, and thermal spray coating. Thermal spraying can produce hydrophobic ceramic coatings with good hydrophobicity due to the high flame temperature.

Bai and colleagues [4], for example, employed oxy-fuel thermal spraying to create a hydrophobic $CeO_2$ coating, to fabricate hydrophobic $TiO_2$ coatings, Sharifi and colleagues [5] used air plasma spraying (APS). Furthermore, O.S. Asiq Rahman and colleagues created hydrophobic $CeO_2$ coatings using an atmospheric plasma-spray technique and tested them for wetting. Following the fabrication of the coating, they measured the water contact angle (WCA) $\theta$ using equation (11.7):

$$\theta = 2\,arc\,\tan\left(\frac{h}{r}\right) \tag{11.7}$$

where $h$ and $r$ are the droplet's height and base radius.

They discovered that the fabricated hydrophobic coating had a high water contact angle (>150°), which they attributed to hierarchical microstructures; Figure 11.6 shows the existence of microsized islands and nanograin clusters [6]. They studied WCA over time to see how the microstructure affects the WCA of the coating. They discovered that the WCA increases and decreases over time as a result of consecutive pinning and depinning caused by the presence of local energy barriers that prevent the water droplet contact line from advancing smoothly.

They also looked into how water droplets behaved during pinning and depinning [6]. WCA rises during water addition because droplet height rises, while radius remains constant because droplets become pinned at the island edges, introducing an energy barrier for the advancing liquid front. After time $t = 2.6$ s, the droplet contact line deepens/stretches and travels farther, wetting the other side of the island, as illustrated in Figure 11.7a. In comparison to the previously pinned stage, the droplet's height is reduced while its radius is raised during depinning, resulting in a decrease in the advancing contact angle. The water droplet is also expected to penetrate the larger scale grooves between the micro-size islands, but not the smaller scale grooves between the nanograins, as illustrated in Figure 11.7b.

**FIGURE 11.6** FE-SEM surface morphology of (a) $CeO_2$ coating and (b) corresponding magnified image.

*Source*: Reprinted with permission from [6].

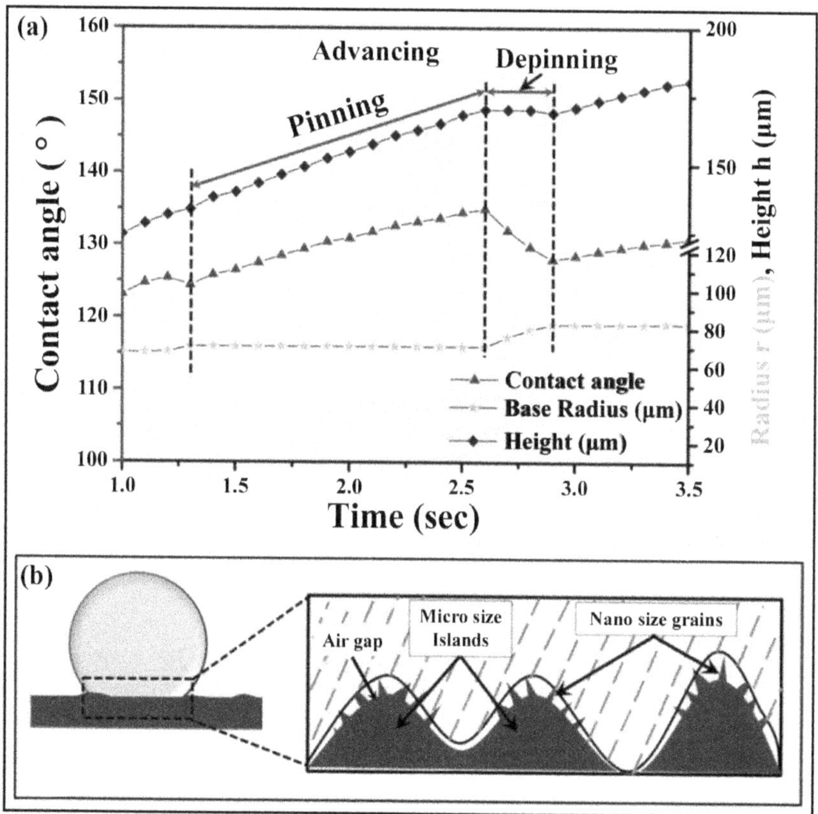


**FIGURE 11.7** (a) Water droplet behavior during pinning-depinning; (b) a visualization of a water droplet in contact with a hierarchical micro-structured surface.

*Source*: Reprinted with permission from [6].

Furthermore, hydrophilic surfaces can be converted to hydrophobic surfaces. Li and colleagues [7] prepared metallic (mixed powders of Fe, Cr, and Ni) coatings; they exhibited hydrophilic nature, and the coatings were then converted to hydrophobic surfaces by spraying a dodecanoic acid acetone solution.

## REFERENCES

[1] Watson, G. S., Gellender, M., & Watson, J. A. (2014). Self-propulsion of dew drops on lotus leaves: A potential mechanism for self cleaning. *Biofouling*, *30*(4), 427–434.

[2] Birjandi, F. C., & Sargolzaei, J. (2014). Super-non-wettable surfaces: A review. *Colloids and Surfaces A: Physicochemical and Engineering Aspects*, *448*, 93–106.

[3] Hu, L., Song, X., Shan, X., Zhao, X., Guo, F., & Xiao, P. (2019). Visible light-activated self-recovery hydrophobic $CeO_2$/black $TiO_2$ coating prepared using air plasma spraying. *ACS Applied Materials & Interfaces*, *11*(40), 37209–37215.
[4] Bai, M., Kazi, H., Zhang, X., Liu, J., & Hussain, T. (2018). Robust hydrophobic surfaces from suspension HVOF thermal sprayed rare-earth oxide ceramics coatings. *Scientific Reports*, *8*(1), 1–8.
[5] Sharifi, N., Pugh, M., Moreau, C., & Dolatabadi, A. (2016). Developing hydrophobic and superhydrophobic $TiO_2$ coatings by plasma spraying. *Surface and Coatings Technology*, *289*, 29–36.
[6] Rahman, O. A., Mukherjee, B., Priyadershini, S., Gunjan, M. R., Raj, R., Aruna, S. T., & Keshri, A. K. (2020). Investigating the wetting phenomena and fabrication of sticky, para-hydrophobic cerium oxide coating. *Journal of the European Ceramic Society*, *40*(15), 5749–5757.
[7] Li, Z., Zheng, Y., & Cui, L. (2012). Preparation of metallic coatings with reversibly switchable wettability based on plasma spraying technology. *Journal of Coatings Technology and Research*, *9*(5), 579–587.

## 11.4 CASE STUDY 4: CORROSION-RESISTANT Zn-Al AND Al COATINGS

Steels have long been universally embraced materials [1] for a spectrum of applications including construction, agriculture, power transmission and automotive due to high strength, toughness, hardness, weldability and low cost [2]. However, in prolonged exposure to high thermal stress, mechanical load, relative humidity, pH, corrosive gases and electrolytes ($O_2$, $H^+$, $OH^-$, $Cl^-$, $Br^-$, $F^-$, $SO_4^{2-}$, $SO_2$, $NO_x$, $Na^+$, $K^+$, $Ca^{2+}$, etc.) [3] in gas turbines, boilers, waste incineration, fluidized bed combustion chambers, coal gasification plants, marine transport, offshore developments and so forth, premature degradation of steel structures is inevitable [4]. The subsequent material losses (red rust of ferrous oxide) and economic downfall, especially in coastal/marine environments, enforces a colossal threat to the iron and steelmaking industries. A benchmark study conducted from 1999 to 2001 under the Department of Transportation Federal Highway Administration (FHWA) and NACE International estimated an economic loss of ~$276 billion (3.1% of the 1998 GDP) per year caused due to corrosion in the United States alone, with additional indirect costs of ~$551.4 billion or more [5]. Alloyed steel reinforced with corrosion-resistant Ni, oxidation resistant Cr and others can offer excellent corrosion resistance, but their performance often depends on the alloying elements, testing environment and specimen zone (submerged/tidal/splash) under usage [6]. The robustness of an engineering module in terms of their corrosion protection capability is often disregarded at 5% surface coverage with corrosion by-products [7]. Hence, anticorrosive coatings play the first line of defense and are of prevalent research importance to (1) mitigate high repairing cost and unforeseen downtime and (2) enhance overall structural integrity and aesthetics of steel components.

### 11.4.1 Description of Concept

Paint or powder-based barrier coatings comprising layers—primer, build and finish coats—although these offer economical alternatives, application of these are only limited to structures of low or moderate corrosivity, as they (1) often spall off, further leading to aggravated corrosion and (2) emit volatile organic compounds (VOCs). Electroplating (or electroless plating) and sherardizing are attempted for small components. Cathodic protection of steels via sacrificial anode of Zn, Al, Mg and others, extensively employed for structural steel, is of interest for the current section. The negative standard reduction potential of Zn and Al in the electrochemical series (shown below) duly validates the anti-corrosion traits only for oxide free surfaces measured at standard condition (at 298 K, 1 atm, and with 1 M solution concentrations).

$$H_2 \rightarrow 2H^+ + 2e^-,\ E° \text{ (vs. SHE)} = 0.00\text{ V}$$

$$Zn \rightarrow Zn^{2+} + 2e^-,\ E° \text{ (vs. SHE)} = 0.762\text{ V}$$

$$Al \rightarrow Al^{3+} + 2e^-,\ E° \text{ (vs. SHE)} = 1.662\text{ V}$$

Nonetheless, galvanic series might come handy at predicting corrosion characteristics in complex electrolyte such as seawater.

Centuries-old hot-dip galvanization is a type of cathodic protection against galvanic corrosion in which Zn-alloy uniform coating is formed via metallurgical bonds in molten Zn bath (at ~450 °C). Although known to exhibit sufficient anti-corrosion properties, hot-dip galvanization (1) is only confined to on-site accessibility, (2) entails negative impacts on humans in large industrial sites, and (3) infringes environmental regulations.

Thermal spray has been in high demand for almost all OEMs [8, 9] with (1) no VOCs emission and (2) additional advantages on anti-slip and wear-resistance insinuated by improving the hardness and friction behavior. The popularity of thermal spray for depositing corrosion-resistant coating for modern engineering components demanding even-higher operational temperatures and fuel efficiency is of no surprise [7]. Often, the different approaches are combined; for example, barrier coating is followed by a galvanized layer in duplex protection. Corrosion susceptivity of a thermally sprayed coating emanating from inaccessible complex areas and discontinuities (e.g., crevices, pits, fissures, sharp edges) is recurrently arrested by employing a thin organic layer (i.e., sealing). Al imparts both atmospheric and solution-immersion corrosion [10] by forming continuous, adherent, insoluble $Al_2O_3.3H_2O$ passivation film but provides no resistance to galvanic corrosion. Zn enhances durability and functionality of machine parts by effectively mitigating corrosion through a sacrificial layer offering threefold mechanisms: (1) galvanic protection via barrier action, (2) cathodic protection via acting as a sacrificial anode providing, and (3) passivation through corrosion products [11, 12, 13]. Complying the combined advantages of Al and Zn consisting of primary Zn phase ($\gamma$ phase) and eutectic Al phase ($\alpha$-rich phase), Zn-Al alloys exhibit superplasticity [14], low cost, mechanical rigidity and outstanding corrosion protection [15], which is reported to be governed by the Zn/Al weight ratio and thickness. Zn-50wt%Al composition attests the best corrosion properties among various Zn-Al alloys, even after 10.5 years of marine and atmospheric exposure [16, 17]. A new avenue in this area is to apply varnish/epoxy/phenolic polymer-based sealant, such as zinc-epoxy-urethane or polysiloxane, immediately after depositing the coating for extending service life [18] through blocking the pores restricting further entry of the corrosive media at the interface.

### 11.4.2 Zn-Al and Al Thermally Sprayed Anti-Corrosion Coatings

A possible corrosion resistance mechanism of Zn-Al based coatings is schematically represented later in this sub-chapter (Figure 11.10b). Studies by H. Lee on arc-sprayed Al coating concluded significantly enhanced corrosion protection for mild steel in 3.5 wt% NaCl solution [19] and SAE J2334 solution (0.5 wt% NaCl, 0.1 and 0.1 wt% $CaCl_2$ 0.075 wt% $NaHCO_3$) [20] more so, at longer exposure time, due to the formation of dense, thick and uniform barrier film that interconnected the pores/defects.

Zn-5wt%Al (Galfan) and Galvalume ($Zn_{43}Al_{55}Si_{1.6}$ in wt%) [21], with enhanced corrosion performance compared to hot-dip galvanized coatings [22], find applications in electric motor housings, automotive components, architectural parts, pipelines and so on [23] in marine and industrial atmospheres. Zhang et al. [24] identified ZnO, $ZnAl_2O_4$ and $Al_2O_3$ as corrosion products, which subsequently were converted into $Zn_6Al_2(OH)_{16}CO_3.4H_2O$, $Zn_2Al(OH)_6Cl.2H_2O$ and/or $Zn_5Cl_2(OH)_8.H_2O$, along with the formation of $Na_4Zn_4SO_4(OH)_6Cl_2.6H_2O$ in Galfan after five years of non-sheltered marine exposure but not in laboratory exposures.

The study by Li [25] confirmed corrosion amalgamation products in Zn-Al (22%–30% Al, trace % of Si) from co-precipitation in dynamic aerated seawater immersion, showing broad XRD peaks of basic carbonate, $Zn_4CO_3(OH)_6.H_2O$, $Zn_6Al_2CO_3(OH)_6.4H_2O$, $Mg_6Al_2CO_3(OH)_6.4H_2O$, $Cl^-$, $Zn_5(OH)_8Cl_2.H_2O$, $OH^-$, $Mg_2Al(OH)_7$ and so on. XRD peaks got sharper with increasing immersion time (up to 18 months) due to amplified crystallinity and grain size. Microcrystals were formed with crystalline state with long-range order and amorphous state with short-range order with gelatinous sedimentation of $Al(OH)_3$ intruding $Zn^{2+}$ and $Mg^{2+}$ ion adsorption. Morphologies of some of the corrosion products are shown in Figure 11.8.

Yang et al. [26] correlated the decreased corrosion rate (hence the pitting damage) following an order Zn-4wt%Al ($I_{corr}$: 3.97 × 10–$^6$, corroded depth was ~2 μm) > Zn-8wt%Al ($I_{corr}$: 3.02 × 10–$^6$) > Zn-12wt%Al ($I_{corr}$: 2.34 × 10–$^6$) > Zn-16wt%Al ($I_{corr}$: 2.06 × 10–$^6$) upon increasing Al % content to the Zn-Al phases distribution in immersion tests in simulated acid rain, and Al-rich phase being prone to preferential attack [27]. The main corrosion products, ZnO and $Zn_6Al_2(OH)_{16}CO_3.4H_2O$

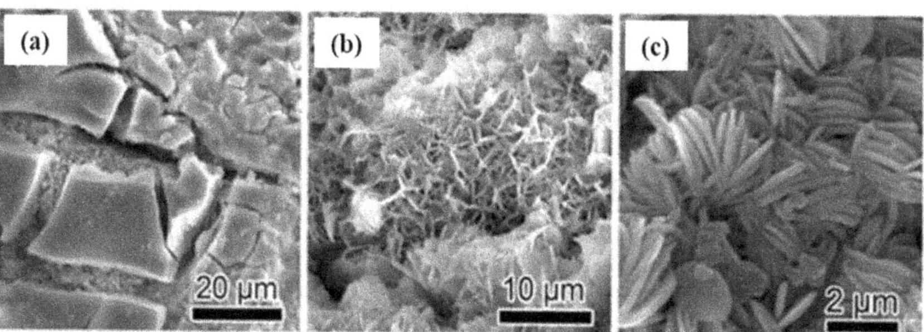


**FIGURE 11.8** The SEM microstructures of corrosion products such as (a) $Zn_6Al_2(OH)_{16}CO_3.4H_2O$, (b) $Zn_5(OH)_8Cl_2.H_2O$, (c) $Zn_5(OH)_6(CO_3)_2$ on Zn-70vol%Al cold-sprayed coating after a 200-day salt spray test.

*Source*: Open access under creative common attribution license [28].

for Zn-4wt%Al, however, did not show in XRD analysis, which could be attributed to the inadequate $Al^{3+}$ concentration prompting the dissolution reaction following equation (11.8) and no precipitation of $Zn_6Al_2(OH)_{16}CO_3.4H_2O$, unlike the Zn-Al alloys with Al wt% higher than 8%.

$$6Zn^{2+} + 2Al^{3+} + 16OH^- + CO_3^{2-} + 4H_2O \leftrightarrow Zn_6Al_2\left(OH\right)_{16}CO_3.4H_2O \quad (11.8)$$

Sugimura et al. [29] performed immersion tests on arc-sprayed Zn-Al-Si ($Zn_{79.5}Al_{19.5}Si_{1.0}$ in mass %) coating consisting of three distinct zones of (1) Zn, Al (co-associated with Si), and Zn-Al identifying $Al_{0.403}Zn_{0.59}$ phase; (2) with $SiO_2$/$(C_5O_2H_8)n$ (mass ratio 1:1) sealer; and (3) with an outer layer of acrylic resin paint $(C_5O_2H_8)n/Mg_3Si_4O_{10}(OH)_2$ (mass ratio 5:3). $E_{corr}$ for whole of arc-sprayed Zn-Al-Si + sealer + paint coated specimen, with scratch, was noted as −750 mV and −880 mV, respectively, in 50 ppm and 200 ppm $SO_4^{2-}$ ion versus Ag/AgCl after 522 days; versus −650 mV versus Ag/AgCl in 50 ppm and 100 ppm $Cl^-$ ion for 60–70 days (Figure 11.9a). However, open circuit potential (OCP) tended to increase with prolonged immersion time for the specimens coated with only Zn-Al-Si layer which showed no corrosion of steel underneath (Figure 11.9b). Corrosion products such as $Zn(OH)_2$, ZnO, and basic zinc salts (BZS): $Zn_5(OH)_6(CO_3)_2$, and $Zn_{12}(CO_3)_3(SO_4)(OH)_{16}$ (trace amount in 50 ppm $SO_4^{2-}$) provided the barrier action in $SO_4^{2-}$ ions. Cathodic/oxygen reduction and metal dissolution were inhibited in the chloride solutions due to (1) the formation of $Zn_5(OH)_6(CO_3)_2$, layered double hydroxides (LDH) such as $Zn_{0.70}Al_{0.30}(OH)_2(CO_3)_{0.15}.xH_2O$ (more stable than BZS at pH>12 [30]); and (2) the absence of ZnO (zincite), unlike in $SO_4^{2-}$ ions, which causes high corrosion rate due to good electrical conductivity. The dependence on type and concentration of ions [31] in the open circuit potential (OCP) evolution, rather than the cathodic protection of Zn, inferred the barrier action of corrosion products for providing the long-term corrosion prevention.

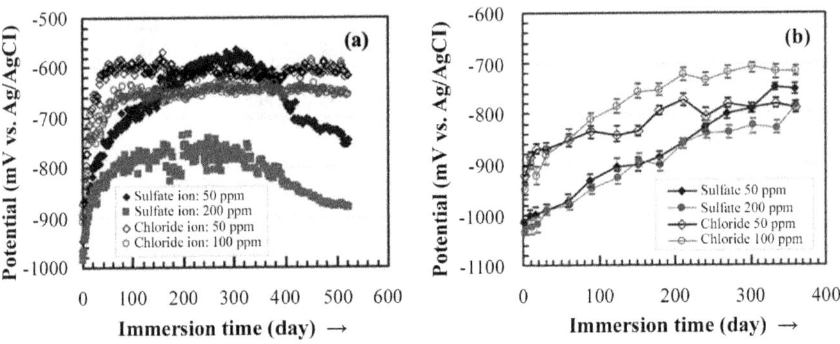


**FIGURE 11.9** OCP evolution in immersion test of specimens coated with (a) arc-sprayed Zn-Al-Si + sealer + paint, and (b) only with arc-sprayed Zn-Al-Si at room temperature in various $Cl^-$ and $SO_4^{2-}$ solutions.

*Source*: Reprinted with permission from [29].

Arc-sprayed Zn-31.50 vol% Al [32] ($E_{corr}$ of −1.123 V vs. Ag/AgCl, $I_{corr}$ of 7.19 μA/cm$^2$, corrosion rate of 120.40 μm/year) showed enhanced potentiodynamic polarization parameters in Tafel extrapolation compared to the bare steel ($E_{corr}$ of −0.950 V vs. Ag/AgCl, $I_{corr}$ of 20.10 μA/cm$^2$, corrosion rate of 233.17 μm/year) with OCP value of −0.904 V vs. Ag/AgCl in 3.5 wt% NaCl even after 29 days, attributed to the compact, adherent and uniform passive oxide film formation.

Selective corrosion through dissolution and subsequent co-precipitation of the zinc-rich inter-dendritic phases of Zn-55wt%Al coating [22] resulted in $NaZn_4(SO_4)Cl(OH)_6.6H_2O$ near the anodic region upon prolonged exposure of 1–3 months in a marine environment [33]. Hydroxycarbonate produced $Zn_5(OH)_8Cl_2.H_2O$, whereas co-precipitation of $Zn^{2+}$ and $Al^{3+}$ with hydroxyl ions formed hydrotalcite $Zn_{0.71}Al_{0.29}(OH)_2(CO_3)_{0.145}.xH_2O$ in the cathodic regions at a basic pH [34].

Pronounced corrosion resistance in accelerated salt spray test in $Cl^-$ over 2000 h attested by twin-wire arc-sprayed Zn-15wt% compared to Al [35] was explained by the microstructural differences, such as (1) dense coating but oxides of gray patches in Al and (2) white rust with less porous microstructure in Zn-15wt%Al.

Katayama et al. [36] performed field exposure tests in coastal environments for about 33 years. Zn-30wt%Al-based coating revealed adequate corrosion performance attributed to selective dissolution of Zn to $Zn_6Al_2(OH)_{16}CO_3.4H_2O$ [25] confirming individual Zn and Al metal XRD peaks demonstrating an oxidized outer layer with a few cracks, an inner layer with flaking, and an interface with a marginal amount of Cl and S. The corrosion protection of Zn-30wt%Al was affected by porosity [37] of thick corrosion products, current distribution in the cracks, and the surface roughness [38]. On the other hand, the Al coatings exhibited much larger corrosion resistance than un-weathered Al coatings upon penetration of seawater electrolytes owing to the presence of oxide layer (with elemental oxygen at the top surface and interfaces) which did not affect the corrosion under the atmospheric condition. Further, the corrosion resistance order: flame-sprayed Zn-15wt%Al with wash primer (70–130 μm) > flame-sprayed Zn-15wt%Al with vinyl sealer (150 μm) > flame-sprayed Al (150 μm) > continuous hot-dip Al/13Si (40 μm) > continuous hot-dip Zn/5Al (50 μm)> continuous hot-dip 55Al/Zn (20 μm) obtained from a 24-month study in an aggressive, windy atmosphere (at corrosion-erosion rate of ~1.4 mm/year for ASTM 1029 steel) by De Rincón et al. [39] confirmed the influence of composition, coating deposition technique, as well as thickness in determining the anti-corrosion performance.

The marine field test inspection by Kuroda et al. [40] revealed (1) spotty red rust-substrate corrosion with minor swelling/exfoliation after 7 years and affected substrate after 10 years for flame-sprayed Al (175 μm); (2) localized corrosion damage till 18 years for arc-sprayed Al (175 μm and 400 μm); (3) no significant corrosion damage for arc-sprayed and flame-sprayed Al with epoxy sealing (175 μm) at the splash zone after 18 years. Although the flame-sprayed Al (175 μm) + primer + polyurethane paint (PU) (of 3 mm) showed no significant damage, epoxy primer with calcium plumbate + epoxy resin paint (3 layers of 100 μm) + PU (100 μm) application on arc-sprayed Al (400 μm) showed accelerated degradation accompanying localized sea-contaminated spotty white rust in the splash zone. Flame-sprayed Zn-13 wt% Al (175 μm), revealing (1) no substantial corrosion damage with epoxy sealing and (2) slightly localized rust without epoxy sealing, suggested adequate corrosion resistance compared to analogs Al coatings.

This study agrees with the observation from a 19-year exposure test conducted by the American Welding Society, concluding adequate corrosion protection of steel in severe seawater and marine and industrial atmosphere via Al coating (80–150 μm) irrespective of the presence of sealing [41].

Although the precise corrosion mechanism of Zn-Al coatings is not yet well formulated, but the feasible reaction scheme of bare iron structure and that of in presence of Zn-Al coatings is schematically presented in Figure 11.10a–b. To conclude, the mechanism of corrosion products (listed in Figure 11.10c) formation in Zn-Al alloys depends on (1) the alloy composition, (2) pH, for example, the rate of $Al_2O_3$ and $Al(OH)_3$ dissolution to $Al(OH)_4^-$ is much higher in alkaline pH than the Zn counterparts forming thermodynamically stable ZnO and/or $Zn(OH)_2$ further converting to $Zn(OH)_3^-$ and $Zn(OH)_4^{2-}$ [42]; (3) presence of oxygen as that can alter the dissolution rates; and (iv) exposure time, for example, Zn-rich phases might become consumed, while pitting corrosion of Al phases might come into the picture at the extended exposure duration. Overall, Zn-Al-based coatings on steel structure can provide adequate corrosion resistance along with anti-wear and anti-skid properties.

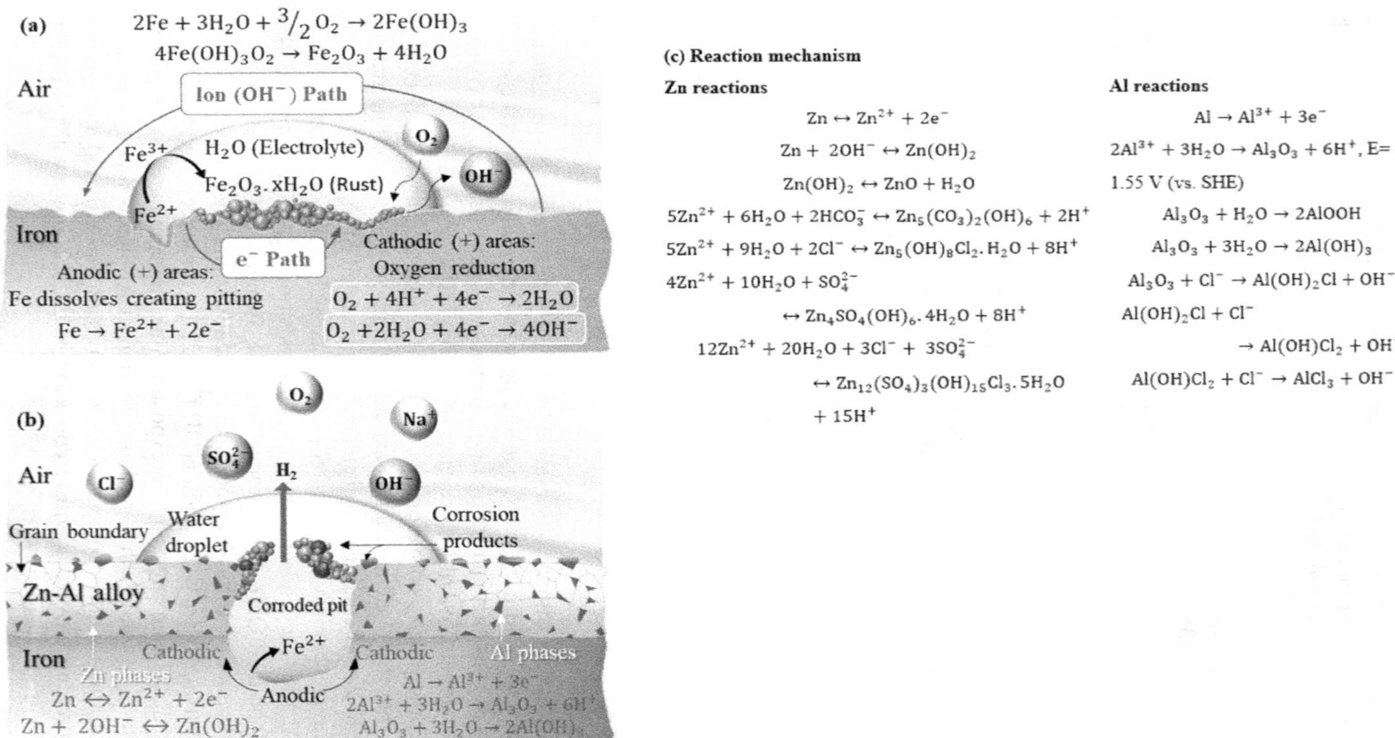


**FIGURE 11.10** (a) Corrosion reactions in iron structures and (b) corrosion protection of iron components through Zn-Al coatings depicting Al < 8 wt%; (c) possible reaction mechanisms of Zn and Al phases in $Cl^-$ and $SO_4^{2-}$ ions environment.

*Source*: Moumita Mistri.

## REFERENCES

[1] Z.L. Zhang, T. Bell, Structure and corrosion resistance of plasma nitrided stainless steel, *Surf. Eng.* 1 (1985) 131–136. https://doi.org/10.1179/sur.1985.1.2.131.

[2] M. Manna, M. Dutta, A.N. Bhagat, Microstructure and electrochemical performance evaluation of Zn, Zn-5 wt% Al and Zn-20 wt% Al alloy coated steels, *J. Mater. Eng. Perform.* 30 (2021) 627–637. https://doi.org/10.1007/S11665-020-05359-8/TABLES/5.

[3] C.G. Soares, Y. Garbatov, A. Zayed, G. Wang, Influence of environmental factors on corrosion of ship structures in marine atmosphere, *Corros. Sci.* 51 (2009) 2014–2026. https://doi.org/10.1016/j.corsci.2009.05.028.

[4] K. Zen, Corrosion and life cycle management of port structures, in: *Corros. Sci.*, Pergamon, 2005: pp. 2353–2360. https://doi.org/10.1016/j.corsci.2005.04.003.

[5] G.H. Koch, M.P.H. Brongers, N.G. Thompson, Y.P. Virmani, J.H. Payer, Cost of corrosion in the United States, in: *Handb. Environ. Degrad. Mater.*, William Andrew Publishing, 2005: pp. 3–24. https://doi.org/10.1016/B978-081551500-5.50003-3.

[6] B. Hou, Y. Li, Y. Li, J. Zhang, Effect of alloy elements on the anti-corrosion properties of low alloy steel, *Bull. Mater. Sci.* 23 (2000) 189–192. https://doi.org/10.1007/BF02719908.

[7] S. Kuroda, J. Kawakita, M. Takemoto, An 18-year exposure test of thermal-sprayed Zn, Al, and Zn-Al coatings in marine environment, *Corrosion.* 62 (2006) 635–647. https://doi.org/10.5006/1.3280677.

[8] Q. Jiang, Q. Miao, W. Liang, F. Ying, F. Tong, Y. Xu, B. Ren, Z. Yao, P. Zhang, Corrosion behavior of arc sprayed Al-Zn-Si-RE coatings on mild steel in 3.5wt% NaCl solution, *Electrochim. Acta.* 115 (2014) 644–656. https://doi.org/10.1016/j.electacta.2013.09.156.

[9] I.A. Gorlach, A new method for thermal spraying of Zn-Al coatings, *Thin Solid Films.* 517 (2009) 5270–5273. https://doi.org/10.1016/j.tsf.2009.03.174.

[10] Y. Xiao, X. Jiang, Y. Xiao, L. Ma, Research on Zn-Al15 thermal spray metal coating and its organic painting composite system protection performance, in: *Procedia Eng.*, No longer published by Elsevier, 2012: pp. 1644–1653. https://doi.org/10.1016/j.proeng.2011.12.632.

[11] D.P. Schmidt, B.A. Shaw, E. Sikora, W.W. Shaw, Corrosion protection assessment of barrier properties of several zinc-containing coating systems on steel in artificial seawater, *Corrosion.* 62 (2006) 323–339. https://doi.org/10.5006/1.3280665.

[12] H.N. McMurray, G. Parry, B.D. Jeffs, Corrosion resistance of Zn-Al alloy coated steels investigated using electrochemical impedance spectroscopy, *Ironmak. Steelmak.* 25 (1998) 210–215.

[13] S. Fujita, D. Mizuno, Corrosion and corrosion test methods of zinc coated steel sheets on automobiles, *Corros. Sci.* 49 (2007) 211–219. https://doi.org/10.1016/J.CORSCI.2006.05.034.

[14] T.K. Ha, J.R. Son, W.B. Lee, C.G. Park, Y.W. Chang, Superplastic deformation of a fine-grained Zn-0.3wt% Al alloy at room temperature, *Mater. Sci. Eng. A.* 307 (2001) 98–106. https://doi.org/10.1016/S0921-5093(00)01952-3.

[15] B. Wang, Z.W. Lai, C.B. Jiang, Study of the corrosion protection properties of Al-Zn films synthesized by IBAD, *J. Mater. Process. Technol.* 74 (1998) 122–125. https://doi.org/10.1016/S0924-0136(97)80135-5.

[16] R. Kain, E. Baker, Marine atmospheric corrosion museum report on the performance of thermal spray coatings on steel, in: *Test. Met. Inorg. Coatings*, ASTM

International, 100 Barr Harbor Drive, PO Box C700, West Conshohocken, PA 19428–2959, 2008: pp. 211–224. https://doi.org/10.1520/STP20039S.
[17] T.P. Hoar, O. Radovici, Zinc-aluminium sprayed coatings, *Trans. IMF.* 42 (1964) 211–222. https://doi.org/10.1080/00202967.1964.11869929.
[18] O. Salas, O. Troconis De Rincón, D. Rojas, A. Tosaya, N. Romero, M. Sánchez, W. Campos, Six-year evaluation of thermal-sprayed coating of Zn/Al in tropical marine environments, *Int. J. Corros.* 2012 (2012). https://doi.org/10.1155/2012/318279.
[19] H.S. Lee, J.K. Singh, J.H. Park, Pore blocking characteristics of corrosion products formed on Aluminum coating produced by arc thermal metal spray process in 3.5 wt% NaCl solution, *Constr. Build. Mater.* 113 (2016) 905–916. https://doi.org/10.1016/J.CONBUILDMAT.2016.03.135.
[20] H.-S. Lee, J. Singh, M. Ismail, C. Bhattacharya, Corrosion resistance properties of aluminum coating applied by arc thermal metal spray in SAE J2334 solution with exposure periods, *Metals (Basel).* 6 (2016) 55. https://doi.org/10.3390/met6030055.
[21] A.Q. Vu, B. Vuillemin, R. Oltra, C. Allély, Cut-edge corrosion of a Zn-55Al-coated steel: A comparison between sulphate and chloride solutions, *Corros. Sci.* 53 (2011) 3016–3025. https://doi.org/10.1016/j.corsci.2011.05.048.
[22] D. Persson, D. Thierry, N. LeBozec, Corrosion product formation on Zn55Al coated steel upon exposure in a marine atmosphere, *Corros. Sci.* 53 (2011) 720–726. https://doi.org/10.1016/j.corsci.2010.11.004.
[23] X. Zhang, T.N. Vu, P. Volovitch, C. Leygraf, K. Ogle, I.O. Wallinder, The initial release of zinc and aluminum from non-treated Galvalume and the formation of corrosion products in chloride containing media, *Appl. Surf. Sci.* 258 (2012) 4351–4359. https://doi.org/10.1016/j.apsusc.2011.12.112.
[24] X. Zhang, C. Leygraf, I. Odnevall Wallinder, Atmospheric corrosion of Galfan coatings on steel in chloride-rich environments, *Corros. Sci.* 73 (2013) 62–71. https://doi.org/10.1016/j.corsci.2013.03.025.
[25] Y. Li, Formation of nano-crystalline corrosion products on Zn-Al alloy coating exposed to seawater, *Corros. Sci.* 43 (2001) 1793–1800. https://doi.org/10.1016/S0010-938X(00)00169-4.
[26] L. Yang, Y. Zhang, X. Zeng, Z. Song, Corrosion behaviour of superplastic Zn-Al alloys in simulated acid rain, *Corros. Sci.* 59 (2012) 229–237. https://doi.org/10.1016/J.CORSCI.2012.03.013.
[27] X.G. Zhang, *Corrosion and Electrochemistry of Zinc*, Springer, 1996. https://doi.org/10.1007/978-1-4757-9877-7.
[28] Z. Zhao, J. Tang, N.U.H. Tariq, J. Wang, X. Cui, T. Xiong, Microstructure and corrosion behavior of cold-sprayed Zn-Al composite coating, *Coatings* 10 (2020) 931. https://doi.org/10.3390/COATINGS10100931.
[29] S. Sugimura, J. Liao, Long-term corrosion protection of arc spray Zn-Al-Si coating system in dilute chloride solutions and sulfate solutions, *Surf. Coatings Technol.* 302 (2016) 398–409. https://doi.org/10.1016/j.surfcoat.2016.06.042.
[30] M. Salgueiro Azevedo, C. Allély, K. Ogle, P. Volovitch, Corrosion mechanisms of Zn(Mg,Al) coated steel: 2. The effect of Mg and Al alloying on the formation and properties of corrosion products in different electrolytes, *Corros. Sci.* 90 (2015) 482–490. https://doi.org/10.1016/j.corsci.2014.07.042.
[31] A. Macias, C. Andrade, The behaviour of galvanized steel in chloride-containing alkaline solutions-I. The influence of the cation, *Corros. Sci.* 30 (1990) 393–407. https://doi.org/10.1016/0010-938X(90)90046-8.

[32] H.-S. Lee, J.K. Singh, M.A. Ismail, C. Bhattacharya, A.H. Seikh, N. Alharthi, R.R. Hussain, Corrosion mechanism and kinetics of Al-Zn coating deposited by arc thermal spraying process in saline solution at prolong exposure periods, *Sci. Rep.* 9 (2019) 3399. https://doi.org/10.1038/s41598-019-39943-3.
[33] I. Odnevall, C. Leygraf, Reaction sequences in atmospheric corrosion of zinc, *ASTM Spec. Tech. Publ.* (1995) 215–229. https://doi.org/10.1520/stp14921s.
[34] J.T. Kloprogge, L. Hickey, R.L. Frost, The effects of synthesis pH and hydrothermal treatment on the formation of zinc aluminum hydrotalcites, *J. Solid State Chem.* 177 (2004) 4047–4057. https://doi.org/10.1016/j.jssc.2004.07.010.
[35] A. Gulec, O. Cevher, A. Turk, F. Ustel, F. Yilmaz, Accelerated corrosion behaviors of Zn, Al and Zn/15Al coatings on a steel surface, *Mater. Technol.* 45 (2011) 477–482.
[36] H. Katayama, S. Kuroda, Long-term atmospheric corrosion properties of thermally sprayed Zn, Al and Zn-Al coatings exposed in a coastal area, *Corros. Sci.* 76 (2013) 35–41. https://doi.org/10.1016/j.corsci.2013.05.021.
[37] A. Lasia, Impedance of porous electrodes, *J. Electroanal. Chem.* 397 (1995) 27–33. https://doi.org/10.1016/0022-0728(95)04177-5.
[38] L. Nyikos, T. Pajkossy, Fractal dimension and fractional power frequency-dependent impedance of blocking electrodes, *Electrochim. Acta.* 30 (1985) 1533–1540. https://doi.org/10.1016/0013-4686(85)80016-5.
[39] O. de Rincón, A. Rincón, M. Sánchez, N. Romero, O. Salas, R. Delgado, B. López, J. Uruchurtu, M. Marroco, Z. Panosian, Evaluating Zn, Al and Al-Zn coatings on carbon steel in a special atmosphere, *Constr. Build. Mater.* 23 (2009) 1465–1471. https://doi.org/10.1016/j.conbuildmat.2008.07.002.
[40] S. Kuroda, J. Kawakita, M. Takemoto, An 18-year exposure test of thermal-sprayed Zn, Al, and Zn-Al coatings in marine environment, *CORROSION.* 62 (2006) 635–647. https://doi.org/10.5006/1.3280677.
[41] American Welding Society [AWS] Committee on Thermal Spraying, *Corrosion tests of flame-sprayed coated steel; 19-year report—Catalog—UW-Madison Libraries*, Miami, FL, 1974. https://search.library.wisc.edu/catalog/999822360002121 (accessed January 20, 2022).
[42] T.N. Vu, P. Volovitch, K. Ogle, The effect of pH on the selective dissolution of Zn and Al from Zn-Al coatings on steel, *Corros. Sci.* 67 (2013) 42–49. https://doi.org/10.1016/j.corsci.2012.09.042.

## 11.5 CASE STUDY 5: $Al_2O_3$-BASED THERMAL BARRIER COATINGS: INFLUENCE OF RETAINED HIGH-TEMPERATURE $ZrO_2$ AND CARBON NANOTUBES

This section of the chapter summarizes the results to illustrate how the high-temperature phases retained during atmospheric plasma spraying [1] are effectively utilized to enhance properties of thermal barrier coatings (TBCs). Here, utilization of retained phases in tuning the mechanical properties, especially fracture toughness of the aluminum oxide ($Al_2O_3$) and yttria stabilized zirconia (YSZ) based nanocomposite coatings with carbon nanotube (CNT).

Conventionally, YSZ serves as the top thermally insulated layer due to the primary prerequisite for the TBC system, such as thermal insulation or thermal expansion coefficient. But relatively poor mechanical (hardness, elastic modulus and fracture toughness) and tribological properties lead to failure of coatings during their operation. So, $Al_2O_3$-based nanocomposite [2–5] is proposed to improve the structural stability of the TBCs. Though the carbon nanotube (CNT) possess high thermal conductivity (~3000 $Wm^{-1}K^{-1}$ along tube axis), the atmospheric plasma-spraying technique is exploited to produce the CNT-reinforced composite coating with low thermal conductivity [1, 6]. The idea to use $Al_2O_3$-YSZ-CNT is to utilize the harder $Al_2O_3$ matrix, thermally insulate YSZ, and CNT to enhance the fracture toughness. The amount of different phases in coatings can significantly change, that is attributed to stress related to CNTs owing to its higher thermal conductivity. The fracture toughness is theoretically calculated to observe the contributions of different phases and CNTs in $Al_2O_3$-based TBC.

### 11.5.1 Retention of Phases

The composite powders (Table 11.2) are synthesized using sprayed atmospheric plasma sprayed (APS) on Inconel 718. The coatings contain a high amount of $ZrO_2$ phase (as monoclinic and tetragonal form) in comparison to that of initial composite powders and it shows maximum for CNT-reinforced coatings (Table 11.2). The X-ray diffraction pattern of the composite coatings shows the signature for the high amounts of *t*-$ZrO_2$ and *c*-$ZrO_2$ which develops due to stress generation at the $ZrO_2$ interfaces due to high thermal conducting of CNTs that encourages heat dissipation from $ZrO_2$ to the surrounding phases and assists retention of high temperature phases. APS-A3YZC and APS-A8YZC coatings retained the ~38 vol% and ~76 vol% cubic $ZrO_2$, respectively, which is maximum in comparison to all other APS coatings.

The retention of CNTs in the composite coatings are confirmed with the Raman spectra of the APS coatings. Though the CNT is highly unstable above 600 °C in the oxidized environment, the low residence time of CNT in the plasma plume assured its stability. The state of CNT (graphitization and defects) after APS is quantified by calculating the ratio of *D*-band and *G*-band intensity ($I_D$ and $I_G$, respectively) of the Raman spectra. The Raman spectra for starting powders (that contains CNT) have shown a high $I_D/I_G$ (0.71–0.87) than that of its counterpart. The highest $I_D/I_G$ (0.91) of pristine CNT assures the relatively pure form of CNT. Decrease in the $I_D/I_G$ ratio of CNT reveals that the CNT is structurally damaged (~4.5–21.9%), while dispersing it in the composite powder. Further, the state of CNT (graphitization and defects) in the coating indicates the disordering and defect density increased up to ~9.9%–35.2%, with $I_D/I_G$ ratio of ~0.59–0.82 of the Raman spectra. It also confirms the structural damage of CNTs that leads to the graphitization. *G*-band peak shift with respect to pristine CNT indicates the accumulation of compressive stress in the CNT. Stress encouraged to alter the interatomic distance and the vibration frequency of the G-band. It is worth noticing that the shift in *G*-band in the case of APS-AC is comparatively higher in comparison to that of YSZ reinforced composite, indicating stress relaxation. Also, high stress is responsible for the increment in the amount of high-temperature $t$-$ZrO_2$ and $c$-$ZrO_2$ phases in the coatings that contains CNT. Further, the structural stability of the CNT ensures its contribution in improving the mechanical and wear resistance of the coatings. So, the contribution of CNT and other retained phases in the mechanical properties (especially fracture toughness) of the TBCs is delineated with modified fractal model [7].

### 11.5.2 Contribution of Retained Phases on Fracture Toughness of the APS Coatings

The effect of transformed $ZrO_2$ and retained CNTs on indentation fracture toughness ($K_{IC}$) of the coatings are measured. The fracture toughness increased with the reinforcement of YSZ and CNTs up to ~51.6% enhancement for A3YZC coating (5.90 MPa.m$^{1/2}$). This increase in fracture toughness is due to increase in retained tetragonal $ZrO_2$ and CNT, via transformation toughening mechanism, and in the case of CNTs via a crack deflection, CNT bridging, and CNT pull-out mechanisms (see Figure 11.11b–e).

The mismatch in coefficient of thermal expansion in coatings leads to form residual stresses, which helps to arrest further crack propagation, owing to increase the fracture toughness. The fractured APS-AC (Figure 11.11b) and APS-A8YZC (Figure 11.11c) coating supports the existence of CNT pull-out and bridging mechanism (shown in Figure 11.11a). Further, the crack closure due to creation of stress persuaded deformation bands are observed in TEM micrograph of APS-A3YZC coating (Figure 11.11d–e).

The contribution of $ZrO_2$ phases, retained martensitic and CNT in the improvement of fracture toughness ($K_{IC,\,Th}$) of $Al_2O_3$ with reinforcements (YSZ and CNT) are theoretically calculated via modified fractal model, as in equation (11.9). The values of $K_{IC,\,Th}$ (Table 11.2) are directly proportional to its interfacial energy of the crack surface ($\gamma_s$), elastic modulus, grain shape and quantity of phases in the $Al_2O_3$ matrix with random distributed phases. Further, a transformation-toughening term $K_{IC,\,Tr}$ is introduced in model due to the martensitic transformation in the composites; see equation (11.9).

$$K_{IC,\ Thoretical(th)=}\left\{K_{IC}^{ALO}-1.07*\left(2E\gamma_{sALO}^{ALO-ALO}\right)^{0.5}*1+1.07*V_f^{ALO}\right. \tag{11.9}$$

$$\left.*\left(2E\gamma_{sCYZx}^{ALO-ALO}\right)^{0.5}\right\}+\left\{V_f^{YZ}*\left(1.35*\left(2E\gamma_{sCYZx}^{ALO-YZ}\right)^{0.5}-A*E_{YZ}\right)\right.$$

$$\left.+K_{IC,\ Tr}\right\}+\left\{V_f^{CNT}*\left(1.35*\left(2E\gamma_{sCYZx}^{ALO-CNT}\right)^{0.5}-B*E_{CNT}\right)\right\}$$

$$\text{where, } K_{IC,\,Tranformation\ toughening(Tr)}=\frac{\eta E e_T V_f^t h^{0.5}}{(1-\nu)} \tag{11.10}$$

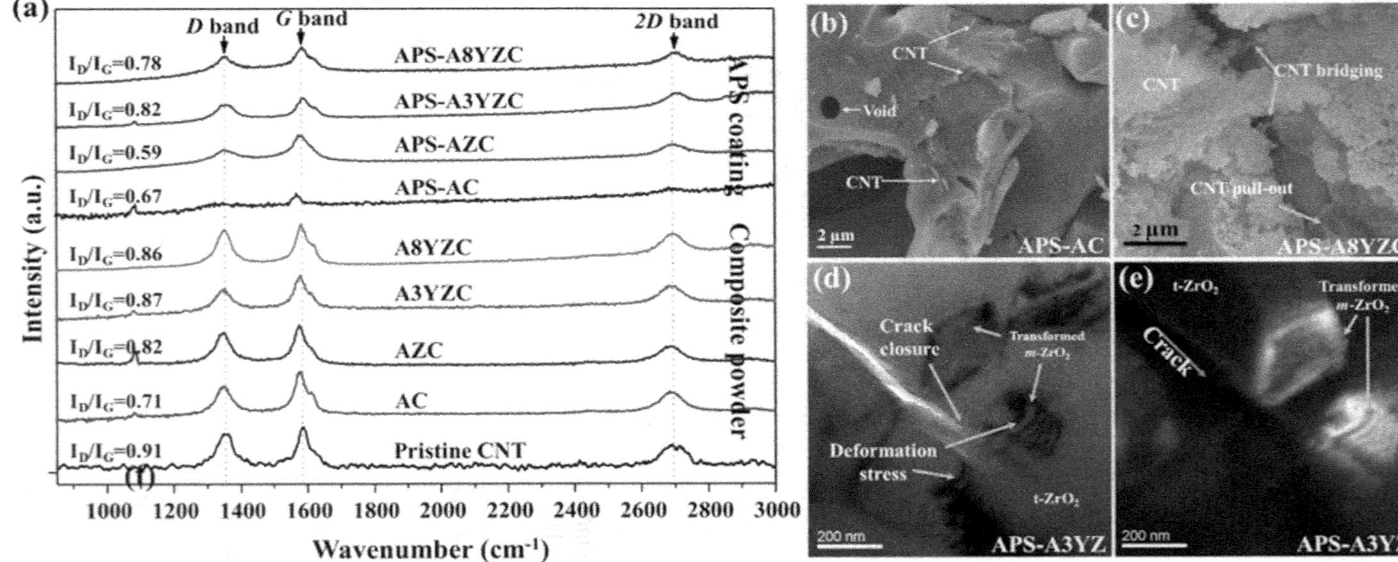


**FIGURE 11.11** (a) Raman spectra of CNT-reinforced $Al_2O_3$-based nanocomposite and coating. Fractured SEM micrograph representing the phenomenon of CNT pull-out and bridging in APS coating of (b) AC and (c) A8YZC. TEM micrograph of APS-A3YZ represents (d) bright field image and (e) dark field of corresponding TEM image presenting deformation stresses which transform tetragonal $ZrO_2$.

*Source*: Reprinted with permission from [1].

where $E_{CNT}$, $E_{YZ}$ and $V_f^{CNT}$, $V_f^{YZ}$ are the elastic modulus and volume fraction of CNT and YSZ, respectively. Interfacial energy ($\gamma_s$) is considered in range of 13.3–19, 23–39 and 2.2–2.8 J/m$^2$ for $Al_2O_3$-CNT, $Al_2O_3$-YSZ and $Al_2O_3$-$Al_2O_3$ interfaces, respectively. And $\eta$, $E$, $e_T$, $V_f^t$, $h$ and $v$ are the transformation zone factor, elastic modulus, dilatational strain, volume fraction of tetragonal $ZrO_2$, phase transformation zone and Poisson's ratio of the coating, respectively.

$V_f^{CNT}$ is the volume fraction of CNT and YSZ, $V_f^t$ is the volume fraction of martensitic tetragonal $ZrO_2$ phase, $K_{IC}$ is the experimentally calculated fracture toughness, $K_{IC,th}$ is the theoretically determined fracture toughness, A is taken as $2.18 \times 10^{-6}$ and B is taken as $1.72 \times 10^{-6}$.

The CNT-CNT, YSZ-YSZ and CNT-YSZ interfaces are not considered due to relatively lower fractions compared to $Al_2O_3$-YSZ and $Al_2O_3$-CNT. The theoretical fracture toughness ($K_{IC,\ Th}$, Table 11.2) were determined via equations (11.9) and (11.10) for all coatings. First, it may be noted that CNT reinforcement alone in $Al_2O_3$ does not provide substantial toughening (up to ~10% toughening). But, YSZ addition has substantially rendered toughening to $Al_2O_3$ composites. There, again, a substantial toughening is not observed with CNT addition in their corresponding composites (i.e., up to 15% toughening only). An increase in the fracture toughness with addition of $ZrO_2$ and 8YSZ (in $Al_2O_3$) is attributed to relatively higher elastic modulus and toughness of zirconia compared to that of $Al_2O_3$.

Toughening arises in CNT reinforced composites owing to CNT pull-out, bridging and crack deflection. It is also realized that CNT pull out (Figure 11.11b–c) leads to toughening mechanisms, and this is feasible due to extraordinary elastic modulus of CNT. Highest roughening is exhibited with synergistic combination of transformation toughening in 3YSZ-CNT reinforcement in $Al_2O_3$ (Table 11.2) owing to $t$-$ZrO_2$ transformation. Such a contribution is additionally confirmed via delineation of contributing toughening effects via isolating CNT and YSZ. The overestimation of fracture toughness in YSZ-reinforced composite is due to the contribution of the residual stress, pore shape and non-homogeneous distribution of reinforcement CNT in $Al_2O_3$ matrix. Synergistically, intrinsic properties of CNT and YSZ and externally induced phase transformation improves fracture toughness of $Al_2O_3$ reinforced with YSZ- and CNT-based TBCs.

In conclusion, $Al_2O_3$ reinforced 20 wt% of YSZ coatings with and without CNTs (4 wt%) were deposited via the atmospheric plasma spraying technique. A relatively low residence time of CNT in the plasma plume assures the retention of CNT in the deposited coating. The average value of fracture toughness was enhanced by ~51% (compared to that of $Al_2O_3$), and the maximum value was observed to be 5.90 MPa.m$^{1/2}$ for CNT- and 3YSZ-reinforced composite coatings, which is attributed to retention of CNT (4 wt%) and high temperature transformable-tetragonal $t$-$ZrO_2$ (~26%) phase. In this regard, the contributions of CNT (via crack bridging and pull-out phenomenon) and $t$-$ZrO_2$ (via transformation toughening) are 12% and 15%, respectively, towards improving the fracture toughness. The analytically calculated fracture toughness increases with YSZ and CNT addition commensurate with the experimental values. The overestimation

**TABLE 11.2**
**Experimentally and Theoretically Calculated Fracture Toughness of the Nanocomposite Coating**

| Composition and Sample ID | APS Coating Without CNT | | | Composition and Sample ID | APS Coating Reinforced with 4 wt% CNT | | | |
|---|---|---|---|---|---|---|---|---|
| | $V_f^t$ | Fracture Toughness ($MPa.m^{1/2}$) | | | $V_f^{CNT}$ | $V_f^t$ | Fracture Toughness ($MPa.m^{1/2}$) | |
| | | $K_{IC}$ | $K_{IC,th}$ | | | | $K_{IC}$ | $K_{IC,th}$ |
| $Al_2O_3$ (APS-ALO) | 0 | ~3.67–4.11 | ~3.73–4.05 | 4% CNT (APS-AC) | 7.30 | 0 | ~4.15–4.43 | ~3.87–4.20 |
| 20wt% $ZrO_2$ (APS-AZ) | 0 | ~4.51–5.31 | ~4.53–5.26 | 20% $ZrO_{2}$-4% CNT (APS-AZC) | 7.75 | 0 | ~5.03–5.59 | ~4.79–5.78 |
| 20wt% 3YSZ (APS-A3YZ) | 7.01 | ~4.82–5.62 | ~5.32–7.18 | 20% 3YSZ-4% CNT (APS-A3YZC) | 7.83 | 8.63 | ~5.59–6.21 | ~5.54–7.20 |
| 20wt% 8YSZ (APS-A8YZ) | 0 | ~4.96–5.18 | ~4.55–5.61 | 20% 8YSZ-4% CNT (APS-A8YZC) | 7.83 | 0 | ~5.24–5.42 | ~5.04–6.11 |

of fracture toughness values is attributed to assumption of uniform distribution of different phases in the $Al_2O_3$ matrix during atmospheric plasma spraying. So, on the basis of fracture toughness, synergy of CNT and 3YSZ reinforcement in $Al_2O_3$ is a suitable choice as TBC for high-temperature gas turbine applications.

## REFERENCES


[1] Ariharan S, Nisar A, Balaji N, Aruna ST, Balani K (2017) Carbon nanotubes stabilize high temperature phase and toughen $Al_2O_3$-based thermal barrier coatings. *Compos Part B Eng* 124:76–87

[2] Balani K, Bakshi SR, Chen Y, Laha T, Agarwal A (2007) Role of powder treatment and carbon nanotube dispersion in the fracture toughening of plasma-sprayed aluminum oxide-carbon nanotube nanocomposite. *J Nanosci Nanotechnol.* 7:3553–3562

[3] Balani K, Harimkar SP, Keshri A, Chen Y, Dahotre NB, Agarwal A (2008) Multiscale wear of plasma-sprayed carbon-nanotube-reinforced aluminum oxide nanocomposite coating. *Acta Mater* 56:5984–5994

[4] Ariharan S, Balani K (2021) Fretting wear behaviour and frictional force mapping of $Al_2O_3$ based thermal barrier coatings. *Int J Refract Met Hard Mater* 98:105525

[5] Pakseresht AH, Saremi M, Omidvar H, Alizadeh M (2019) Micro-structural study and wear resistance of thermal barrier coating reinforced by alumina whisker. *Surf Coatings Technol* 366:338–348

[6] Ariharan S, Wangaskar B, Xavier V, Venkateswaran T, Balani K (2019) Process induced alignment of carbon nanotube decreases longitudinal thermal conductivity of $Al_2O_3$ based porous composites. *Ceram Int* 45:18951–18964

[7] Rishabh A, Joshi MR, Balani K (2010) Fractal model for estimating fracture toughness of carbon nanotube reinforced aluminum oxide. *J Appl Phys* 107:1–7


## 11.6 CASE STUDY 6: ULTRA-HIGH-TEMPERATURE CERAMICS

The use of carbon/carbon (C/C) composites in thermal protection systems for hypersonic re-entry vehicles in nose cones and leading edges is facilitated owing to their low weight, higher resistance to thermal shock and commendable ablation resistance at extreme temperatures. However, the severe working conditions during re-entry, wherein the vehicle can experience high temperatures (> 2000 °C), high air pressure and sharp impact particle erosion, makes them vulnerable to oxidation. This condition sustains for several seconds to a few minutes. However, C/C composites starts to oxidize at ~600 °C and under high-speed gas jet undergo ablation and particle erosion, restricting their application in this field. Hence, enormous efforts have been undertaken to improve their resistance to ablation for higher-temperature application and in atmospheres containing oxygen. In this regard, enormous amount of research has been conducted in recent years with the intention of applying ultra-high-temperature ceramics (UHTCs) to C/C composites, essentially due to their high melting temperatures (>3000 °C) and excellent thermal properties [1, 2]. UHTCs are usually defined as materials having melting temperatures beyond 3000 °C, including borides, carbides and nitrides of some transition metals like Zr, Hf, Ti and Ta [3]. Coating approaches are established to protect C/C composites because of the simplicity in their fabrication and integrity with the base material (substrate). UHTCs have been investigated extensively for the purpose of developing thermal protection system (TPS) for space vehicles.

Tului et al. [4] prepared $ZrB_2$ + SiC (25 wt%) composite coating on graphite plates using inert gas plasma spraying (IPS). In a first phase, tests were conducted on small samples (i.e., laboratory scale samples) for which coatings were mechanically separated from the substrate and subjected to spark erosion for machining into (4 × 1.5 × 22 $mm^3$) specimens followed by exposure to 1900 °C in air for 120 s using flame heating (acetylene + oxygen) followed by rapid cooling in water (thermal shock) or natural cooling in air (thermal cycling). In the second phase, validation of the developed material is done under simulated operative conditions, testing demonstrative components in a plasma wind tunnel (PWT). For this purpose, a graphite cone of base diameter 125 mm and height of 175 mm with tip curvature radius of 5 mm as shown in Figure 11.12a was used. Graphite was enveloped in a 5 mm thick layer of C/SiC, and on it a UHTC film was deposited via plasma spraying. High temperature exposure in the first phase resulted in a decrease of ultimate tensile strength (from 41 ± 6.6 MPa to 32 ± 8.1 MPa) and fracture strength (from 81 ± 6.1 MPa to 60 ± 10.4 MPa), and an improvement in Young's modulus (from 47 ± 7.8 MPa to 76 ± 12.2 MPa), which could be due to enhanced surface defects caused by the formation of an oxidation layer and also due to the densification of material induced by sintering effects. A total of 9 and 24 cycles of exposure for water cooling and air cooling, respectively, resulted in failure. Phase composition modifications due to oxidation tended to form sort of exfoliation layers on sample surface. After each cooling cycle, such layers detached from the sample surface and brittle fractures (typical of ceramic materials) were observed after nine cycles, when the specimen section reduced to around half the original. For air cooling tests, no brittle fractures were observed, and this phenomenon extended until the last cycle to failure causing complete material consumption. Visual inspection after preliminary plasma test did not show any coating detachment from the substrate (Figure 11.12b). However, change in color was observed in the thermally most stressed part of the sample, from pale gray to dark gray.

In another study [5], they fabricated $ZrB_2$ + SiC (20 vol%) + $MoSi_2$ (10 vol%) composite (ZSM) coating on graphite plates using controlled atmosphere plasma spraying in Ar gas at 1200 mbar, under high-pressure plasma spraying (HPPS) mode, employing a F4MB torch and a plasma gas mixture of Ar (55 SLPM) and $H_2$ (13 SLPM). The coatings were mechanically separated from the substrate and subjected to spark erosion for machining into (45 × 4 × 3 $mm^3$) specimens for carrying out the mechanical characterization tests. For mechanical testing, a four-point bending test was performed at room temperature and at higher temperatures of 500 °C, 1000 °C and 1500 °C using ASTM: C1211–02 (2008). Increase in elastic modulus from 108 ± 6.3 GPa at room temperature to 205 ± 0.6 GPa at 1000 °C with a significant drop to 84 ± 4.5 GPa at 1500 °C was recorded. The flexural strength showed same trend with 115.1 ± 6.5 GPa at room temperature, which increased to 201.9 ± 70.1 GPa at 1000 °C and dropped to 144.3 ± 4.0 GPa at 1500 °C. It was observed that mechanical properties

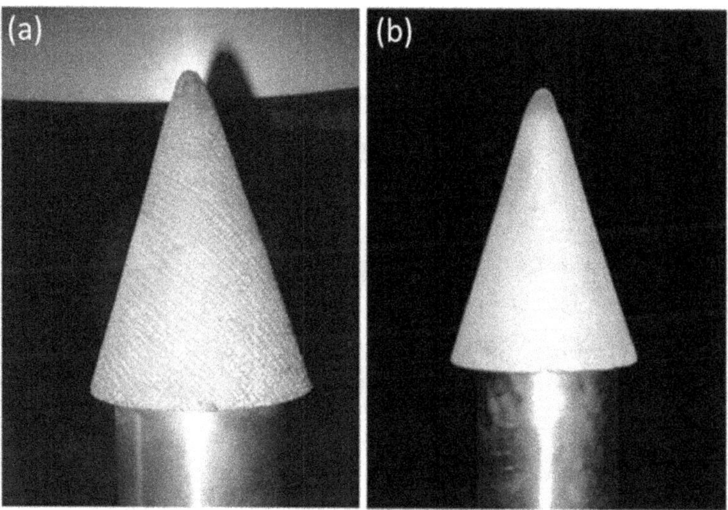


**FIGURE 11.12** Test article of $ZrB_2$-SiC composite (a) before plasma test and (b) after plasma test.

*Source*: Reprinted with permission from [4].

improve till 1000 °C and then decrease at 1500 °C. Main factor ascribed to the stiffness decrease was the intersplat and intrasplat cracking. The specimen tested at 1500 °C exhibited a denser microstructure, with some residual porosity but a higher splat cohesion due to the sintering effect of the high temperature treatment. This yielded a relatively high modulus of elasticity at 1500 °C, where stiffness reduction due to high temperature is partially compensated by the sintering effect. Tested coatings resulted in brittle behavior up to 1000 °C. However, at 1500 °C, ZSM exhibited a significant plastic deformation. Silica scale with a $ZrO_2$ layer was observed on the surface. $ZrO_2$ was observed to provide sealing effect, avoiding the presence of open cracks on the surface. Thus, sintering and the healing of surface defects by oxide formation probably provided the strengthening mechanisms at 1000 °C, improving the mechanical properties till 1000 °C. However, at 1500 °C, though good mechanical properties retained, but plastic deformation was evident.

In another study [1], for enhanced ablation resistance of the C/C composites, SiC buffer layer on C/C composites was coated by a thick ZrC layer using vacuum plasma spraying. An HVOF system equipped with a diamond jet (DJ) gun (DJH2700 gun; Oerlikon Metco, US) was utilized for carrying out the test. To produce a supersonic flame, it is accelerated through convergent/divergent nozzle. The specimen installed at 60 mm from the nozzle was subjected to a 30 s ablation test, as shown in Figure 11.13.

During the ablation test on ZrC-coated sample (Figure 11.14a), a layer formed (Figure 11.14b), which completely protected the C/C composite from thermal

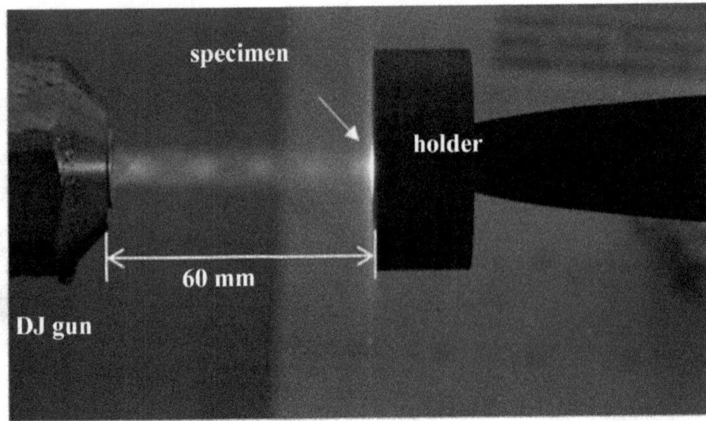


**FIGURE 11.13** Test specimen mounted on holder for arc-jet exposure damage tolerance.

*Source*: Open reprints from [1].

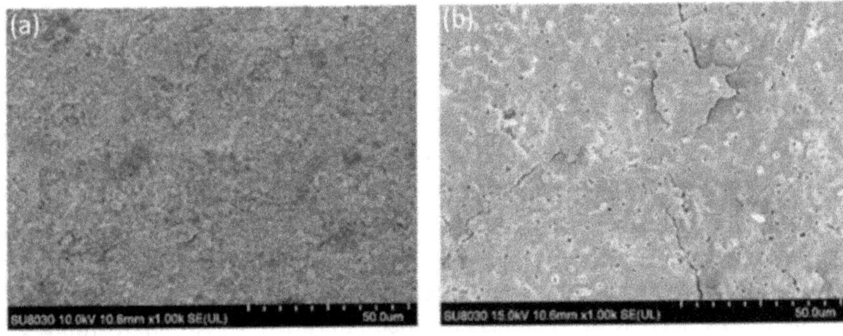


**FIGURE 11.14** Surface microstructure of ZrC-coated sample (a) before the ablation test and (b) after the ablation test.

*Source*: Open reprints from [1].

oxidation. Though cracks and pores were apparent on the surface of the coating layer after the ablation test (Figure 11.14b), no delamination occurred at the coating layer interface.

Phase analysis, which was carried out by XRD and EDS, revealed the transfer of $ZrO_2$ from surface into the C/C composites. Thus, the coating approach using VPS effectively protected a C/C composite in an ablation environment. Moreover, the absence of any detached coating between the ZrC and substrate provides an indication of good adherence between them. Thus, UHTC compounds can be considered as the potential candidates for use in the extreme environments owing to their thermal and chemical stability. Though use of monolithic UHTC

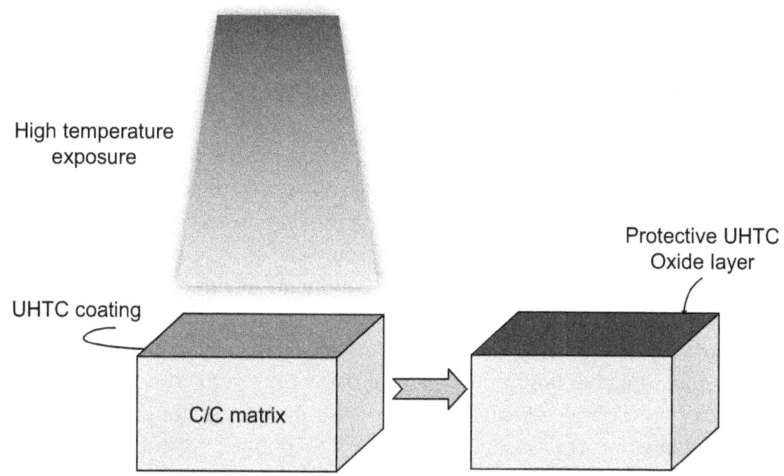


**FIGURE 11.15** Schematic of UHTC coating forming protective oxide scale, which provides protection to the C/C matrix.

ceramics in large and hot structures could be limited by their brittleness and their difficulty of manufacturing into large parts, a feasible alternate is the employment of UHTC coatings to guarantee an overall good reliability of the structure. From the case studies discussed above, it can be concluded that UHTC plasma-sprayed coatings possess good mechanical properties essentially due to sintering effects at higher temperatures and healing of defects and have this capability of withstanding high-temperature oxidizing conditions. Figure 11.15 presents a simple schematic of UHTC coating providing protection to the base material against high temperature exposure.

## REFERENCES

[1] B. R. Kang, H. S. Kim, P. Y. Oh, J. M. Lee, H. I. Lee, and S. M. Hong, "Characteristics of ZrC barrier coating on SiC-coated carbon/carbon composite developed by thermal spray process," *Materials*, vol. 12, no. 5, p. 747, 2019.

[2] E. M. Alosime, M. S. Alsuhybani, and M. S. Almeataq, "The oxidation behavior of $ZrB_2$-SiC ceramic composites fabricated by plasma spray process," *Materials*, vol. 14, no. 2, p. 392, 2021.

[3] A. Nisar, R. Hassan, A. Agarwal, and K. Balani, "Ultra-high temperature ceramics: Aspiration to overcome challenges in thermal protection systems," *Ceramics International*, vol. 48, no. 7, p. 8852–8881, 2021.

[4] M. Tului, G. Marino, and T. Valente, "Plasma spray deposition of ultra high temperature ceramics," *Surface Coatings Technology*, vol. 201, no. 5, pp. 2103–2108, 2006.

[5] G. Pulci, M. Tului, J. Tirillò, F. Marra, S. Lionetti, and T. Valente, "High temperature mechanical behavior of UHTC coatings for thermal protection of re-entry vehicles," *Journal of Thermal Spray Technology*, vol. 20, no. 1, pp. 139–144, 2011.

## 11.7 CASE STUDY 7: THERMAL SPRAYING AS AN ALTERNATIVE TO CHROME PLATING

Hard chrome plating is a technique in which chromium is electroplated onto a metallic object. For over a century, these coatings have been used in automotive components like pistons and valves. They gained popularity due to their superior wear and corrosion resistance. But there are several environmental and health hazards associated with the waste produced by the galvanic industry, which has motivated the search for adopting such alternatives.

Hexavalent chromium is present in paint pigments, metal plating (hard chrome plating), wood preservatives and so forth. Chromium leaks from industry can also contaminate soil and ground water. Exposure to Cr(VI) can lead to several health problems, including but not limited to respiratory tract problems, pneumonia, tracheobronchitis and lung cancer. The increased awareness about environmental damage and health risks associated with Cr(VI) motivated the search for "clean coatings" [1].

Some popular and commercially viable surface coating alternatives include physical vapor deposition (PVD), chemical vapor deposition (CVD) and thermal spray. It is very important that the alternative coating meets the physical requirements like wear, erosion and corrosion resistance, good surface finish and so on, at least as effective as chrome plating [2].

### 11.7.1 Why Thermal Spray?

Thermal spray, especially high-velocity oxygen fuel (HVOF) and plasma spray, has emerged as one of the leading technologies to substitute hard chromium plating because they do not generate hazardous wastes, generate high-performance coating, are quicker than PVD and CVD, and are cost-effective.

These techniques can be used to spray metals as well as ceramics. Now depending on the application, a suitable material can be used; for example, a composite coating deposited using HVOF in Figure 11.16a shows a better wear resistance than chromium plating. Commonly used coating materials include Cr-Ni, Ni-Mo, WC-12Co, WC-10Co-4Cr, NiCrBSi, AISI316L, WC-CrC-Ni and CrC-NiCr.

Legg et al. [2] studied the cost involved in coating a cylinder with outer diameter 125 mm and height 300 mm, as shown in Figure 11.16b. The study compared the overall cost required to coat the cylinder by chrome plating, HVOF and PVD. It compared the cost required for preheating, coating, finishing, utilities and so on, as it is clear from the graph, HVOF is much cheaper than PVD and chrome plating [2].

### 11.7.2 Wear-Resistant Coatings

The most popular commercial application of hard chrome plating is to form hard, wear-resistant coatings. CrC-NiCr and WC coated using HVOF are some of the most widely used and rather superior wear-resistant alternatives. The quality

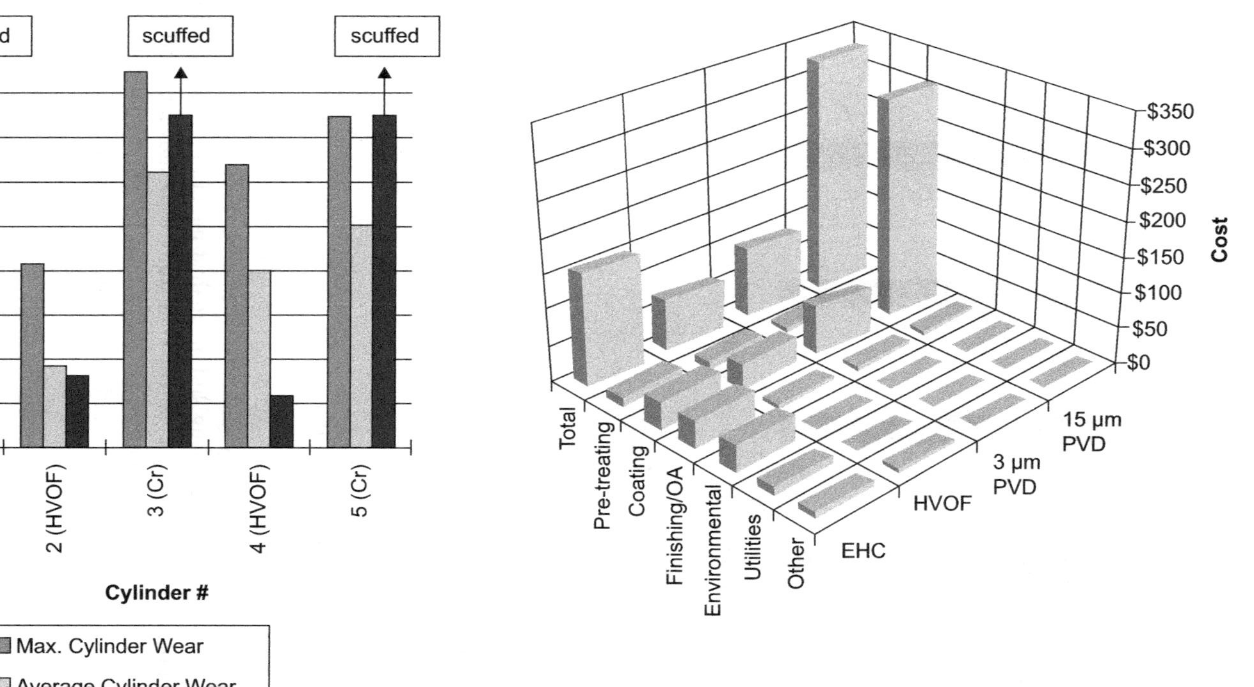


**FIGURE 11.16** (a) Result of engine tests on hard chrome and HVOF $Cr_3C_2$-composite-coated piston rings; (b) cost factor comparison of various techniques used to coat a cylinder with outer diameter 125 mm and height 300 mm.

*Source*: Open reprints from [2].

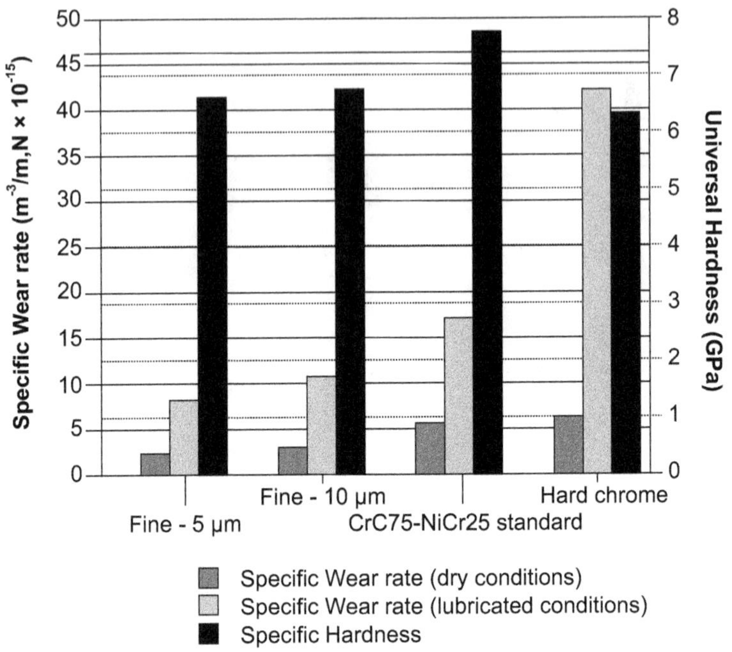


**FIGURE 11.17** Specific wear rate and universal hardness of different coatings.

*Source*: Reprinted with permission from [3].

of coating depends on thermal spray and powder parameters. For example, a study by Picas et al. showed that CrC75-NiCr25 is more wear resistant than hard chrome plating. It also illustrates that powder parameters like size influence the coating performance [3].

Another study shows the sliding wear behavior of different WC-Co coated on a ring using HVOF, plasma spray and chrome plating. Figure 11.17 shows sliding wear test results for coated ring sliding against metal block. HVOF in general shows good sliding wear resistance. This study also shows the effect of coating composition on performance [4]. These wear resistant coatings are commercially used in piston, valves, turbine shafts, aluminum bearings and gears.

## REFERENCES


[1] C. Pellerin, S.M. Booker, Reflections on hexavalent chromium: health hazards of an industrial heavyweight, *Environmental Health Perspectives* 108(9) (2000) A402–A407.

[2] K. Legg, M. Graham, P. Chang, F. Rastagar, A. Gonzales, B. Sartwell, The replacement of electroplating, *Surface and Coatings Technology* 81(1) (1996) 99–105.

[3] J. Picas, A. Forn, G. Matthäus, HVOF coatings as an alternative to hard chrome for pistons and valves, *Wear* 261(5–6) (2006) 477–484.
[4] A. Savarimuthu, I. Megat, H. Taber, J. Shadley, E. Rybicki, W. Emery, J. Nuse, D. Somerville, Sliding wear behavior as a criterion for replacement of chromium electroplate by tungsten carbide (WC) thermal spray coatings in aircraft applications, *ITSC 2000*, ASM International, 2000, pp. 1095–1104.

## 11.8 CASE STUDY 8: THERMALLY SPRAYED CYTOCOMPATIBLE HYDROXYAPATITE COATINGS

The number of patients undergoing total joint arthroplasty is increasing every year. This can be attributed to surgical advancements, obesity, financial feasibility, aging population, lifestyle changes, and so on. But implants have a limited life, and therefore the amount of revision total joint arthroplasty has also increased. The requirement of revision arthroplasty depends on a number of factors like age, level of activity and quality of implant. As the demand for primary and revision arthroplasties is only going to increase, engineers and scientists are trying to make the implants more cost-effective and durable. One way to increase the life and mechanical stability of the implant is to modify the implant surface by increasing the surface roughness and/or coating the substrate with a bioactive material. Another important biomedical application, dental implants, which involve surface modification for improved osseointegration and functionality, just like prosthesis are ever increasing in frequency.

Implants are usually made of metals like titanium and stainless steel. Coating of a material like hydroxyapatite (HA) therefore promotes osseointegration which reduces the risk of implant failure thereby increasing implant life and reducing requirement of revision surgery. When implants fail because of a variety of reasons like infections, aseptic loosening and instability/dislocation, a revision total knee arthroplasty (TKA) or total hip arthroplasty (THA) needs to be performed. Also, the life of an implant depends on a number of factors like the age of the patient, the level of activity and the quality of primary implant. Surface modification is one of the most crucial, cost-effective developments in the field of medical implants which can be used to increase implant life. After the implant is inserted into the patient's body, it is the surface that interacts with the bodily fluids and tissues. Hence, surface roughness and chemistry play an important role in osseointegration of the implant. It is very important for its acceptance and long-term functionality.

### 11.8.1 Thermal-Sprayed Hydroxyapatite-Based Coatings

It is very important to choose the right material and deposition technique to coat implants. The chemistry and physical properties of the coating influences the acceptance of implant by the body, recovery of the patient and the time required to resume normal activity.

$Ca_{10}(PO_4)_6OH_2$, known as hydroxyapatite, is the most important bio-ceramic, also known as bone mineral, because its calcium-to-phosphate ratio is the same as that of human bone (Ca:P = 1.67). It is highly bioactive, and it helps for bone growth [1]. HA is highly biocompatible to the surrounding tissue. Despite good biocompatibility and its bioactive nature, its applications are limited. HA cannot be used for load-bearing applications because of its poor mechanical properties and fracture toughness (0.8–1.3 $MPa.m^{-1/2}$). To improve the mechanical properties of HA, various type of reinforcements, including carbon nanotubes (CNT), graphene, alumina ($Al_2O_3$), yttria stabilized zirconia and titanium alloy, have been added into the HA matrix [2]. Two different types of coatings (HA, HA-CNT) through plasma spray were produced over Ti6Al4V substrate (see Figure 11.18). Both coatings were adhered to the surface, and no delamination was found in any case. Variation was observed in the thickness of coating for HA (150 μm) and for HA-CNT (110 μm). In vitro studies (cell culture) were performed over HA-CNT plasma-sprayed coating with human osteoblasts hFOB 1.19. It was observed that cells grow at the CNT surface (see Figure 11.19), which ensures that CNT is non-toxic to surrounding tissue [3].

Two most advanced thermal spraying techniques are solution precursor plasma spraying (SPPS) and suspension plasma spraying (SPS), because in these techniques liquid feedstock is used in place of the solid powder feedstock, which produces a coating of submicron thickness [4]. High adhesion strength (40 MPa) and micrometer thick coating (<50 μm) of HA can be deposited by axial suspension plasma spraying (ASPS). For suspension plasma spraying (SPS) three different types of solutions were prepared: ASPS-A (water + 23 wt% HA), ASPS-B (water + 13 wt% HA) and ASPS-C (water + 80 wt% HAp + 20 wt% bioglass). SEM micrographs of these coatings are shown in Figure 11.20.

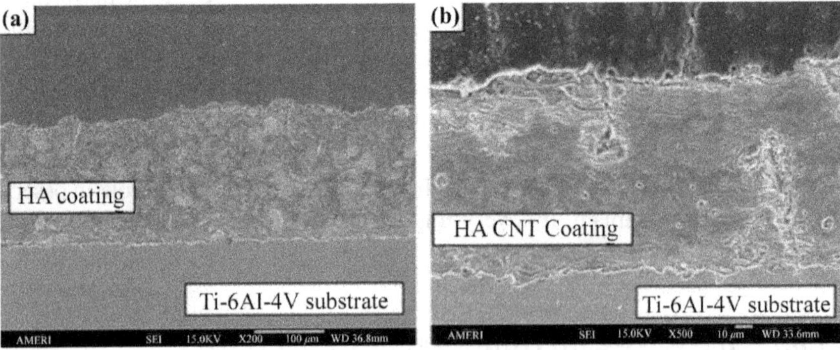


**FIGURE 11.18** Scanning electron microscopy image of plasma-sprayed (a) HA and (b) HA-CNT.

*Source*: Reprinted with permission from [3].

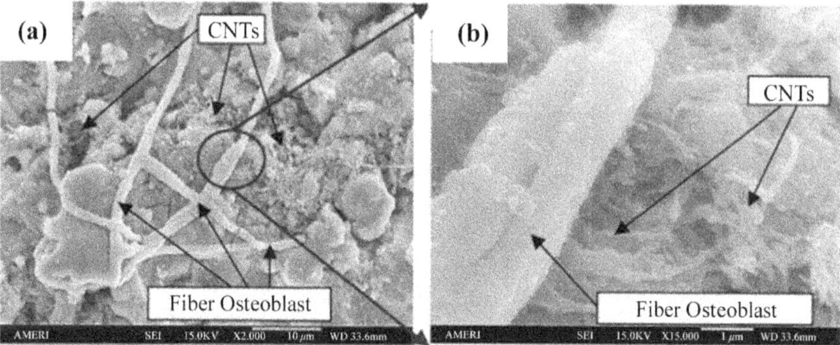


**FIGURE 11.19** Scanning electron microscopy image shows cell growth (a) at CNTs and (b) near CNTs in thermally sprayed hydroxyapatite-based coatings.

*Source*: Reprinted with permission from [3].

Microstructural analysis, shown in Figure 11.20, was performed to understand the features of the coating, and it was observed that delamination and larger cracks were found in ASP (axial plasma spraying), whereas in ASPS finer cracks were observed and there was no delamination. ASPS-B in Figure 11.20c shows the minimum thickness of coating <50 μm. Adhesion strength of these coatings was measured by adhesion test where adhesive agent was used to join top and bottom counterparts after that tensile load was applied and the ratio between the load at which rupture occurred and surface area of coating was calculated. The adhesion strengths of the coatings were 29 ± 3 MPa (for APS), 30 ± 4 MPa (for ASPS-A), 42 ± 10 MPa (for ASPS-B) and 15 ± 3 MPa (for ASPS-C) [5].

Simulated body fluid was used to study the tribological behavior of HA-CNT-based plasma-sprayed coatings [6]. Wear performance on coatings was performed using pin-on-disc wear tester at the load of 8.8 N, where zirconia pin (50 mm long and 3 mm diameter) was used as a counter body. HA-CNT coating applied by the plasma-spray technique showed a lower wear rate (38.92 g/m$^2$) compared to that of HA (60.15 g/m$^2$) coating. This is because of the lubricating nature of CNT. Figure 11.21a shows the bar chart of wear rate for three different samples of base, and Figure 11.21b displays the cumulative weight loss of each sample with time.

CNT as a reinforcement provide enhanced fracture toughness due to the CNTs bridging (Figure 11.22a) and stretching (Figure 11.22b), biocompatibility, and superior wear resistance. CNTs bridging between HA splats restrict the motion of these splats, which further helps to reduce the weight loss and wear rate [6]. A coating of a few micrometers thickness (50–200 μm) can be obtained by thermal spray technique.

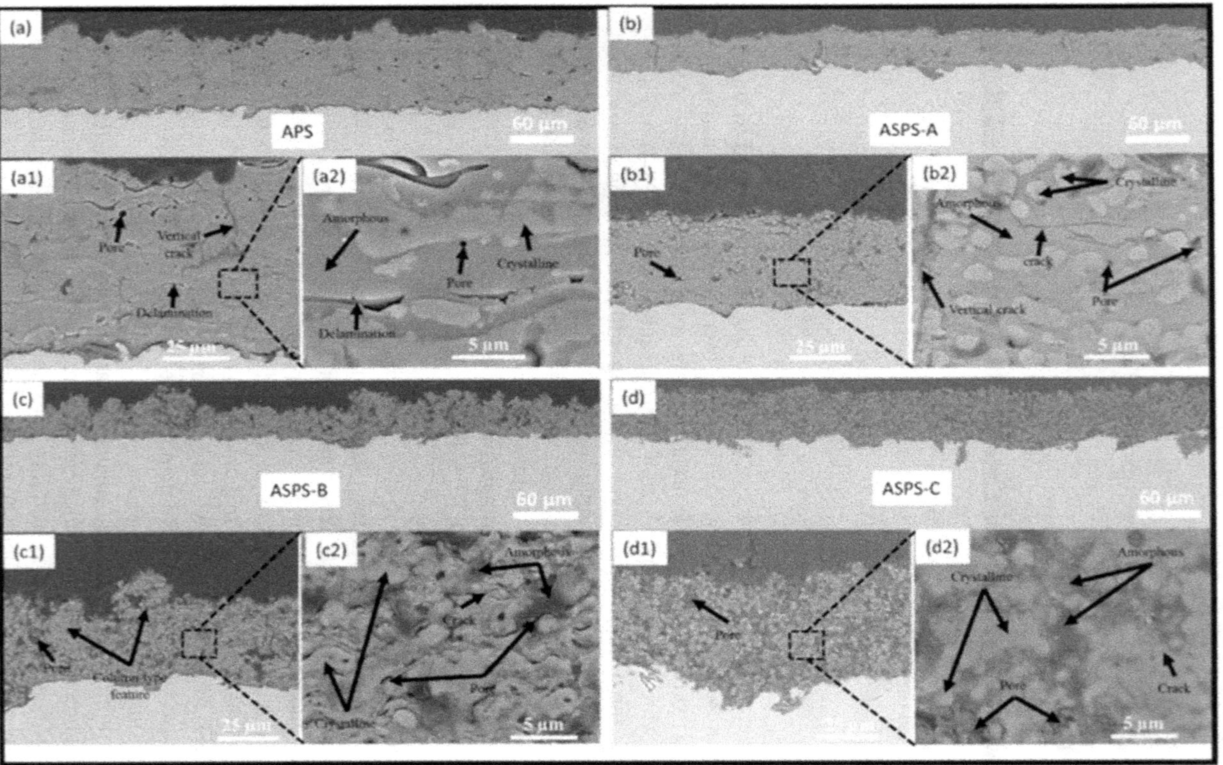


**FIGURE 11.20** Scanning electron microscopy image: (a) APS, (a1) and (a2) reveal large cracks and delamination in the coating; (b) ASPS-A, (b1) and (b2) reveal finer cracks and no delamination in the coating; (c) ASPS-B, (c1) and (c2) highlight crystalline and amorphous regions in the coating; and (d) ASPS-C, (d1) and (d2) highlight crystalline and amorphous regions in the coating.

*Source*: Reprinted with permission from [5].

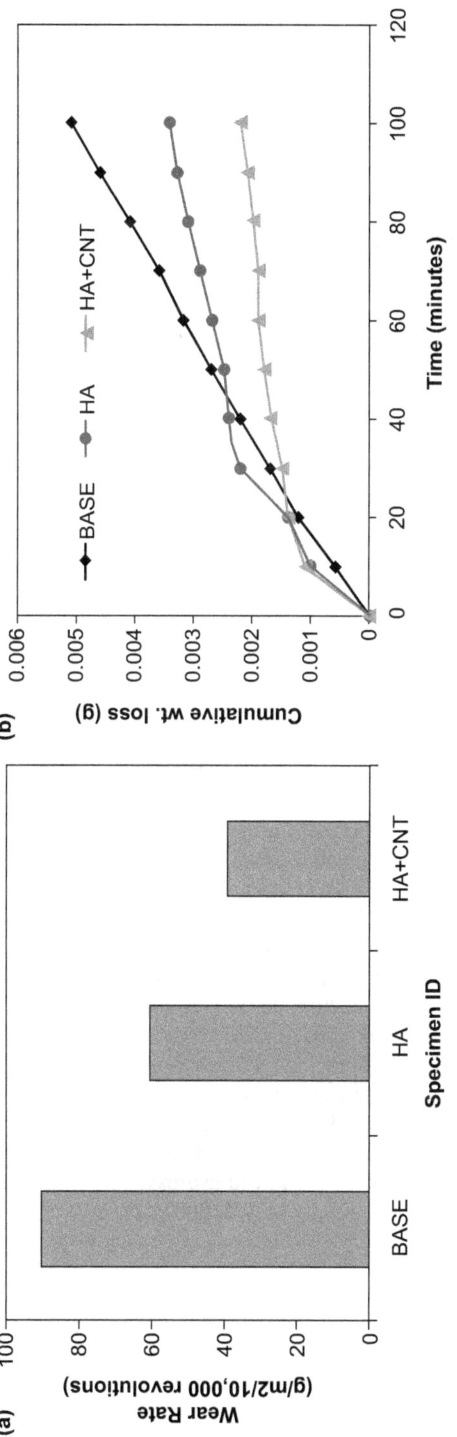


**FIGURE 11.21** Plots of (a) wear rate and (b) weight loss.

*Source*: Reprinted with permission from [6].

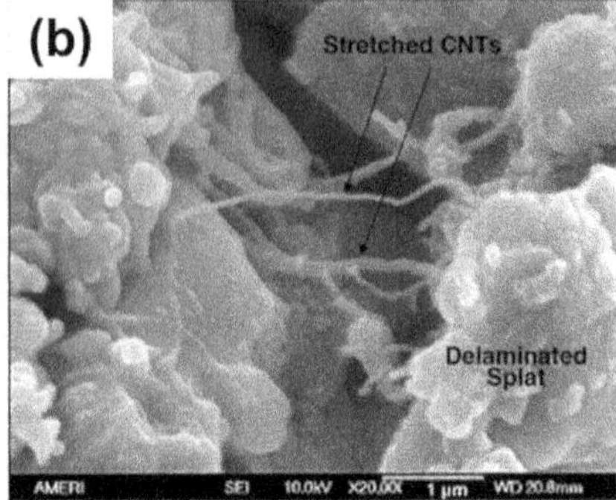


**FIGURE 11.22** SEM images showing behavior of CNTs as reinforcement in plasma-sprayed hydroxyapatite coatings: (a) CNTs bridging; (b) stretched CNTs.

*Source*: Reprinted with permission from [6].

## REFERENCES

[1] D. Kim, Y. Han, J. H. Lee, I. Kang, B. Jang, and S. Kim, "Characterization of Multiwalled Carbon Nanotube-Reinforced Hydroxyapatite Composites Consolidated by Spark Plasma Sintering," *BioMed. Res. Int.*, vol. 2014, 2014, doi:10.1155/2014/768254.

[2] Y. Meng, W. Qiang, and J. Pang, "Preparation and Characterization of Mechanical Properties of Carbon Nanotube Reinforced Hydroxyapatite Composites Consolidated by Spark Plasma Sintering," *IOP Conf. Ser.: Mater. Sci. Eng.*, 2017, doi:10.1088/1757–899X/231/1/012164.

[3] K. Balani *et al.*, "Plasma-sprayed Carbon Nanotube Reinforced Hydroxyapatite Coatings and Their Interaction with Human Osteoblasts in Vitro," *Biomaterials*, vol. 28, pp. 618–624, 2007, doi:10.1016/j.biomaterials.2006.09.013.

[4] A. Ganvir, N. Curry, and N. Markocsan, "Characterization of Thermal Barrier Coatings Produced by Various Thermal Spray Techniques Using Solid Powder, Suspension, and Solution Precursor Feedstock Material," *Int. J. Appl. Ceram. Technol.*, vol. 332, pp. 324–332, 2016, doi:10.1111/ijac.12472.

[5] A. Ganvir, S. Nagar, N. Markocsan, and K. Balani, "Deposition of Hydroxyapatite Coatings by Axial Plasma Spraying : Influence of Feedstock Characteristics on Coating Microstructure, Phase Content and Mechanical Properties," *J. Eur. Ceram. Soc.*, vol. 41, no. 8, pp. 4637–4649, 2021, doi:10.1016/j.jeurceramsoc.2021.02.050.

[6] K. Balani, Y. Chen, S. P. Harimkar, and N. B. Dahotre, "Tribological behavior of plasma-sprayed carbon nanotube-reinforced hydroxyapatite coating in physiological solution," *Acta Biomater.*, vol. 3, pp. 944–951, 2007, doi:10.1016/j.actbio.2007.06.001.

# Index

Note: Page numbers in **bold** indicates tables on the corresponding page.

For Product Safety Concerns and Information please contact our EU representative GPSR@taylorandfrancis.com
Taylor & Francis Verlag GmbH, Kaufingerstraße 24, 80331 München, Germany

www.ingramcontent.com/pod-product-compliance
Lightning Source LLC
Chambersburg PA
CBHW060806220326
41598CB00022B/2547

* 9 7 8 1 0 3 2 3 4 4 0 2 7 *